Introduction to Inverse Problems in Imaging
Second Edition

Mario Bertero
Patrizia Boccacci
Christine De Mol

CRC Press
Taylor & Francis Group
Boca Raton London New York

CRC Press is an imprint of the
Taylor & Francis Group, an **informa** business

Second edition published 2022
by CRC Press
6000 Broken Sound Parkway NW, Suite 300, Boca Raton, FL 33487-2742

and by CRC Press
2 Park Square, Milton Park, Abingdon, Oxon, OX14 4RN

© 2022 Taylor & Francis Group, LLC

First edition published by CRC Press 1998

CRC Press is an imprint of Taylor & Francis Group, LLC

ISBN: 978-0-367-47005-0 (hbk)
ISBN: 978-0-367-46786-9 (pbk)
ISBN: 978-1-003-03275-5 (ebk)

DOI: 10.1201/9781003032755

Typeset in LM Roman
by KnowledgeWorks Global Ltd.

Introduction to Inverse Problems in Imaging

Second Edition

To Marisa

Contents

Part I: Image Deconvolution

Part II: Linear Inverse Problems

Part III: Statistical Methods

Preface to the first edition

This book arises from a series of lectures in image restoration and tomography we teach to students in computer science and physics at the University of Genova. It is a course of approximately fifty one-hour lectures, supplemented by about fifty hours of training with the computer, devoted to the implementation and validation of the principal methods presented in the lectures.

The mathematical background common to these students consists of first courses in mathematical analysis, geometry, linear algebra, probability theory and Fourier analysis. This is also the mathematical background we assume for our readers.

The selection of the methods and applications is partly subjective, because determined by our research experience, and partly objective, because dictated by the need of presenting efficient methods which are also easily implementable. Therefore the book not only does not pretend completeness but is based on a strategy of incompleteness. In spite of this choice, which concerns also references, we hope that the book will provide the reader with the appropriate background for a clear understanding of the essence of inverse problems (ill-posedness and its cure) and, consequently, for an intelligent assessment of the rapidly and, sometimes, chaotically, growing literature on these problems.

We wish to thank our friends and colleagues Paola Brianzi, Piero Calvini, and Carlo Pontiggia, who have read and commented on preliminary drafts of the book. In particular, we are indebted to Christine De Mol, from the University of Brussels, for an accurate and critical reading of the manuscript and for several suggestions intended to improve the clarity of the presentation. We also express our gratitude to Laura Opisso for her help during the preparation of the Latex file.

For illustrating some of the methods described in the book, we used images (both synthetic and real) from the Hubble Space Telescope. These images were produced by the Space Telescope Science Institute, with support operated by the Association of Universities for Research in Astronomy, Inc., from NASA contract NAS5-26555, and were reproduced with permission from AURA/STScI.

The manuscript of the book was completed while M. Bertero was with the Department of Physics of the University of Genova.

M. B. and P. B.

Preface to the second edition

More than twenty years after the first publication of this book, the time has come for a new edition. The first edition mainly focused on the foundations of the theory and on the basic methods for the solution of linear inverse problems as well as on their physical and mathematical motivations. This remains true for the present edition. Meanwhile, however, within this well-established context, considerable progress has been made concerning the development of new methods aiming at the reconstruction of specific classes of objects. We believe that within the huge literature on the subject, we can at least single out three classes of such newly developed methods: edge-preserving methods, sparsity-enforcing methods and statistical methods with a focus on methods for Poisson data. Therefore we thought that this material had to be included in the book and these topics are now covered in three dedicated chapters.

This second edition derives from a reorganization and expansion of the content of the first edition. Due to the necessity of maintaining a reasonable size for the printed book, the chapter on preliminary mathematical tools (except for some parts which are now included in the relevant chapters) and the third part with mathematical appendices, which were included in the first edition, have been removed. However, their content with additional and auxiliary mathematical results is included in the online material, which is provided with the present edition. Moreover, the general structure of the book has been modified and consists now of three parts. The first part is essentially unchanged and contains the detailed study of the problem of image deconvolution, for which the tools of Fourier analysis greatly simplifies the discussion of the peculiarities of inverse problems and the presentation of the first regularization methods developed for the solution of ill-posed problems. The second part is devoted to more sophisticated linear inverse problems and also contains the edge-preserving and sparsity-enforcing regularization approaches. Finally, the third part is completely devoted to statistical methods, with a focus on Poisson data but with also hints on other possible kinds of noise.

Let us stress the fact that the introduction in the Tikhonov theory of non-differentiable penalties has stimulated in recent years extensive research on optimization methods for the design of efficient iterative methods able to minimize the relevant functionals. We deliberately decided not to cover these developments which are quite technical and therefore beyond the scope of the present book. Moreover, as in the first edition, we chose to focus on linear inverse problems. Nonlinear inverse problems have also been the subject of a lot of recent research, including for practical applications. However, although the basic principles of ill-posedness and regularization are the same as for linear problems, their treatment requires more advanced mathematical tools and problem-specific methods that go beyond this introductory textbook on inverse imaging problems.

Author Bios

Mario Bertero, born in 1938, is Professor Emeritus of the University of Genova. He received the advanced degree in Physics cum laude in 1960 and the *libera docenza* in Theoretical Physics in 1968. In 1981, he became a Professor of Mathematics and in 1997 a Professor of Imaging Science at the University of Genova. He retired in 2010 from teaching (but not from research). From 1990 to 1994, he was the Editor-in-chief of the journal *Inverse Problems*, published by the Institute of Physics, and is currently a member of the International Advisory Panel of the same journal. He organized several workshops and conferences on Inverse Problems and he was the promoter and Chair of the conference *Applied Inverse Problems: Theory and Practice* held in Montecatini (Italy), in June 2001, the first of a successful series organized every two years in Europe and North America.

He is the author or co-author of about 150 articles, co-editor, with E. R. Pike, of the book *Inverse Problem in Scattering and Imaging* and co-author with P. Boccacci and V. Ruggiero of the book *Inverse Imaging with Poisson Data: From cells to galaxies*. His first research activity concerned potential scattering in quantum theory. Around 1975, he became interested in the theory and applications of inverse and ill-posed problems. He developed methods based on the singular value decomposition of linear operators for the solution of inverse problems in various domains of applied science. In recent years, he mainly contributed to the study of astronomical imaging and, more generally, to the study and analysis of inverse problems with Poisson data.

He is a member of the Institute of Physics and of the 'Accademia Ligure di Scienze ed Arti'.

Patrizia Boccacci, born in 1955, received the advanced degree in Physics from the University of Genova, Italy, in 1980. From 1981 to 1983, she worked on a project of multi-directional holographic interferometry, funded by ESA (European Space Agency). From 1985 to 2000, she was technical research manager with the Department of Physics (DIFI) of the University of Genova, Italy. Since May 2004, she is an Associate Professor at the Department of

Informatics, Bioengineering, Robotics and System Engineering (DIBRIS) of the University of Genova.

She is a co-author of several research articles and of two books. Her main research interest is related to numerical methods for the solution of inverse ill-posed problem and to their applications, such as optical tomography for the investigation of crystal growth in micro-gravity, confocal microscopy, particle sizing, seismic tomography and astrophysics. More recently, she has been mainly interested in image restoration problems related to the LBT (Large Binocular Telescope) project and in medical imaging.

Christine De Mol, born in 1954, holds a Ph.D. in Physics (1979) and a habilitation degree in Mathematical Physics (1992) from the 'Université libre de Bruxelles' (ULB). Since 1975, she has held several research positions with the Belgian National Fund for Scientific Research (FNRS) that she left in 1998 as an Honorary Research Director to become a full-time Professor (Emeritus since 2019) at ULB, where she is affiliated to both the Mathematics Department and the European Centre for Advanced Research in Economics and Statistics (ECARES). She has held several visiting positions at the Universities of London, Rome, Montpellier, Paris-Sud, Genoa, Bonn and St.Gallen. In 2012 and 2013, she served as the elected Chair of the SIAM Activity Group on Imaging Science.

She is the author or co-author of more than eighty publications. Her research interests in applied mathematics include inverse problems, sparsity-enforcing regularization theory, wavelets and applications, learning theory, portfolio theory, as well as analysis and forecast-ing of time series data.

Acronyms

AEM	Alternating Extragradient Method
AO	Adaptive Optics
AREMF	Average Relative Error Magnification Factor
BV	Bounded Variation
CCD	Charge-Coupled Device
CG	Conjugate Gradient
COC	Circle of Confusion
COSTAR	Corrective Optics Space Telescope Axial Replacement
CSLM	Confocal Scanning Laser Microscope/Microscopy
CT	Computed Tomography
dMRI	diffusion Magnetic Resonance Imaging
DFT	Discrete Fourier Transform
DTI	Diffusion Tensor Imaging
ELT	Extremely Large Telescope
EM	Expectation Maximization
EM-RL	Expectation Maximization - Richardson Lucy
ESO	European Southern Observatory
FFAH	Far-Field Acoustic Holography
FBP	Filtered Back-Projection
FFT	Fast Fourier Transform
FISTA	Fast Iterative Shrinkage-Thresholding Algorithm
FLAO	First Light Adaptive Optics
FOF	Fiber Orientation Function
FOM	Figure of Merit
FT	Fourier Transform
FWHM	Full Width at Half Maximum
GCV	Generalized Cross Validation
HARDI	High Angular Resolution Diffusion Imaging
HST	Hubble Space Telescope
ISRA	Image Space Restoration Algorithm
ISTA	Iterative Soft Thresholding Algorithm
KKT	Karush-Kuhn-Tucker
KL	Kullback-Leibler
LBT	Large Binocular Telescope
LBTI	Large Binocular Telescope Interferometer
LBTO	Large Binocular Telescope Observatory
l.h.s.	left hand side
MAP	Maximum a Posteriori
ML	Maximum Likelihood
ML-EM	Maximum Likelihood - Expectation Maximization
MM	Majorization-Minorization
MRI	Magnetic Resonance Imaging
MSE	Mean Squared Error
MSEIF	Mean Square Error Improvement Factor
NFAH	Near-Field Acoustic Holography

OSEM	Ordered Subset Expectation Maximization
OSL	One-Step Late
OTF	Optical Transfer Function
PET	Positron Emission Tomography
PSF	Point Spread Function
PSNR	Peak Signal-to-Noise Ratio
PSWF	Prolate Spheroidal Wave Functions
r.h.s.	right hand side
RL	Richardson-Lucy
RMSE	Root Mean Square Error
ROF	Rudin-Osher-Fatemi
RON	Read-out Noise
r.v.	random variable
SAR	Synthetic Aperture Radar
SGP	Scaled Gradient Projection
SIRT	Simultaneous Iterative Reconstruction Technique
SNOM	Scanning Near-field Optical Microscope/Microscopy
SNR	Signal-to-Noise Ratio
SPECT	Single Photon Emission Computed Tomography
STED	Stimulated Emission Depletion Microscopy
SVD	Singular Value Decomposition
TF	Transfer Function
TMT	Thirty Meter Telescope
TV	Total Variation
VLT	Very Large Telescope
WM	White Matter

1

Introduction

The first clinical machine for the detection of head tumors, based on a new technique called X-ray *computed tomography* (CT), was installed in 1971 at the Atkinson Morley's Hospital, Wimbledon, Great Britain. The announcement of this machine, by G. H. Hounsfield at the 1972 British Institute of Radiology annual conference [163], has been considered as the greatest achievement in radiology since the discovery of X-rays by W. C. Roengten in 1895. In 1979 G. H. Hounsfield shared with A. Cormack the Nobel Prize for Physiology and Medicine. As stated by Hounsfield, the CT computer-generated images were obtained by a specially designed algorithm since no viable method was found in the existing literature. Anyway, these images were presumably the first example of images obtained by solving a mathematical problem in the class of the so-called inverse and ill-posed problems, first discussed by the Russian mathematician A. N. Tikhonov in 1943 [282].

The first workshop completely devoted to these problems was organized in 1971, the same year as the beginning of tomography, by L. Colin at the Ames Research Center, Moffet Field, California [70]. Among the topics considered were inversion methods applied to passive and active atmospheric sounding, ionospheric sounding, particle scattering, electromagnetic scattering and seismology. Subsequently, important advances, also driven by the success of tomography, were made both in the theory and in the practice of inverse problems. A typical mathematical property of these problems, the so-called ill-posedness, was understood and methods for overcoming the difficulties due to this property were developed, mainly along the lines indicated by A. N. Tikhonov [284]. Applications to imaging have been especially impressive; in particular, the applications to diagnostic medicine, inspired by CT, allowed the development of new imaging techniques such as SPECT (*Single Photon Emission Computed Tomography*), PET (*Positron Emission Tomography*) and MRI (*Magnetic Resonance Imaging*). The development of new imaging modalities with applications to diagnostic medicine still stimulates nowadays research in this important and expanding field.

1.1 What is an inverse problem?

From the point of view of a mathematician, the concept of inverse problem has a certain degree of ambiguity which is well illustrated by a frequently quoted statement by J. B. Keller [179]: 'We call two problems *inverses* of one another if the formulation of each involves all or part of the solution of the other. Often, for historical reasons, one of the two problems has been studied extensively for some time, while the other is newer and not so well understood.

DOI: 10.1201/9781003032755-1

In such cases, the former is called the *direct problem*, while the latter is called the *inverse problem'*.

In any domain of mathematical physics, one finds problems satisfying the requirements stated by Keller. In general, these problems are related by a sort of duality in the sense that one problem can be obtained from the other by exchanging the role of the data and that of the unknowns: the data of one problem are the unknowns of the other and conversely. As a consequence of this duality, it may seem arbitrary to decide which is the direct and which is the inverse problem.

For a physicist, however, the situation is quite different because the two problems are not on the same level: one of them, and precisely the one called the direct problem, is considered as more fundamental than the other and, for this reason, is also much more investigated. In other words, the historical reasons mentioned by Keller are basically physical reasons.

With reference to physics, one can say that a direct problem is a problem oriented along a cause-effect sequence or, also, a problem which consists in computing the consequences of given causes; then, the corresponding inverse problem is associated with the reversal of the cause-effect sequence and consists in finding the unknown causes of known consequences [289]. It follows that the definition of a direct-inverse pair must be based on well-established physical laws, which specify what are the causes and what the effects and provide the equations relating the effects to the causes. It also follows that for each domain of physics (mechanics, astronomy, wave propagation, heat conduction, geophysics, etc.), it is necessary to specify the direct problems typical of that domain as well as the corresponding inverse problems. A few examples may clarify these statements.

In classical mechanics, a direct problem is, for instance, the computation of the trajectories of particles from the knowledge of the forces. Then the inverse problem is the determination of the forces from the knowledge of the trajectories. From this point of view, Newton not only stated the basic laws of mechanics, and therefore the basic equations of the direct problem, but also solved the first inverse problem when he determined the gravitation force from the Kepler laws describing the trajectories of the planets.

Other examples, however, are more appropriate for the modern applications of inverse methods. In scattering and diffraction theory, the direct problem is the computation of the scattered (or diffracted) waves from the knowledge of the sources and obstacles, while the inverse problem consists in the determination of the obstacles from the knowledge of the sources and of the scattered (or diffracted) waves. Inverse problems of this kind are fundamental for various methods of non-destructive evaluation (including medical imaging), which consist in sounding an object by means of a suitable radiation. Another example of direct problem in wave-propagation theory is the computation of the field radiated by a given source, for instance, the radiation pattern of a given antenna; then the inverse problem is the determination of the source from the knowledge of the radiated field (in this case, the determination of the current distribution in the antenna from the knowledge of the radiation pattern). Analogously in potential theory, which is basic in geodesy, the direct problem is the determination of the potential generated by a known mass distribution, while the inverse problem is the determination of the mass distribution from the values of the potential, and so on.

Other examples come from instrumental physics, i.e. the physics of instruments such as electronic devices and imaging systems. Here the direct problem is the computation of the output of the instrument (the image), being given the input (the object) and the characteristics of the instrument (impulse response function, etc.). Then the inverse problem is the identification of the input of a given instrument from the knowledge of the output. The first part of this book is devoted to an important particular case of this inverse problem.

As we stated above, a direct problem is a problem oriented along a cause-effect sequence; it is also a problem directed towards a loss of information: its solution defines a transition from a physical quantity with a certain information content to another quantity with smaller information content. This property, which is common to most direct problems, will be reformulated in a more precise way in the next section. In general, it implies that the solution is much smoother than the data: the image provided by a bandlimited system is smoother than the corresponding object, the scattered wave due to an obstacle is smooth even if the obstacle is rough, and so on. Here we briefly discuss a simple example where this property has a nice physical interpretation.

In the case of heat propagation, let us consider the direct problem of computing the temperature distribution at a time $t > 0$, being given the temperature distribution at $t = 0$ (plus additional boundary conditions). A simplified version of the problem is the following: to solve in the domain $\mathcal{D} = \{0 \leq x \leq a; t > 0\}$ the one-dimensional equation of the heat conduction

$$\frac{\partial^2 u}{\partial x^2} = \frac{1}{D}\frac{\partial u}{\partial t} , \tag{1.1}$$

(the positive constant D measures the thermal conductivity), being given the following initial and boundary conditions

$$u(x,0) = f(x) , \quad u(0,t) = u(a,t) = 0 . \tag{1.2}$$

This is our direct problem, the cause being the given temperature distribution $f(x)$ at $t = 0$ and the corresponding effect being the temperature distribution $u(x,t)$ at the time $t > 0$. The problem can be easily solved by means of Fourier series expansions. If we expand the data function as follows

$$f(x) = \sum_{n=1}^{\infty} f_n \sin\left(\frac{n\pi}{a}x\right) \tag{1.3}$$

with

$$f_n = \frac{2}{a}\int_0^a f(x)\sin\left(\frac{n\pi}{a}x\right)dx , \tag{1.4}$$

then the solution of the problem is given by

$$u(x,t) = \sum_{n=1}^{\infty} f_n e^{-D(n\pi/a)^2 t}\sin\left(\frac{n\pi}{a}x\right) . \tag{1.5}$$

Let us assume that the initial state is known with a certain precision ε, in the sense that only the Fourier coefficients f_n, such that $|f_n| > \varepsilon$, are known. Since the Fourier coefficients f_n tend to zero when $n \to \infty$, it follows that only a finite number of Fourier coefficients, let us say N_ε, is known. This is the information content of the data of the direct problem, or also of the initial state of the physical system. Then, if the final state is the temperature distribution at the time $t = T$, i.e. $u(x,T)$, its number of Fourier coefficients greater than ε is much smaller than N_ε as an effect of the decaying factor $\exp[-D(n\pi/a)^2 T]$. We conclude that the information content of the solution is much smaller than the information content of the data. This result is related to the well-known fact that the entropy of the system increases for increasing time.

The corresponding inverse problem is the problem of determining the temperature distribution at $t = 0$, being given the temperature distribution at the time $t = T > 0$. Therefore the data function is now given by $u(x,T)$, while the unknown function is $f(x)$. From the previous discussion of the direct problem, it follows that, if we know the data with precision ε, then it will be difficult (maybe impossible) to obtain $f(x)$ with the same precision because some information has been lost in the natural evolution of the system. In other words, there exist many functions $f(x)$ which correspond, within the precision ε, to the given $u(x,T)$.

This conceptual difficulty is common to most inverse problems because, by solving these problems, we would like to accomplish a transformation that should correspond to a gain of information. It provides the explanation of a typical mathematical property of inverse problems which is known as *ill-posedness*. This point will be considered in the next section. Here we only observe as an interesting fact that the different physical interpretations of the two problems, direct and inverse, are associated with different mathematical properties.

The example previously discussed is an example of a linear problem and, in this book, we will consider only linear problems. The reason is twofold: first, linear problems, possibly deriving from the linearization of nonlinear ones, are still nowadays the most frequently met in the applications; second, a well-developed and unified mathematical framework and efficient numerical algorithms for their solution are available.

Undoubtedly, most inverse problems are basically nonlinear and very interesting both from the mathematical and from the practical point of view. We would like to recall that the first inverse problems, which were the object of elegant and deep mathematical investigations, were in fact nonlinear, namely, the *inverse Sturm-Liouville problem* and the *inverse scattering problem in quantum theory*.

The first dates back to Lord Rayleigh who, in describing the vibrations of strings of variable density, discussed the possibility of deriving the density distribution from the frequencies of vibration. Important contributions to this problem have been made by outstanding mathematicians, such as Levinson, Marchenko and Krein. An easy-to-read survey of these results is given in a famous paper by M. Kac [174].

As concerns the second problem, it was originated by the attempt of determining the unknown nuclear forces from the measured scattering data. Basic results in this problem are due, for instance, to Levinson, Jost, Kohn, Gel'fand, Levitan, Marchenko and many others. A valuable and complete presentation of these results is the book by Chadan and Sabatier [61]. Among books devoted to inverse scattering, we would also like to mention the book by Colton and Kress [72] on inverse problems in acoustic and electromagnetism.

Let us mention that in recent years research on specific nonlinear inverse problems has been rapidly growing, in connection with applications such as *impedance tomography* [68], [49], [291], *optical tomography* [10], [11], or seismic inversion [276], to single out a few of them.

1.2   What is an ill-posed problem?

In the previous section, we mentioned that a typical property of inverse problems is ill-posedness, a property which is opposite to that of well-posedness. We now comment on these concepts.

The basic concept of *well-posed problem* was introduced by the French mathematician Jacques Hadamard in a paper on boundary-value problems for partial differential equations and their physical interpretation [143]. In this first formulation, a problem is called well posed when its solution is unique and exists for arbitrary data. In subsequent work [144] Hadamard emphasizes the requirement of continuous dependence of the solution on the data, claiming that a solution that varies considerably for a small variation of the data is not really a solution in the physical sense. Indeed, since physical data are never known exactly, this would imply that the solution is not known at all.

From an analysis of several cases, Hadamard concludes that, essentially, problems motivated by physical reality are well posed. An example is provided by the initial value problem, also called the Cauchy problem, for the D'Alembert equation, which is basic in the description of wave propagation

$$\frac{\partial^2 u}{\partial x^2} - \frac{1}{c^2}\frac{\partial^2 u}{\partial t^2} = 0 \qquad (1.6)$$

where c is the wave velocity. If we consider, for instance, the following initial data at $t = 0$

$$u(x,0) = f(x) \ , \quad \frac{\partial u}{\partial t}(x,0) = 0 \ , \tag{1.7}$$

then there exists a unique solution given by

$$u(x,t) = \frac{1}{2}\left[f(x-ct) + f(x+ct)\right] \ . \tag{1.8}$$

This solution exists for any continuous function $f(x)$. Moreover, it is obvious that a small variation of $f(x)$ produces a small variation of $u(x,t)$, as seen from equation (1.8).

Another example of a well-posed problem is provided by the forward problem of the heat equation, briefly analyzed in the previous section. The solution, as given by equation (1.5), clearly exists, is unique and depends continuously on the data function $f(x)$.

The previous problems are well posed and, of course, are basic in the description of physical phenomena. They are examples of direct problems. An impressive example of an ill-posed problem and, in particular, of non-continuous dependence on the data, was also provided by Hadamard [143]. This problem which, at that time, was deprived of physical motivations, is the Cauchy problem for the Laplace equation in two variables

$$\frac{\partial^2 u}{\partial x^2} + \frac{\partial^2 u}{\partial y^2} = 0 \ . \tag{1.9}$$

If we consider the following Cauchy data at $y = 0$

$$u(x,0) = \frac{1}{n}\cos(nx) \ , \quad \frac{\partial u}{\partial y}(x,0) = 0 \ , \tag{1.10}$$

then the unique solution of equation (1.9), satisfying the conditions (1.10), is given by

$$u(x,y) = \frac{1}{n}\cos(nx)\cosh(ny) \ . \tag{1.11}$$

The factor $\cos(nx)$ produces an oscillation of the surface representing the solution of the problem. When n is sufficiently large, this oscillation is imperceptible near $y = 0$ but becomes enormous at any given finite distance from the x-axis. More precisely, when $n \to \infty$, the data of the problem tend to zero but, for any finite value of y, the solution tends to infinity.

This is now a classical example illustrating the effects produced by a non-continuous dependence of the solution on the data. If the oscillating function (1.10) describes the experimental errors affecting the data of the problem, then the error propagation from the data to the solution is described by equation (1.11) and its effect is so dramatic that the solution corresponding to real data is deprived of physical meaning. Moreover, it is also possible to show that the solution does not exist for arbitrary data but only for data with some specific analyticity property. Let us observe that an inverse problem of electrocardiography considered in recent years [71], i.e. the reconstruction of the epicardial potential from body surface maps, can be formulated precisely as a Cauchy problem for an elliptic equation, i.e. a generalization of the Laplace equation.

Another example of non-continuous dependence of the solution on the data is provided by the backward problem for the heat equation, i.e. the inverse problem discussed in the previous section. If we consider the following data at the time $t = T$

$$u(x,T) = \frac{1}{n}\sin\left(\frac{n\pi}{a}x\right) \tag{1.12}$$

then the solution of the heat equation (1.1) is given by

$$u(x,t) = \frac{1}{n} \sin\left(\frac{n\pi}{a}x\right) e^{D(n\pi/a)^2(T-t)} . \tag{1.13}$$

For $t < T$ and, in particular, for $t = 0$ we find again that, when $n \to \infty$, the data function tends to zero while the solution tends to infinity. Since this pathology is clearly due to the exponential factor which, as discussed in the previous section, is responsible of the loss of information in the solution of the direct problem, we find a link between this loss of information and the lack of continuity in the solution of the inverse problem.

A problem satisfying the requirements of uniqueness, existence and continuity is now called *well posed* (in the sense of Hadamard), even if the complete formulation in terms of the three requirements was first given by R. Courant [75]. The problems which are not well posed are called *ill posed* or also *incorrectly posed* or *improperly posed*. Therefore an ill-posed problem is a problem whose solution is not unique or does not exist for arbitrary data or does not depend continuously on the data.

The conviction of Hadamard that problems motivated by physical reality had to be well posed is essentially generated by the physics of the nineteenth century. The requirements of existence, uniqueness and continuity of the solution correspond to the ideal of a unique, complete and stable determination of the physical events. As a consequence of this point of view, ill-posed problems were considered, for many years, as mathematical anomalies and were not seriously investigated. The discovery of the ill-posedness of inverse problems has completely modified this conception.

The previous observations and considerations can now justify the following general statement: a direct problem, i.e. a problem oriented along a cause-effect sequence, is well posed while the corresponding inverse problem, which implies a reversal of the cause-effect sequence, is ill posed. This statement, however, is meaningful only if we provide a suitable mathematical setting for the description of direct and inverse problems. Before doing so, let us remark that we are mainly considering problems of imaging, and therefore, we use a language appropriate to these problems.

The first point is to define the class of the objects to be imaged, which are described by suitable functions with certain properties. In this class, we also need a *distance*, in order to establish when two objects are close to each other and when they are not. In such a way our class of objects takes the structure of a *metric space* of functions. We denote this space by $\mathcal{X}$ and we call it the *object space*.

The second point is to solve the direct problem, i.e. to compute, for each object, the corresponding image which can be called the computed image or the *noise-free image*. Since the direct problem is well posed, to each object we associate one, and only one, image. As we already remarked, this image may be rather smooth as a consequence of the fact that its information content is smaller than the information content of the corresponding object. This property of smoothness, however, may not be true for the measured images, also called *noisy images*, because they correspond to some noise-free image corrupted by the noise affecting the measurement process.

Therefore the third point is to define the class of images in such a way that it contains both the noise-free and noisy images. It is convenient to introduce a distance also in this class so that it is also a metric space. We denote it by $\mathcal{Y}$ and we call it the *image space*. In conclusion, the solution of the direct problem defines a mapping (operator), denoted by A, which transforms any object of the space $\mathcal{X}$ into a noise-free image of the space $\mathcal{Y}$. This operator is continuous, i.e. the images of two close objects are also close, because the direct problem is well posed. The set of the noise-free images is usually called, in mathematics, the *range* of the operator A, and, as follows from our previous remark, this range does not coincide with the image space $\mathcal{Y}$ because this space contains also the noisy images.

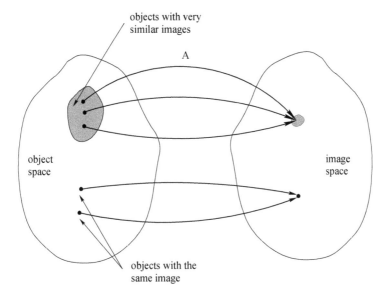

FIGURE 1.1 Schematic representation of the relationship between objects and images. The dashed subsets in $\mathcal{X}$ and $\mathcal{Y}$ illustrate the loss of information due to the imaging process.

By means of this mathematical scheme, it is possible to describe the loss of information which, as we said, is typical in the solution of the direct problem. It has two consequences which are described in a different way from the mathematical point of view, even if their practical effects are very close. First, it is possible that two, or even more, objects have exactly the same image. In the case of a linear operator, this is related to the existence of objects whose image is exactly zero. These objects will be called *invisible objects*. Then, given any object of the space $\mathcal{X}$, if we add to it an invisible object, we obtain a new object which has exactly the same image. Second, and this fact is much more general than the previous one, it may be possible that two very distant objects have images that are very close. In other words, there exist very broad sets of distinct objects such that the corresponding sets of images are very small. All these properties are illustrated in Figure 1.1.

If we consider now the inverse problem, i.e. the problem of determining the object corresponding to a given image, we find that this problem is ill posed as a consequence of the loss of information intrinsic to the solution of the direct one. Indeed, if we have an image corresponding to two distinct objects, the solution of the inverse problem is not unique. If we have a noisy image, which is not in the range of the operator A, then the solution of the inverse problem does not exist. If we have two neighboring images such that the corresponding objects are very distant, then the solution of the inverse problem does not depend continuously on the data.

1.3 How to cure ill-posedness

The property of non-continuous dependence of the solution on the data applies only to ill-posed problems formulated in infinite-dimensional spaces. In practice, one has discrete data, and one has to solve discrete problems. These, however, are obtained by discretizing problems with very bad mathematical properties. What happens in these cases?

If we consider a linear inverse problem, its discrete version is a linear algebraic system, apparently a rather simple mathematical problem. Many methods exist for solving numerically such problem. However, the solution obtained is not necessarily satisfactory. A lively description of some first attempts at data inversion is given by S. Twomey in the preface of his book [290]: 'The crux of the difficulty was that numerical inversions were producing results which were physically unacceptable but were mathematically acceptable (in the sense that *had* they existed they would have given measured values identical or almost identical with what was measured)'. These results 'would be rejected as impossible or ridiculous by the recipient of the computer's answer. And yet the computer was often blamed, even though it had done all that had been asked of it'. … 'Were it possible for computers to have ulcers or neuroses there is little doubt that most of those on which early numerical inversion attempts were made would have required both afflictions'.

The explanation can be found if one has in mind the examples discussed in the previous section, where small oscillating data produce large oscillating solutions. In any inverse problem, data are always affected by noise which can be viewed as a small randomly oscillating function. Therefore the solution method amplifies the noise producing a large and wildly oscillating function that completely hides the physical solution corresponding to the noise-free data. This property holds true also for the discrete version of the ill-posed problem. Then one says that the corresponding linear algebraic system is *ill conditioned*: even if the solution exists and is unique, it is completely corrupted by a small error on the data.

In conclusion, we have the following situation: we can compute one and only one solution of our algebraic system, but this solution is unacceptable for the reasons indicated above; the physically acceptable solution we are looking for is not a solution of the problem but only an approximate solution in the sense that it does reproduce the data not exactly but only within the experimental errors. However, if we look for approximate solutions, we find that they constitute a set that is extremely broad and contains completely different functions, a consequence of the loss of information in the direct problem, as we discussed in the previous sections. Then the question arises: how can we choose the good ones?

We can state now what we consider to be the 'golden rule' for solving inverse problems that are ill posed: *search for approximate solutions satisfying additional constraints coming from the physics of the problem*. This point can be clarified in the framework of the mathematical model introduced in the previous section. The set of the approximate solutions corresponding to the same data function is just the set of the objects with images close to the measured one (the two sets are the dashed sets represented in Figure 1.1). The set of the objects is too broad, as a consequence of the loss of information due to the imaging process. Therefore we need some additional information to compensate for this loss. This information, which is also called *a priori* or *prior information*, is additional in the sense that it cannot be derived from the image or from the properties of the mapping A, which describes the imaging process, but instead expresses some expected physical properties of the object. Its role is to reduce the set of objects compatible with the given image or also to discriminate between interesting objects and spurious objects, generated by uncontrolled propagation of the noise affecting the image.

The most simple form of additional information is that the object cannot be too large. This implies a constraint that consists in an upper bound on the object itself, or on its intensity, or on its energy, etc. Another kind of additional information may be that the object is smooth so that, for instance, its derivatives must be smaller than a certain quantity. Moreover, it may be known that the object must be nonnegative or that it must be different from zero only inside a given bounded region. A quite different kind of additional information may consist of statistical properties of the objects. In this case, one assumes that the objects to be restored are realizations of a random process with a known probability distribution. This can be a way of expressing our previous experience about the possible objects to be restored. Even if complete knowledge of the probability distribution is not available, a partial knowledge of statistical properties consisting, for instance, in the knowledge of the expectation values and covariance matrices may be useful.

The idea of using prescribed bounds to produce approximate and stable solutions was introduced by C. Pucci in the case of the Cauchy problem for the Laplace equation [237], i.e. the first example of ill-posed problem discussed by Hadamard. The constraint of positivity was used by F. John in the solution of the heat equation for preceding time [171], i.e. the problem previously considered of determining the temperature distribution at the time $t = 0$ from the knowledge of the temperature distribution at the time $t = T > 0$. A general version of similar ideas was formulated independently by V. K. Ivanov [169]. His method and the method of D. L. Phillips for Fredholm integral equations of the first kind [230] were the first examples of *regularization methods* for the solution of ill-posed problems. The theory of these methods was formulated by A. N. Tikhonov one year later [283].

The principle of regularization methods is to use the additional information explicitly, at the start, to construct families of approximate solutions, i.e. of objects compatible with the given image. These methods are now one of the most powerful tools for the solution of inverse problems, another one being provided by the so-called Bayesian methods, where the additional information used is of statistical nature.

1.4   An outline of the book

The book by Tikhonov and Arsenin [284], published in 1977 (the Russian edition appeared in 1974 and the French translation in 1976), was the first book on regularization theory for ill-posed problems. Subsequently, many other books did appear on the subject, and we can only list a few of them. Some are written by mathematicians for mathematicians, as the book by Groetsch [141], which is not too technical and contains a commented list of references, and the reference book by Engl, Hanke and Neubauer [114], which provides a valuable, thorough and well-organized presentation of the most important mathematical results on regularization theory. In this vein, we can also mention the more recent books by Kirsch [180], Hanke [146] and Scherzer et al. [249]. On inverse problems in imaging, let us single out the encyclopedic references [20] and [248], as well as [72], [66] and [214]. There are also books oriented towards more specific fields such as [76], [266] and [265] on inverse problems in astronomy, [276] and [43] on inverse problems in geophysics and [156], [218] and [219] on tomography. For statistical approaches, we can refer to [175] and [278] and for the numerical aspects related to inverse problems to [299] and [149]. Nonlinear inverse problems are considered in [177], [249] and [214] and inverse problems for partial differential equations in [166].

The main purpose of our book, which is not addressed to mathematicians (but we hope that it could nevertheless be useful also to mathematicians), is to introduce the reader to the basic ideas and methods for the solution of inverse and ill-posed problems even if she/he does not master the mathematics of the functional treatment of operator equations. To this aim, we thought that it would be useful to focus on a few significant examples, such as image deconvolution and tomography, because the basic mathematical tool which is needed in these cases is common to anyone working in applied science and, in particular, in signal and image processing, namely Fourier analysis.

The book is divided into three parts: the first one is essentially dedicated to image deconvolution; the second one to more general linear inverse problems and more abstract regularization methods, including edge-preserving and sparsity-enforcing methods; and finally the third one to statistical methods, including a presentation of different noise models and a detailed discussion of the case of Poisson data.

As concerns the first part, which consists of four chapters, it is completely devoted to the problem of image deconvolution, which is a particular example of the more general problem of image reconstruction. Other – more or less equivalent – names for image deconvolution are image deblurring, image enhancement, image restoration, etc. The motivation is that in such problems, we meet all the conceptual difficulties of more general inverse problems but the treatment only requires the knowledge of Fourier analysis. Hence it is possible to introduce in a natural and easy way the most common regularization methods and to easily compute all the relevant quantities.

Accordingly, in Chapter 2, we describe several examples of space-invariant imaging systems described by convolution operators. In such a case, the information loss due to the imaging process can be described in terms of the transfer function of the system.

In Chapter 3, the ill-posedness of the problem of image deconvolution is investigated, as well as the relationship between the ill-posedness of the problem formulated in function spaces and the ill-conditioning of the corresponding discrete problem. Moreover, the notions of least-squares solution and of generalized solution are introduced.

The concept of constrained least-squares solution is used for introducing in Chapter 4 the Tikhonov regularization method. The behavior of the regularized solutions as a function of the regularization parameter is investigated and a property called *semi-convergence* is emphasized. Thanks to this property, the user knows that there must exist an optimal value of the regularization parameter, even if its determination may be difficult. Moreover, the general concept of a regularization method is illustrated by giving other examples of such methods. The concept of global point spread function is also introduced as a tool for comparing the performances of different linear regularization algorithms. Finally, a short description of various methods for the choice of the regularization parameter is given.

In Chapter 5, we discuss an interesting property of a few iterative methods converging to least-squares solutions: they can be viewed as regularization methods, the role of the regularization parameter being played by the inverse of the number of iterations. Indeed, the semi-convergence property, which holds true again, implies the existence of an optimal value of this number of iterations. With the exception of the so-called Landweber method, all methods are nonlinear and therefore they provide examples of nonlinear regularization methods for linear inverse problems. In particular, the projected Landweber method is discussed as a method that approximates the solutions of constrained least-squares problems in the case of convex constraints (positivity, etc.). In special cases, this method can provide very accurate solutions even if it is not very efficient from the computational point of view. The second part of the book, which consists of five chapters, is devoted to linear imaging systems that are not described by convolution operators but instead by linear operators that can be represented by means of a singular value decomposition. This representation is essentially a generalization of the well-known representation of a symmetric matrix in terms of its eigenvalues and eigenvectors (spectral representation).

In Chapter 6 examples of imaging systems described by linear operators other than convolutions are discussed with particular attention to tomographic systems (both for transmission and emission tomography).

Chapter 7 is devoted to the singular value decomposition (SVD). It is first derived in the case of a matrix and in the case of a semi-discrete imaging operator. The case of integral operators is also discussed and a rather simple derivation of the singular value decomposition of the Radon transform, the basic operator in tomography, is also given.

In the first four sections of Chapter 8, the inversion methods introduced in the case of image deconvolution are revisited for the case of imaging operators represented by a singular value decomposition. Since the numerical application of this technique may be difficult in the case of large images, due to the large number of singular values and singular functions to be computed, the SVD is used as an analytical tool for getting explicit expressions of the relevant quantities as in the case of convolution operators. Moreover, the method of truncated SVD is introduced in Section 8.3. In the last two sections, we first discuss the two related problems of super-resolution by data inversion and of out-of-band extrapolation, describing an old and simple iterative method for solving this problem. Next, we describe an improved version of the filtered back-projection, the basic method for CT. For these last problems, Fourier-based methods can still be used.

The next two chapters discuss methods that are still in the framework of regularization theory but require to go beyond the familiar framework of Hilbert spaces common to the methods presented in the previous chapters.

In Chapter 9, we describe the so-called total-variation method, which uses a penalty allowing the reconstruction of images with edges, hence removing the Gibbs artifacts discussed in Chapter 4. We give a simple introduction to the method and to its main properties. Since the regularizing penalty is not differentiable, we also consider a simple differentiable approximation of it, which allows for more efficient numerical methods. With several numerical simulations both for denoising and deblurring, we illustrate the efficacy of the method (as well as the new artifacts it may generate) and we compare the results to those obtained with the differentiable approximation.

In Chapter 10, we present sparsity-enforcing regularization methods which use penalties different from the usual quadratic ones. We also derive an iterative soft thresholding algorithm generalizing the Landweber scheme and allowing to compute the corresponding solutions. Several applications enforcing the sparsity of the solution on different bases are described and analyzed by means of numerical simulations.

The third part of the book is devoted to statistical (or probabilistic) methods and consists of three chapters. Only the case of discrete data is considered.

Chapter 11 contains, in a sense, the foundations of the methods. We first discuss different noise models which lead to different discrepancy functions, possibly replacing the least-squares functional related to the special case of the so-called additive Gaussian noise. Similarly we introduce the concept of priors which, in a first instance, replace the regularization functionals and we focus on the so-called Gibbs priors, which have an exponential structure, whose exponent is in general selected among usual regularization functionals. Then we give a general presentation of maximum likelihood and Bayesian methods in the context of linear inverse problems.

Chapter 12 is devoted to the case of Gaussian additive noise which, in the case of a covariance matrix which is not a multiple of the identity matrix, leads to weighted least-squares methods. Moreover, an iterative algorithm converging to nonnegative least-squares solutions is discussed.

In Chapter 13 the problem of inverse problems with Poisson data is approached by deriving from the negative logarithm of the likelihood a discrepancy function which is a divergence instead of a metric distance. It is called the generalized Kullback-Leibler (KL) divergence. The classic ML-EM method for the minimization of this discrepancy function is introduced and its properties are discussed; moreover, it is shown that the Bayesian methods lead to a penalization of the KL-divergence and simple methods of minimization are discussed. The effect of a total variation penalty is investigated by means of numerical simulations.

Finally, Chapter 14 contains some conclusions and comparisons between different methods as well as some suggestions for simulation programs.

In addition, online supplementary material is available with the book. It contains mathematical prerequisites and additional developments on some specific topics. This material is organized in two parts: Appendix A containing basic definitions and elements of functional analysis, and Appendix B containing results of numerical analysis. The purpose is to provide a tool such that the reader is not obliged to search frequently for the needed results in mathematical textbooks. In the printed book, we do not make explicit reference to this online material, which is nevertheless at the disposal of the reader as an introduction to or a refresher on some basic concepts and useful mathematical properties.

I

Image Deconvolution

2

Examples of image blurring

An image is a signal carrying information about a physical object and, in general, provides a degraded representation of the object. Two sources of degradation can be considered: the process of image formation and the process of image recording.

The degradation due to the process of image formation is usually denoted by *blurring* and is a sort of bandlimiting of the object. In the case of aerial photographs, for instance, the blurring is due to relative motion between the camera and the ground, to aberrations of the optical components of the camera and, finally, to atmospheric turbulence. All these different kinds of blurring will be described in this chapter.

The degradation introduced by the recording process is usually denoted by *noise* and is due to measurement errors, counting errors, etc.

Blurring is a deterministic process and, in most cases, one has a sufficiently accurate mathematical model for its description. On the other hand, noise is a statistical process, so that the noise affecting a particular image is not known. One can, at most, assume a knowledge of the statistical properties of the process.

In this chapter, we first discuss blurring and noise in general terms and then we provide several examples of blurring. These examples are given both to show the large number of applications of the particular image reconstruction problem known as *deconvolution* and to provide models that are useful for testing the methods discussed in this book.

2.1   Blurring and noise

We will use, for convenience, the language of optical image formation, even if some of the examples discussed in the following are not derived from optics.

In optical image formation, an unknown spatial radiance distribution, which will be called the *object* and denoted by $f^{(0)}(\mathbf{x})$ (most frequently a function of two variables, $\mathbf{x} = \{x_1, x_2\}$, which can be a 2D mapping, or projection, of a 3D scene), produces a radiance distribution in the image domain of the optical system (most frequently a plane). This distribution will

DOI: 10.1201/9781003032755-2

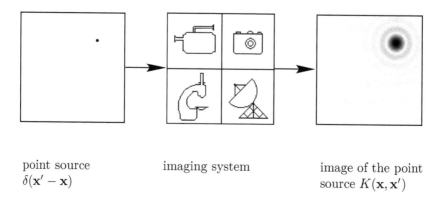

point source   imaging system   image of the point
$\delta(\mathbf{x}' - \mathbf{x})$ source $K(\mathbf{x}, \mathbf{x}')$

FIGURE 2.1   Schematic representation of the point spread function.

be called, for a reason which will become clear shortly, the *noise-free image* of $f^{(0)}(\mathbf{x})$ and denoted by $g^{(0)}(\mathbf{x})$.

A fundamental property of optical instruments, on which the description provided in this section is based, is that the image of a point, more precisely of a point source, is not a point but a small spot with a given radiance distribution. Such a phenomenon is typical of instruments that can be modeled by physical optics such as microscopes and telescopes, or of optical systems affected by aberration. An example is the human eye when affected by myopia, a big problem for many people including two of the authors of this book.

If the point source, located at a given point $\mathbf{x}$ of the object domain, has an intensity distribution given by $f^{(0)}(\mathbf{x}') = \delta(\mathbf{x}' - \mathbf{x})$ (where $\delta(\mathbf{x})$ denotes the Dirac delta distribution concentrated at the origin) then its image is given by a function of $\mathbf{x}'$ which we denote by $K(\mathbf{x}, \mathbf{x}')$. In general, this function of $\mathbf{x}'$ has a peak in $\mathbf{x}' = \mathbf{x}$, surrounded by a sort of little cloud with some structured distribution. A schematic representation of this process is given in Figure 2.1. The function $K(\mathbf{x}, \mathbf{x}')$ is called, for an obvious reason, the *point spread function* (PSF) of the imaging system. The effect of the PSF is called *blurring*. Since it represents the response to a delta distribution, the PSF is also called *impulse response* of the system.

To get the effect of blurring on the noise-free image $g^{(0)}(\mathbf{x})$, which is called a blurred version of the object $f^{(0)}(\mathbf{x})$, we can proceed as follows. Let us denote by A the imaging operator which transforms the object into the noise-free image and let us assume that it is *linear*. This assumption implies that if we have an object consisting of two point sources located in the points $\mathbf{x}_1$ and $\mathbf{x}_2$, with respective intensities f_1 and f_2, i.e. $f^{(0)}(\mathbf{x}') = f_1\,\delta(\mathbf{x}' - \mathbf{x}_1) + f_2\,\delta(\mathbf{x}' - \mathbf{x}_2)$, then its image is given by $g^{(0)}(\mathbf{x}') = f_1 K(\mathbf{x}_1, \mathbf{x}') + f_2 K(\mathbf{x}_2, \mathbf{x}')$. Therefore, if one writes a generic object $f^{(0)}(\mathbf{x})$ as a linear superposition of point sources, i.e. as

$$f^{(0)}(\mathbf{x}) = \int \delta(\mathbf{x} - \mathbf{x}')f^{(0)}(\mathbf{x}')\,d\mathbf{x}'\ , \tag{2.1}$$

then, from the previous remark, one easily derives that the corresponding blurred and noise-free image $g^{(0)}$ of $f^{(0)}$ is given by

$$g^{(0)}(\mathbf{x}) = \int K(\mathbf{x}, \mathbf{x}')f^{(0)}(\mathbf{x}')\,d\mathbf{x}'\ . \tag{2.2}$$

In principle, the PSF can be obtained by solving the so-called *direct problem* (or else *forward problem*) associated with the imaging process, namely the computation of an image associated to a given object. In the case of an optical system, for instance, this means computing

object
$f^{(0)}(\mathbf{x})$

imaging system
$K(\mathbf{x}, \mathbf{x}')$

noise-free image
$g^{(0)}(\mathbf{x})$

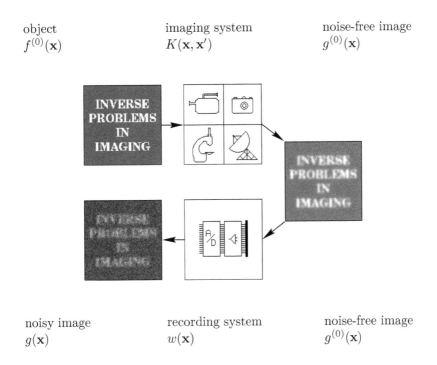

noisy image
$g(\mathbf{x})$

recording system
$w(\mathbf{x})$

noise-free image
$g^{(0)}(\mathbf{x})$

FIGURE 2.2 Schematic representation of the formation of the noisy image g.

the propagation of light from a point source in the object domain to points in the image domain through the elements of the system (lenses, mirrors, etc.). If no exact or approximate solution of the direct problem is available, the PSF could be measured by generating a point source, by moving the point source in the object domain and by recording the images produced by the system for the various positions of the source. Once the PSF is known, one can insert the result into the basic equation (2.2).

A second fundamental point is that, in order to measure the blurred image of $f^{(0)}(\mathbf{x})$, a recording system is placed in the image domain, for instance, an array of detectors. However the effect of the recording process is to corrupt the noise-free image $g^{(0)}(\mathbf{x})$ by a random process which is usually called *noise*; accordingly, the recorded image, denoted by $g(\mathbf{x})$, is called the *noisy image*. Since the process is random, its effect is unknown and, when we are lucky, we know at least its statistical properties. It is obvious that this is a quite difficult and intriguing problem in the formulation and treatment of image reconstruction. In Section 11.1, we discuss a few relevant models of noise perturbations while in Chapters 12 and 13, we treat image reconstruction respectively in the case of additive Gaussian noise and in the case of Poisson noise, the two cases which are the most relevant in the applications.

In Figure 2.2, we give a schematic representation where the two processes, imaging and detection, are clearly separated. This separation is assumed for the sake of simplicity even if, in some cases, some form of interaction exists between the two processes.

In the first two parts of the book, which are essentially based on the approach introduced by the Russian mathematician Andrei Tikhonov and known as *regularization theory* [283], we use a simplified model of noise perturbation. This model, in general, is not realistic but is certainly useful when nothing is known about the noise properties. In this model the contribution of the detection system consists in the addition of an unknown function, *independent of the noise-free image*, so that the noisy image can be written in the following form

$$g(\mathbf{x}) = \int K(\mathbf{x}, \mathbf{x}') f^{(0)}(\mathbf{x}') \, d\mathbf{x}' + w(\mathbf{x}) \tag{2.3}$$

where $w(\mathbf{x})$ represents the unknown noise term. In some instances, one further assumes that $w(\mathbf{x})$ is a bounded function with respect to some suitable norm (in general the L^2-norm – see later) and that a bound ε on this norm is known, so that

$$\|w\| \leq \varepsilon \, . \tag{2.4}$$

More precisely, as shown in Chapter 12, the usual regularization theory is based on the tacit assumption that $w(\mathbf{x})$ is the realization of a Gaussian random process with zero mean and given variance σ^2; in such a case, one should consider the expected value of the squared norm of $w(\mathbf{x})$ and remark that ε^2 is related to the variance of the noise.

It is important to remark that looking only at the value of ε in equation (2.4) is not very informative for estimating how much noise corrupts the noise-free image; ε is not necessarily a small quantity in the mathematical sense, i.e. a quantity smaller or much smaller than 1. Indeed, in several applications, it can be a large or extremely large quantity but still smaller than the signal or image. For this reason, an important quantity called *Signal-to-Noise Ratio* (SNR) is introduced in signal and image processing. A very frequent definition is given in terms of the ratio between the mean value of the image (in the case of noise with zero mean, the mean of the noise-free and that of the noisy image coincide) and the standard deviation of the noise (the square root of the variance σ^2)

$$\mathrm{SNR} = \frac{1}{N^2 \sigma} \sum_{m,n=1}^{N} g_{m,n} \, , \tag{2.5}$$

where we consider an image that has been discretized on a grid of $N \times N$ *pixels* and where $g_{m,n}$ denotes its value on the pixel labeled by m and n. Because many images have a very wide *dynamic range*, the SNR is frequently expressed using a *logarithmic decibel scale* so that, using the definition of *decibel* (dB), the SNR is expressed in dB as follows

$$(\mathrm{SNR})_{\mathrm{dB}} = 20 \log_{10} (\mathrm{SNR}) \, . \tag{2.6}$$

This definition of SNR applies to the whole image but, of course, we could have different SNR values in different regions of the image; an example is shown in Figure 2.3, where we indicate the different SNR in each of the corresponding small squares and on the remaining background. The eye being unable to see noise when the SNR is higher than 40 dB, we can verify on the figure that, since in all cases the noise is visible, the SNR is indeed smaller than 40 dB in all regions. To define the SNR value for the whole image, one can also consider the maximum value or the mean of the values computed on different regions.

As clear from this example, the local version of SNR is useful in the case of noisy versions of a piecewise constant object. In the case of natural objects where, in general, no sufficiently wide region with a constant intensity exists, a frequently used quantity in image processing is the so-called *Peak Signal-to-Noise Ratio* (PSNR). This quantity is typically used in simulations since it is based on the knowledge of the noise-free image $g^{(0)}$; indeed, it is defined by

$$\mathrm{PSNR} = 20 \log_{10} \left(\frac{\max_{m,n} g_{m,n}^{(0)}}{\sqrt{\frac{1}{N^2} \sum_{m,n=1}^{N} |g_{m,n} - g_{m,n}^{(0)}|^2}} \right) \tag{2.7}$$

the denominator in equation (2.7) being the mean squared error (MSE) between the noise-free image and the noisy one. As the SNR, the PSNR is also usually expressed in decibel.

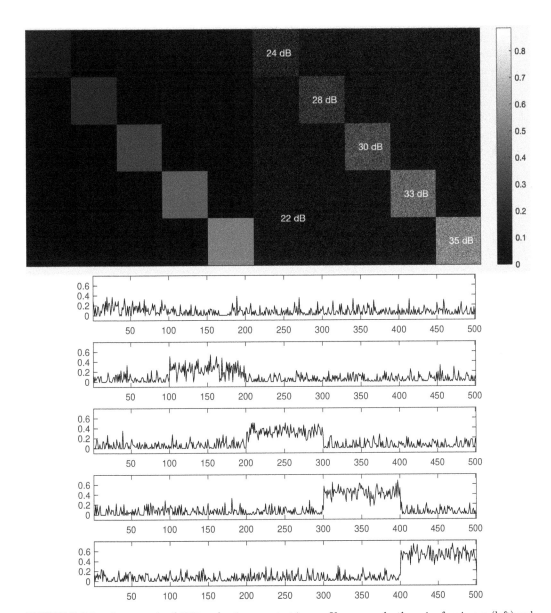

FIGURE 2.3 An example of SNR evaluation on a test image. Upper panels: the noise-free image (left) and the noisy image (right) corrupted by additive Gaussian noise and annotated with the local SNR computed by means of equation (2.6). Lower panel: plots of the noisy image along horizontal lines through the squares.

Let us now refine our previous analysis of the imaging process and focus on the main topic of this first part of the book, namely the study of *imaging systems which are invariant with respect to translations*. This specification means that the PSF appearing in equation (2.2) satisfies the following condition: the image $K(\mathbf{x}, \mathbf{x}_0)$ of a point source located at $\mathbf{x}_0$ is the translated by $\mathbf{x}_0$ of the image $K(\mathbf{x}, \mathbf{0})$ of a point source located at the origin of the object plane, i.e. $K(\mathbf{x}, \mathbf{x}_0) = K(\mathbf{x} - \mathbf{x}_0, \mathbf{0})$. It follows that $K(\mathbf{x}, \mathbf{x}')$ is a function of the difference $\mathbf{x} - \mathbf{x}'$ and we can write $K(\mathbf{x} - \mathbf{x}')$ instead of $K(\mathbf{x}, \mathbf{x}')$.

Therefore such an imaging system, which in optics is also called *isoplanatic*, is characterized by a function that depends on only one space variable $\mathbf{x}$, is denoted by $K(\mathbf{x})$ and is also called PSF, more precisely *space-invariant* PSF. For brevity, in the following, it will be simply called PSF. If the imaging system is not space-invariant, its PSF is said to be *space-variant*; this case is briefly discussed in Section 6.1.

A space-invariant PSF can be determined by detecting the image of a single point source, for instance, located in the center of the object domain. To simplify the analysis, it is assumed that this PSF as well as the object and the image are defined everywhere on $\mathbb{R}^d$, $d = 1, 2, 3$. In addition, $K(\mathbf{x})$ is assumed to satisfy the normalization condition

$$\int_{\mathbb{R}^d} K(\mathbf{x}) \, d\mathbf{x} = 1 \ . \tag{2.8}$$

To summarize, in the case of a space-invariant system equation (2.2) becomes

$$g^{(0)}(\mathbf{x}) = \int_{\mathbb{R}^d} K(\mathbf{x} - \mathbf{x}') f^{(0)}(\mathbf{x}') \, d\mathbf{x}' \tag{2.9}$$

and therefore the noise-free image is the so-called *convolution product* of the PSF $K(\mathbf{x})$ and the object $f^{(0)}$, which we write $g^{(0)} = K * f^{(0)}$. Remark that condition (2.8) implies that the object and the noise-free image have the same integral:

$$\int_{\mathbb{R}^d} g^{(0)}(\mathbf{x}) \, d\mathbf{x} = \int_{\mathbb{R}^d} f^{(0)}(\mathbf{x}) \, d\mathbf{x} \ . \tag{2.10}$$

A remarkable property is that the convolution product is diagonalized by the Fourier Transform (FT). Therefore the treatment of space-invariant systems is greatly simplified by the use of this mathematical tool. In the next section, we briefly summarize its main properties.

2.2   The Fourier Transform

The Fourier Transform (FT) is described in so many textbooks that we just recall here the basic definitions and properties we need in the following. For a review of the fundamental mathematical concepts, see e.g. the books by Papoulis [226] and Bracewell [50].

The FT of a function $f(\mathbf{x})$ depending, in general, on a d-dimensional variable $\mathbf{x} = \{x_1, x_2, \ldots, x_d\}$ (in imaging $d = 1, 2, 3$) is defined as follows

$$\hat{f}(\boldsymbol{\omega}) = \int_{\mathbb{R}^d} e^{-i\boldsymbol{\omega} \cdot \mathbf{x}} f(\mathbf{x}) \, d\mathbf{x} \ , \tag{2.11}$$

where $\boldsymbol{\omega} \cdot \mathbf{x} = \omega_1 x_1 + \omega_2 x_2 + \cdots + \omega_d x_d$ denotes the Euclidean scalar product in $\mathbb{R}^d$ and $d\mathbf{x} = dx_1 dx_2 \ldots dx_d$. If the components of the vector $\mathbf{x}$ have the physical meaning of space variables, then the components of the vector $\boldsymbol{\omega} = \{\omega_1, \omega_2, \ldots, \omega_d\}$ are called *spatial frequencies*. The set of all vectors $\boldsymbol{\omega}$ is also called the *frequency domain* or *Fourier domain*.

The above integral has, of course, to make sense. We will consider two classes of functions where the FT is well defined: (Lebesgue) integrable and square-integrable functions.

A function f is said to be *(absolutely) integrable* if

$$\int_{\mathbb{R}^d} |f(\mathbf{x})| \, d\mathbf{x} \; < \; +\infty \tag{2.12}$$

where the integral is intended in the sense of Lebesgue.

The Fourier transform of an integrable function has the following properties: the function $\hat{f}(\boldsymbol{\omega})$ is continuous and bounded

$$|\hat{f}(\boldsymbol{\omega})| \le \int_{\mathbb{R}^d} |f(\mathbf{x})| \, d\mathbf{x} \; ; \tag{2.13}$$

moreover $|\hat{f}(\boldsymbol{\omega})|$ tends to zero when $|\boldsymbol{\omega}| \to \infty$. The latter property is usually referred to as the *Riemann-Lebesgue theorem* (or *lemma*) [226].

Unfortunately, the FT of an integrable function need not be integrable and hence its FT is not necessarily well defined. However, when it does so, one can establish the following inversion formula

$$f(\mathbf{x}) = \frac{1}{(2\pi)^d} \int_{\mathbb{R}^d} e^{i\mathbf{x}\cdot\boldsymbol{\omega}} \, \hat{f}(\boldsymbol{\omega}) \, d\boldsymbol{\omega} \; . \tag{2.14}$$

REMARK 2.1 *As said, even if $|\hat{f}(\boldsymbol{\omega})| \to 0$ at infinity, the* FT *of an integrable function is not necessarily integrable. Examples of integrable functions whose* FT *is not integrable are provided by the characteristic functions. The characteristic function of a set $\mathcal{D}$, denoted by $\chi_{\mathcal{D}}(\mathbf{x})$, is defined as the function which is equal 1 over the set and to zero elsewhere. In the case of one variable, if $\mathcal{D}$ is the interval $[-\pi, \pi]$, then*

$$\hat{\chi}_{\mathcal{D}}(\omega) = 2\pi \, \mathrm{sinc}(\omega) \; , \tag{2.15}$$

the 'cardinal sine' function $\mathrm{sinc}(\xi)$ being defined by

$$\mathrm{sinc}(\xi) = \frac{\sin(\pi\xi)}{\pi\xi} \; . \tag{2.16}$$

The function $\mathrm{sinc}(\xi)$ is not integrable because the integral of its modulus does not converge. Analogously, in two variables, if $\chi_{\mathcal{D}}(\mathbf{x})$ is the characteristic function of the disc of radius π, then its Fourier transform is given by

$$\hat{\chi}_{\mathcal{D}}(\boldsymbol{\omega}) = 2\pi^2 \, \frac{J_1(\pi|\boldsymbol{\omega}|)}{|\boldsymbol{\omega}|} \tag{2.17}$$

where $J_1(\xi)$ is the Bessel function of order 1. Also in this case, the Fourier transform is not integrable.

However, the characteristic functions considered in this remark are square integrable and their FT's have the same property. Indeed, a function f is said to be *square integrable* if

$$\int_{\mathbb{R}^d} |f(\mathbf{x})|^2 d\mathbf{x} \; < \infty \; . \tag{2.18}$$

For these functions the FT, in general, is not defined by means of the integral (2.11). A rigorous theory requires Lebesgue's theory of measure and integration. Roughly speaking, one defines the FT by restricting the integral (2.11) to the ball of radius R and then by taking the limit when $R \to \infty$. It can be proved that the FT defined in such a way is also square integrable and that the inversion formula (2.14) holds true. Moreover, we have the so-called *generalized Parseval equality*

$$\int_{\mathbb{R}^d} f(\mathbf{x})\, h^*(\mathbf{x})\, d\mathbf{x} = \frac{1}{(2\pi)^d} \int_{\mathbb{R}^d} \hat{f}(\boldsymbol{\omega})\, \hat{h}^*(\boldsymbol{\omega})\, d\boldsymbol{\omega} \ , \qquad (2.19)$$

where * denotes complex conjugation. It reduces for $f = h$ to the *Parseval equality*

$$\int_{\mathbb{R}^d} |f(\mathbf{x})|^2\, d\mathbf{x} = \frac{1}{(2\pi)^d} \int_{\mathbb{R}^d} |\hat{f}(\boldsymbol{\omega})|^2\, d\boldsymbol{\omega} \ . \qquad (2.20)$$

The space of all square-integrable functions on $\mathbb{R}^d$, the so-called L^2-*space* or $L^2(\mathbb{R}^d)$, has a nice geometrical structure. We do not give here a precise definition of this space because this also requires Lebesgue's theory of measure and integration.

The L^2-space is linear because a linear combination of square-integrable functions is still a square-integrable function. Moreover, given two functions f and h of this space, one defines their *scalar product* – or else *inner product* – as follows

$$(f, h) = \int_{\mathbb{R}^d} f(\mathbf{x})\, h^*(\mathbf{x})\, d\mathbf{x} \ . \qquad (2.21)$$

A linear space equipped with a scalar (inner) product is called an *inner product space* or a *pre-Hilbert space*. Then two functions f and h are said to be *orthogonal* if their scalar product is zero. Notice that from equation (2.19) it follows that, if two functions f and h are orthogonal, then also their FT $\hat{f}$ and $\hat{h}$ are orthogonal. It also follows that, up to normalization, the FT defines in $L^2(\mathbb{R}^d)$ a so-called *unitary transform* since it preserves the scalar product. To each function f one can then also associate a *norm* defined as follows

$$\|f\| = \left(\int_{\mathbb{R}^d} |f(\mathbf{x})|^2\, d\mathbf{x} \right)^{1/2} \qquad (2.22)$$

which is related to the scalar product (2.21) by $\|f\| = (f, f)^{1/2}$. A function f is said to be *normalized* if $\|f\| = 1$. Unless specified otherwise, in the following, the notation $\|f\|$ for a function f will refer to its L^2-norm. A set of functions $\{f_1, f_2, \ldots, f_n\}$ is said to be an *orthonormal set* if they satisfy the conditions

$$(f_j, f_k) = \delta_{j,k} \qquad (2.23)$$

where $\delta_{j,k}$ is the *Kronecker symbol*, equal to one when $j = k$ and equal to zero when $j \neq k$. It is also possible to introduce the *distance* of f from h, defined as follows

$$d(f, h) = \|f - h\| \ . \qquad (2.24)$$

This distance is the *mean square error* we commit if we take h as an approximation of f. The ball of center f and radius ε is the set of all functions h satisfying the condition $\|f - h\| \leq \varepsilon$ and therefore is the set of all functions which approximate f with a mean square error that is not greater than ε. We still recall two very useful inequalities which will be often used in the following: the *Schwarz* or *Cauchy-Schwarz inequality* for scalar products

$$|(f, h)| \leq \|f\|\, \|h\| \qquad (2.25)$$

and the *triangular inequality* for norms

$$\|f + h\| \leq \|f\| + \|h\| \ . \qquad (2.26)$$

An inner product space which in addition is complete is called a *Hilbert space*. It means that all Cauchy sequences of elements of the space have a limit that also belongs to the space. *This is the case of the space L^2 which, therefore, is a Hilbert space; it is also separable, a property which implies that there exists in the space a complete and countable orthonormal set.*

Let us now recall the well-known *convolution theorem*, which states that the FT of the convolution product of two functions is given by the product of their FT's, namely

$$(\widehat{f * h})(\boldsymbol{\omega}) = \hat{f}(\boldsymbol{\omega})\,\hat{h}(\boldsymbol{\omega})\,, \tag{2.27}$$

the convolution product being defined as in equation (2.9) by

$$(f * h)(\mathbf{x}) = \int_{\mathbb{R}^d} f(\mathbf{x} - \mathbf{x}')\,h(\mathbf{x}')\,d\mathbf{x}' \tag{2.28}$$

provided the integral exists. We will not state here the general mathematical conditions under which the convolution theorem holds. For our purposes it is sufficient to know that it holds true e.g. when $\hat{f}(\boldsymbol{\omega})$ is bounded and $\hat{h}(\boldsymbol{\omega})$ is square integrable. In such a case $(\widehat{f * h})(\boldsymbol{\omega})$ is square integrable and therefore the function $f * h$ is also square integrable.

The FT of the *point spread function $K(\mathbf{x})$* (PSF), i.e. $\hat{K}(\boldsymbol{\omega})$, is called the *transfer function* (TF) of the imaging system. Notice that the normalization condition (2.8) of the PSF implies $\hat{K}(0) = 1$. Moreover, the integrability of the PSF which we assume implies that the TF is a bounded and continuous function of $\boldsymbol{\omega}$ and that $|\hat{K}(\boldsymbol{\omega})| \to 0$ if $|\boldsymbol{\omega}| \to \infty$, by the Riemann-Lebesgue theorem. This property plays an important role in generating difficulties in the problem of image reconstruction which, from now on, will be called *image deconvolution*.

The two functions PSF and TF are essential in the description of the behavior of the system. The PSF provides the response of the system to any point source wherever it is located. On the other hand, the TF tells us how a signal with a fixed frequency is propagated through the linear system, so that the blurring can be viewed as a sort of frequency filtering. Indeed, in terms of the Fourier transforms, thanks to the convolution theorem, the FT of equation (2.9) with the addition of the noise contribution becomes

$$\hat{g}(\boldsymbol{\omega}) = \hat{K}(\boldsymbol{\omega})\,\hat{f}^{(0)}(\boldsymbol{\omega}) + \hat{w}(\boldsymbol{\omega})\,. \tag{2.29}$$

In several important cases, a space-invariant imaging system is characterized by a *bandlimited* PSF; this means that the support of $\hat{K}(\boldsymbol{\omega})$ is bounded. It will be denoted by $\mathcal{B}$ and is called the *band of the* PSF. The band of the PSF is also called the *band of the imaging system*. In the following, we give examples of blurring which can be modeled by means of a convolution integral as in equation (2.9). Some of them are examples of bandlimited blurring. In other cases, the TF tends to zero for large spatial frequencies so that the corresponding imaging systems are approximately bandlimited.

Let us recall that a basic property of a bandlimited function is that it can be represented, without any loss of information, by means of its values (samples) taken on a countable set of points. For simplicity, we give the result only for the case of functions of one variable, in which case this property is expressed by the well-known *sampling theorem* or *Whittaker-Shannon theorem* [170].

Let $f(x)$ be a bandlimited and square-integrable function of the space variable x, with a band interior to the interval $[-\Omega, \Omega]$. In this case, we call Ω the *bandwidth* of f. Then, according to the sampling theorem, f can be represented as follows

$$f(x) = \sum_{n=-\infty}^{+\infty} f\left(n\frac{\pi}{\Omega}\right)\,\text{sinc}\left[\frac{\Omega}{\pi}\left(x - n\frac{\pi}{\Omega}\right)\right] \tag{2.30}$$

the 'sinc' function being defined by equation (2.16).

The points $x_n = n\pi/\Omega$ are called the *sampling points* and the distance π/Ω is called the *sampling distance* or else the *Nyquist distance*. We stress the fact that the sampling distance is inversely proportional to the bandwidth of f. Moreover, thanks to the symmetry between the direct and inverse Fourier transforms, a similar formula applies to the Fourier transform of a space-limited function, i.e. a function $f(x)$ whose support is interior to some interval $[-X, X]$. Obviously, in such a case, $f(x)$ must be replaced by $\hat{f}(\omega)$ and Ω by X in equation (2.30).

2.3   The Discrete Fourier Transform

We now want to discuss the discretization of equation (2.9). Since this equation can be easily treated by means of the FT, it is quite natural to look for a discrete equation which can be treated by means of the so-called *Discrete Fourier Transform* (DFT), with the advantage that the celebrated *Fast Fourier Transform* (FFT) algorithm can be used for its fast computation. Therefore, we specify here the notation we use for this important transform and, for simplicity, we give the definition and main properties only in the 1D case.

Let $\mathbf{f} = \{f_0, f_1, \ldots, f_{N-1}\}$ be a N-dimensional vector. Then its DFT is the N-dimensional vector $\hat{\mathbf{f}} = \{\hat{f}_0, \hat{f}_1, \ldots, \hat{f}_{N-1}\}$ defined by

$$\hat{f}_m = \sum_{n=0}^{N-1} f_n \exp\left(-i\frac{2\pi}{N}mn\right) . \tag{2.31}$$

The DFT has an inversion formula analogous to that of the FT:

$$f_n = \frac{1}{N}\sum_{m=0}^{N-1} \hat{f}_m \exp\left(i\frac{2\pi}{N}mn\right) . \tag{2.32}$$

We recall that, when necessary, one can choose N as a power of 2, $N = 2^p$, because in such a case one can use the fastest version of the FFT algorithm for the computation of the DFT and of the inverse DFT.

Moreover, the following *generalized Parseval equality* holds true

$$\sum_{n=0}^{N-1} f_n\, h_n^* = \frac{1}{N}\sum_{m=0}^{N-1} \hat{f}_m\, \hat{h}_m^* \tag{2.33}$$

which reduces to

$$\sum_{n=0}^{N-1} |f_n|^2 = \frac{1}{N}\sum_{m=0}^{N-1} |\hat{f}_m|^2 . \tag{2.34}$$

for $\mathbf{h} = \mathbf{f}$.

If we consider now the set of all N-dimensional vectors, it is a linear space where we can introduce the canonical Hermitian scalar product

$$\mathbf{f} \cdot \mathbf{h} = \sum_{n=0}^{N-1} f_n\, h_n^* \tag{2.35}$$

with the associated norm of the vector $\|\mathbf{f}\| = (\mathbf{f}\cdot\mathbf{f})^{1/2}$, i.e.

$$\|\mathbf{f}\| = \left(\sum_{n=0}^{N-1} |f_n|^2\right)^{1/2}, \tag{2.36}$$

and the associated distance of a vector $\mathbf{f}$ from a vector $\mathbf{h}$

$$d(\mathbf{f}, \mathbf{h}) = \|\mathbf{f} - \mathbf{h}\| . \tag{2.37}$$

From equation (2.33) we see that, if two vectors are orthogonal, i.e. if $\mathbf{f} \cdot \mathbf{h} = 0$, then also the corresponding DFT's are orthogonal: $\hat{\mathbf{f}} \cdot \hat{\mathbf{h}} = 0$. It also follows that, up to normalization, the DFT defines a unitary transform.

Let us stress the fact that, if we consider the vectors as elements of a N-dimensional vector space equipped with this canonical scalar product, then the DFT provides the decomposition of a vector with respect to an orthonormal basis. Indeed, let us introduce the vectors $\mathbf{v}_0, \mathbf{v}_1, \ldots, \mathbf{v}_{N-1}$ defined as follows, $(\mathbf{v}_m)_n$ denoting the n-th component of the vector $\mathbf{v}_m$:

$$(\mathbf{v}_m)_n = \frac{1}{\sqrt{N}} \exp\left(i\frac{2\pi}{N}mn\right) , \quad m = 0, \ldots, N-1 , \tag{2.38}$$

which satisfy the following orthonormality conditions

$$\mathbf{v}_m \cdot \mathbf{v}_l = \delta_{m,l} \tag{2.39}$$

as follows from the well-known fact that the sum of all the N-th roots of unity is zero. Then equations (2.31) and (2.32) can be written as follows

$$\hat{f}_m = \sqrt{N} \, \mathbf{f} \cdot \mathbf{v}_m , \quad \mathbf{f} = \frac{1}{\sqrt{N}} \sum_{m=0}^{N-1} \hat{f}_m \mathbf{v}_m . \tag{2.40}$$

Hence we see that the values of the DFT of $\mathbf{f}$ are (except for the factor $\sqrt{N}$) the components of $\mathbf{f}$ with respect to the orthonormal basis $\mathbf{v}_0, \mathbf{v}_1, \ldots, \mathbf{v}_{N-1}$. The inversion formula is simply the representation of $\mathbf{f}$ in terms of these components.

In the theory of DFT, one also considers in many instances the periodic sequence with period N associated with the vector $\mathbf{h}$. Finally, if we consider two vectors of length N, $\mathbf{K}$ and $\mathbf{f}$, with the corresponding periodic sequences of period N, then their *cyclic discrete convolution product* is the vector $\mathbf{g}$ of length N, with the corresponding sequence of period N, defined by

$$g_m = \sum_{n=0}^{N-1} K_{m-n} \, f_n . \tag{2.41}$$

Indeed, it is easy to verify, using the periodicity of K_n, that this equation, considered for any m, defines a periodic sequence. We rewrite this equation as

$$\mathbf{g} = \mathbf{K} * \mathbf{f} \tag{2.42}$$

using a notation similar to the one used for the convolution product (2.27). By taking the DFT of both sides of equation (2.41), one can easily prove the following *cyclic discrete convolution theorem*

$$\hat{g}_m = \hat{K}_m \, \hat{f}_m . \tag{2.43}$$

The extension of the previous results to the 2D and 3D cases is straightforward and left to the reader. A complete discussion of the properties of the DFT can be found in the book [54], which uses a different definition. This alternative definition, however, is useful for understanding the relationship between the DFT, as defined above, and the discretization of the Fourier transform.

Consider again a function of one variable $f(x)$ and assume that it is essentially different from zero inside an interval $[-X, X)$. We subdivide this interval into N subintervals of length $\delta_x =$

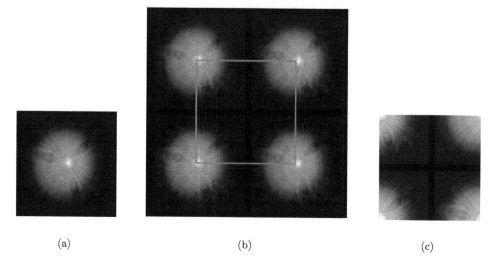

(a)   (b)   (c)

FIGURE 2.4   Relationship between the discretized PSF (60×60) and the shifted PSF: (a) discretisation of the PSF (60×60); (b) piece (120×120) of the translated periodic continuation of the discretized PSF (the square indicates the region of the points corresponding to the shifted array); (c) shifted PSF (60×60), corresponding to the square in (b), to be used in the discrete equation (2.49).

$2X/N$. These intervals are delimited by the points $x_n = n\delta_x$ with $n = -N/2, ..., 0, ..., N/2-1$ and we take the values of the function in the points x_n for approximating the Fourier integral, restricted to $[-X, X)$, by means of a discrete sum. We obtain the following N-point approximation of the Fourier transform of f

$$\hat{f}_N(\omega) = \sum_{n=-N/2}^{N/2-1} f(x_n)\, e^{-i\omega x_n}\, \delta_x \ . \tag{2.44}$$

Now, we want to compute this FT in N points by means of the DFT. To this purpose we consider an interval $[-\Omega, \Omega)$ and, as above, we subdivide the interval into N subintervals of length $\delta_\omega = 2\Omega/N$ and we look for the values of $\hat{f}_N(\omega)$ in the points $\omega_m = m\delta_\omega$ with $m = -N/2, ..., 0, ..., N/2 - 1$. In principle, this approach could be used for a discrete approximation of the inverse Fourier transform.

The important point is that, if we intend to use the DFT for these computations, then the value of Ω is not arbitrary but is fixed. Indeed, if we set ω_m in place of ω in the exponent of equation (2.44), we find $\omega_m x_n = \delta_\omega \delta_x mn$ and therefore to get the usual factor of the DFT the following condition has to be satisfied

$$\delta_x \delta_\omega = \frac{2\pi}{N} \quad \rightarrow \quad \delta_x = \frac{\pi}{\Omega} \ , \quad \delta_\omega = \frac{\pi}{X} \tag{2.45}$$

i.e. we obtain a relationship between X, N and Ω, which implies that the lengths of the subintervals are precisely the sampling distances suggested by the sampling theorem. In other words, we can choose only two of the three quantities X, N, Ω.

If we set $f_n = f(x_n)\delta_x$ and $\hat{f}_m = \hat{f}_N(\omega_m)$, the final result is

$$\hat{f}_m = \sum_{n=-N/2}^{N/2-1} f_n \exp\left(-i\frac{2\pi}{N}mn\right) \ . \tag{2.46}$$

This equation is very similar to equation (2.31) except for the index range. Therefore, for computing equation (2.46) by means of the DFT, we must consider a periodic continuation of the vector appearing in this equation and perform a shift of the components as follows

$$f_n^{(S)} = f_n \ , \quad n = 0, 1, ..., \frac{N}{2} \ ; \quad f_n^{(S)} = f_{N-n} \ , \quad n = \frac{N}{2} + 1, ..., N - 1 \ . \qquad (2.47)$$

The extension to the 2D and 3D cases is obvious. In the 2D case, the result of the shift procedure is illustrated in Figure 2.4.

Let us consider now the discretization of equation (2.9). It contains three functions, $g^{(0)}, K$ and $f^{(0)}$ and we consider the 2D case by assuming that the three functions are essentially zero outside the same square. The three functions can be sampled and replaced by $N \times N$ arrays of their values taken at the points of a regular lattice. These arrays will be denoted respectively by $g_{m,n}^{(0)}$, $f_{k,l}^{(0)}$ and $K_{i,j}$. If we proceed as above, the indices run from $-N/2$ to $N/2 - 1$; therefore, the convolution integral is replaced by the following discrete convolution

$$g_{m,n}^{(0)} = \sum_{k,l=-N/2}^{N/2-1} K_{m-k,n-l} \ f_{k,l}^{(0)} \ . \qquad (2.48)$$

For computing correctly this convolution product by means of the DFT, we should apply the shift operator to the two arrays in the r.h.s. However, the important point is to apply the shift to the PSF because, after multiplication of the two DFT we must perform an inverse DFT. Indeed, if we apply a direct and inverse DFT to an array, we always re-obtain the original one and therefore it is not necessary to apply the shift to $\mathbf{f}^{(0)}$. In conclusion, for computing correctly the r.h.s. of equation (2.48), we must compute by means of the DFT the following convolution product

$$g_{m,n}^{(0)} = \sum_{k,l=0}^{N-1} K_{m-k,n-l}^{(S)} \ f_{k,l}^{(0)} \ . \qquad (2.49)$$

Obviously, if we are interested in the correct and sampled Fourier transform of the r.h.s., then the shift operator must be applied also to the array of the sampled $f^{(0)}$.

2.4   Linear motion blur

The simplest example of blurring is that due to relative motion, during exposure, between the camera and the object being photographed. A simple case is that of a visible object moving across a uniform background (corresponding, for instance, to photographic density zero) in such a way that no part of the object goes out of the image domain during the motion. Examples arise, for instance, when photographing the Earth, Moon and planets by aerial vehicles such as aircrafts and spacecrafts. Examples can also be found in forensic science when potentially useful photographs are sometimes of extremely bad quality due to motion blur.

Let $f^{(0)}(\mathbf{x})$ be the illuminance which would result in the image plane of the camera in the absence of relative motion between the camera and the object (here we assume that the blurring due to the camera is negligible). Nonlinear behavior of the photographic material is neglected so that we can assume that it reacts to successive exposures just by linearly adding their effects. The relative motion can be a translation or a rotation or even a more general one. The blurring is space-invariant only if the object and image planes are parallel and if the motion is a translation.

If we assume that these conditions are satisfied, then the translation motion is described by a time-dependent vector

$$\mathbf{a}(t) = \{a_1(t), a_2(t)\}, \quad 0 \le t \le T , \tag{2.50}$$

where T is the exposure interval and $\mathbf{a}(0) = 0$. This vector defines the path, with respect to a fixed coordinated system in the image plane. If $f_0(\mathbf{x})$ is the object at time $t = 0$, the object at time t is $f_0[\mathbf{x} - \mathbf{a}(t)]$; therefore, from the linearity of the imaging system, at time T the image is given by

$$g^{(0)}(\mathbf{x}) = \frac{1}{T} \int_0^T f^{(0)}(\mathbf{x} - \mathbf{a}(t)) \, dt \tag{2.51}$$

(the factor $1/T$ is introduced as a normalization factor).

This relationship can be written in the form of a convolution product by introducing the PSF

$$K(\mathbf{x}) = \frac{1}{T} \int_0^T \delta(\mathbf{x} - \mathbf{a}(t)) \, dt , \tag{2.52}$$

where $\delta(\mathbf{x})$ is the 2D delta distribution. Then, by means of straightforward computations, one finds that the TF is given by

$$\hat{K}(\boldsymbol{\omega}) = \frac{1}{T} \int_0^T e^{-i\boldsymbol{\omega} \cdot \mathbf{a}(t)} \, dt . \tag{2.53}$$

A very simple form of the TF is obtained in the case of linear motion, i.e.

$$\mathbf{a}(t) = t\mathbf{c} , \tag{2.54}$$

where $\mathbf{c}$ is the constant velocity of the motion. In this case, the integral (2.53) can be computed and the result is

$$\hat{K}(\boldsymbol{\omega}) = \exp\left(-i\frac{T}{2}\mathbf{c} \cdot \boldsymbol{\omega}\right) \mathrm{sinc}\left(\frac{T}{2\pi}\mathbf{c} \cdot \boldsymbol{\omega}\right) , \tag{2.55}$$

the sinc function being defined in equation (2.16). $\hat{K}(\boldsymbol{\omega})$ can also be written in terms of the displacement vector $\mathbf{s} = T\mathbf{c}$ as follows

$$\hat{K}(\boldsymbol{\omega}) = \exp\left(-\frac{i}{2}\mathbf{s} \cdot \boldsymbol{\omega}\right) \mathrm{sinc}\left(\frac{1}{2\pi}\mathbf{s} \cdot \boldsymbol{\omega}\right) . \tag{2.56}$$

As seen from equation (2.56) the TF is zero on the family of parallel straight lines

$$\mathbf{s} \cdot \boldsymbol{\omega} = 2\pi n , \quad n = \pm 1, \pm 2, \ldots \tag{2.57}$$

which are orthogonal to the vector $\mathbf{s}$. The FT of the noise-free image $g^{(0)}(\mathbf{x})$ is also zero on these lines, because we have $\hat{g}^{(0)}(\boldsymbol{\omega}) = \hat{K}(\boldsymbol{\omega})\hat{f}^{(0)}(\boldsymbol{\omega})$. However, this is not true for the FT of the noisy image $\hat{g}(\boldsymbol{\omega})$, as given by equation (2.29). The noise perturbs the lines of zeroes but, if it is not too large, the perturbed lines remain close to the unperturbed ones.

In Figure 2.5, we give an example of an image blurred by a horizontal linear motion and contaminated by the addition of white Gaussian noise. As we see, in this case, the lines of zeroes of $\hat{g}(\boldsymbol{\omega})$ are rather close to straight lines, so that, if the displacement vector $\mathbf{s}$ is not known, these lines of zeroes can provide a reasonable way for estimating the components of $\mathbf{s}$. The problem of fitting straight lines to these lines of zeroes is greatly simplified by the knowledge that the lines are parallel and uniformly separated. Once the fitted straight lines are drawn, the components s_1, s_2 of $\mathbf{s}$ can be easily found.

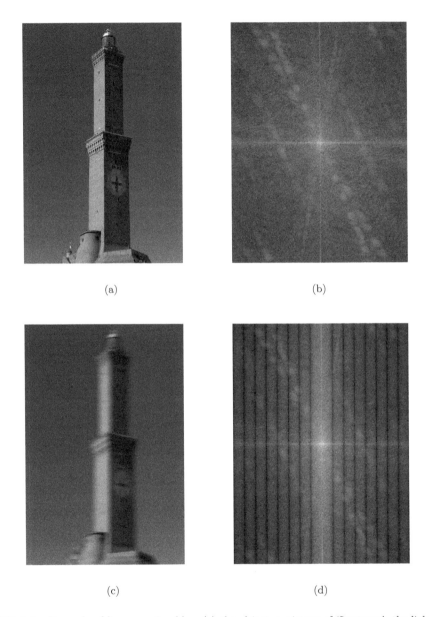

(a)   (b)

(c)   (d)

FIGURE 2.5   Example of linear motion blur: (a) the object: a picture of 'Lanterna', the lighthouse of the Genoa port; (b) the modulus of the FT of the object; (c) the blurred and noisy image; (d) the modulus of the FT of (c).

We briefly comment Figure 2.5 where we consider a case of a horizontal motion. This simplification is not restrictive because, by means of a rotation of the photograph, an arbitrary linear motion can always be reduced to this one. In such a case, the blurring equation (2.51) can be written as a 1D convolution for each horizontal line

$$g^{(0)}(x_1, x_2) = \int_{-\infty}^{\infty} K(x_1 - x_1') f^{(0)}(x_1', x_2) \, dx_1' \ , \tag{2.58}$$

the 1D PSF being given by $(s = cT)$

$$K(x) = \begin{cases} 1/s, & 0 \leq x \leq s \\ 0, & \text{elsewhere} \ . \end{cases} \tag{2.59}$$

The corresponding discrete PSF is given by

$$K_n = \begin{cases} 1/S, & 0 \leq n \leq S - 1 \\ 0, & S < n \leq N - 1 \end{cases} \tag{2.60}$$

where S is the number of pixels corresponding to the displacement distance s, i.e. $S = s/\delta$ if δ is the sampling distance. The DFT of K_n can be easily computed and is given by

$$\hat{K}_m = \frac{1}{S} \exp\left[-i\frac{\pi}{N}(S-1)m\right] \frac{\sin\left(\frac{\pi S}{N}m\right)}{\sin\left(\frac{\pi}{N}m\right)} \ . \tag{2.61}$$

When the ratio N/S is an integer, then $\hat{K}_m$ is zero for $m = Nk/S$ with $k = 1, 2, ..., S - 1$. When the ratio N/S is not an integer, then $\hat{K}_m$ is never zero but can be small for the values of m close to Nk/S.

2.5   Out-of-focus blur

In the case of a lens, an object point is in focus if its distance from the lens, d_0, satisfies the lens conjugation law

$$\frac{1}{d_0} + \frac{1}{d_i} = \frac{1}{d_f} \tag{2.62}$$

where d_i is the distance between the lens and the image plane and d_f is the focal length of the lens. If this condition is satisfied then, according to geometrical optics, the image of a point is again a point (the case where diffraction effects are important is discussed in the next section). If d_0 does not satisfy condition (2.62), then the image of the point, always according to geometrical optics, is a disc, called the *circle of confusion* (COC), whose radius can be computed as a function of d_0, d_i, d_f and of the effective lens diameter [234]. Moreover, as concerns the intensity distribution within the COC, it follows from geometrical optics that it is approximately uniform over the COC, so that the image of an out-of-focus point, located on the optical axis, is given by

$$K(\mathbf{x}) = \frac{1}{\pi D^2} \chi_{\text{coc}}(\mathbf{x}) \tag{2.63}$$

where D is the radius of the COC and $\chi_{\text{coc}}(\mathbf{x})$ its characteristic function. A more accurate computation of $K(\mathbf{x})$ [271] includes the effects of diffraction (discussed in the next section), which are neglected in the present simplified treatment.

When a 3D scene is imaged by a camera some of the points are in focus and other are out-of-focus, so that the blurring of the image is, in general, space-variant. Only in the

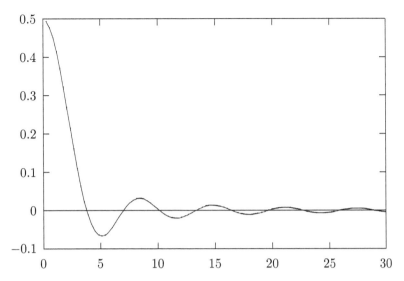

FIGURE 2.6 Plot of the function $J_1(\xi)/\xi$.

following case one has a space-invariant blurring: an out-of-focus flat object whose plane is parallel to the image plane. In such a case, the PSF is given by equation (2.63) and therefore the TF can be expressed in terms of the Bessel function of order 1, $J_1(\xi)$

$$\hat{K}(\boldsymbol{\omega}) = 2 \, \frac{J_1(D|\boldsymbol{\omega}|)}{D|\boldsymbol{\omega}|} \, . \tag{2.64}$$

The function $J_1(\xi)/\xi$ is plotted in Figure 2.6. This function has an infinity of zeroes, $\xi_n = \pi x_n$, $n = 1, 2, 3, ...$, which, for large n, are approximately equispaced, with spacing π. We also recall that the asymptotic behavior of $J_1(\xi)$ for large ξ is given by

$$J_1(\xi) \simeq \sqrt{\frac{2}{\pi \xi}} \cos\left(\xi - \frac{3}{4}\pi\right), \quad \xi \to \infty \, , \tag{2.65}$$

while its asymptotic behavior for small ξ is given by

$$J_1(\xi) \simeq \frac{1}{2} \, \xi \, , \quad \xi \to 0 \, . \tag{2.66}$$

The first three zeroes of $J_1(\xi)$ correspond to $x_1 = 1.220, x_2 = 2.233$ and $x_3 = 3.238$. As we see $x_3 - x_2$ is already very close to 1.

From these properties of the Bessel function $J_1(\xi)$ it follows that the TF of the out-of-focus blur is zero on a family of circumferences with center at the origin and radii

$$\omega_n = \frac{\pi}{D} \, x_n \, ; \quad n = 1, 2, 3, ... \tag{2.67}$$

Even for moderate values of n these circumferences are approximately equispaced, in the sense that $\omega_{n+1} - \omega_n \simeq \pi/D$. Moreover, for large values of $|\boldsymbol{\omega}|$, the TF tends to zero as given by the asymptotic behavior (2.65)

$$\hat{K}(\boldsymbol{\omega}) \simeq \frac{1}{\sqrt{\pi}} \left(\frac{2}{D|\boldsymbol{\omega}|}\right)^{3/2} \cos\left(D|\boldsymbol{\omega}| - \frac{3}{4}\pi\right). \tag{2.68}$$

As in the case of the linear motion blur, the FT of the noise-free image $g^{(0)} = K * f^{(0)}$ is zero on the lines where the TF is zero while this property does not hold for the noisy

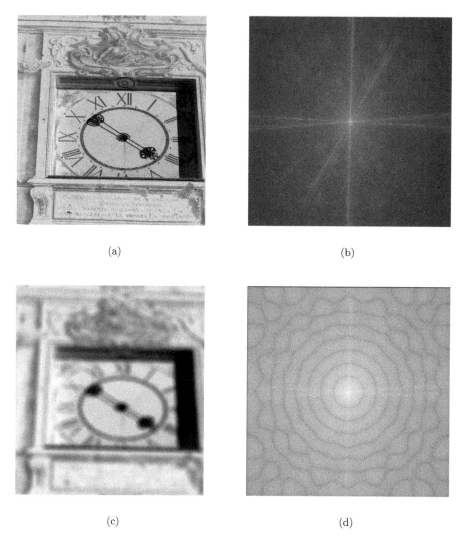

(a)

(b)

(c)

(d)

FIGURE 2.7   Example of out-of-focus blur: (a) the object: a picture of the clock of 'Palazzo San Giorgio' in Genoa; (b) the central part of the modulus of the FT of the object; (c) the blurred and noisy image; (d) the central part of the modulus of the FT of (c).

image because the noise perturbs these lines. In Figure 2.7, we give an example of an image degraded by out-of-focus blur and by the addition of white Gaussian noise. In this example the lines of zeroes of $\hat{g}(\boldsymbol{\omega})$ are rather close to concentric circumferences. This property of $\hat{g}(\boldsymbol{\omega})$ and the distribution of zeroes given in equation (2.67) can be used for estimating the radius of the COC when this radius is not known. Its theoretical determination, on the other hand, would require a precise knowledge of the camera misadjustment and of the object position.

2.6   Diffraction-limited imaging systems

According to geometrical optics, the image of a point source provided by an optical instrument is a perfectly sharp point. However, because of diffraction effects, the image of a point, which is in focus, is not a point but a small light patch, called the *diffraction pattern*. This is precisely the PSF shown in Section 2.1. Optical instruments where diffraction effects are important are also called diffraction-limited imaging systems. Examples are provided by instruments with a very high resolving power (i.e. ability to separate the images of two neighboring points) such as microscopes and telescopes.

Diffraction-limited imaging systems are usually treated by Fourier methods, and the corresponding theory is called Fourier optics [137]. In this theory, an optical system is viewed as a linear system, characterized by its PSF and the corresponding TF. This description is correct in the case of either spatially coherent or spatially incoherent illumination. An optical system is not linear however when the so-called partial-coherence effects are not negligible. The case of incoherent illumination is perhaps the most frequent one even if coherent illumination is often important in microscopy and has gained additional importance since the discovery of the laser.

We assume, for simplicity, that the scalar theory of light diffraction can be used. In this theory, monochromatic light is represented by a scalar function which is usually written as a complex-valued function, called the *complex amplitude* (or phasor) whose modulus and phase are respectively the amplitude and phase of the light disturbance. In the case of *spatially coherent illumination* (for short, coherent illumination) the relative phase of two object points is constant in time, i.e. even if the two phases can vary randomly in time they vary in an identical fashion. In the case of *spatially incoherent illumination* (for short, incoherent illumination) the phases of all points are varying in a statistically independent fashion.

We consider systems producing real (non-virtual) images. Indeed, even in the case of a virtual image, the detection system must contain a lens converting the virtual image to a real one. Then this lens can be considered as the final element of the imaging system. We also assume that the system is *isoplanatic*, i.e. space-invariant. In practice, optical imaging systems are seldom isoplanatic over the whole object field, but it is usually possible to partition the object field into regions within which the system is approximately space-invariant (see Section 6.1). Finally, we assume that the magnification factor of the optical system has been reduced to one by a rescaling of the space variables in the image plane.

(A) *Coherent illumination*

In this case, the system is linear in the complex amplitude. Then the process of image formation is described by an equation like equation (2.2) (equation (2.9) if the system is isoplanatic) where the object $f^{(0)}(\mathbf{x})$ is the complex amplitude of the ideal image predicted by geometrical optics, and $g^{(0)}(\mathbf{x})$ is the complex amplitude of the diffraction-limited image. For monochromatic light with wavelength λ and in the absence of aberrations, i.e. in the case of a system consisting of perfectly corrected lenses, the PSF is, through a change of variable, the inverse FT of the *pupil function*, which is the characteristic function $\chi_{\mathcal{P}}(\mathbf{x})$ of the exit pupil $\mathcal{P}$ of the instrument. Therefore an ideal diffraction-limited system is a bandlimited system which behaves like a perfect low-pass Fourier filter.

The change of variable relating the band $\mathcal{B}$ of the system to the exit pupil $\mathcal{P}$ is the following

$$\boldsymbol{\omega} = \frac{2\pi}{\lambda d_i}\mathbf{x} \ . \tag{2.69}$$

Here $\mathbf{x}$ is a point of $\mathcal{P}$, $\boldsymbol{\omega}$ is the corresponding point of $\mathcal{B}$, λ is the wavelength of the light and d_i is the distance between the image plane and the exit pupil plane. Then the PSF is given by

$$K(\mathbf{x}) = \frac{1}{(2\pi)^2}\int \chi_{\mathcal{B}}(\boldsymbol{\omega})\, e^{i\mathbf{x}\cdot\boldsymbol{\omega}}\, d\boldsymbol{\omega} \ . \tag{2.70}$$

In the case of a square exit pupil of side $2a$ one has

$$K(\mathbf{x}) = \frac{\sin(\Omega x_1)}{\pi x_1}\, \frac{\sin(\Omega x_2)}{\pi x_2} \tag{2.71}$$

where the cut-off frequency Ω is given by (see equation (2.69))

$$\Omega \ = \ \frac{2\pi}{\lambda}\, \frac{a}{d_i} \ , \tag{2.72}$$

while in the case of a circular pupil of radius a one has

$$K(\mathbf{x}) = \frac{\Omega}{2\pi}\, \frac{J_1(\Omega|\mathbf{x}|)}{|\mathbf{x}|} \ , \tag{2.73}$$

where the cut-off frequency is also given by equation (2.72). The properties of the Bessel function $J_1(\xi)$ have been discussed in Section 2.5. The first zero of $K(\mathbf{x})$,

$$\rho_1 = 1.22\, \frac{\pi}{\Omega} = 1.22\, \frac{d_i}{a}\, \frac{\lambda}{2} \ , \tag{2.74}$$

is the famous *Rayleigh resolution distance*, which is usually viewed as a measure of the resolving power of the instrument.

The effect of aberrations can be described by means of a phase distortion over the band of the instrument, so that the TF can be written in the following form

$$\hat{K}(\boldsymbol{\omega}) = \chi_{\mathcal{B}}(\boldsymbol{\omega})\exp[ik\phi(\boldsymbol{\omega})] \tag{2.75}$$

where $k = 2\pi/\lambda$ is the wavenumber of the light radiation. The phase distortion depends on the type of aberration. In the case of a focusing error, the phase is given by

$$\phi(\boldsymbol{\omega}) = \frac{\alpha}{2}\left(\frac{d_i}{k}\right)^2 |\boldsymbol{\omega}|^2 \tag{2.76}$$

where, in the case of a lens,

$$\alpha = \frac{1}{d_0} + \frac{1}{d_i} - \frac{1}{d_f} \ , \tag{2.77}$$

d_0 being the distance between the object plane and the plane of the lens (see equation (2.62)). Equations (2.75) and (2.76) are the starting point for the analysis of the out-of-focus blur in the case of diffraction-limited systems [271].

(B) *Incoherent illumination*

In this case, the system is linear with respect to the light intensity, which is proportional to the square of the modulus of the complex amplitude. In equation (2.9), $f^{(0)}(\mathbf{x})$ represents now the intensity distribution of the ideal image as predicted by geometrical optics while $g^{(0)}(\mathbf{x})$ is the intensity distribution of the diffraction-limited image.

The PSF of the system in the incoherent case is proportional to the square of the modulus of the PSF of the system in the coherent case [137]. If we denote by $K_{\mathrm{coh}}(\mathbf{x})$ the coherent PSF of the system and by $K_{\mathrm{inc}}(\mathbf{x})$ the incoherent PSF of the same system, we have

$$K_{\mathrm{inc}}(\mathbf{x}) = |K_{\mathrm{coh}}(\mathbf{x})|^2 \ . \tag{2.78}$$

Therefore the relationship between the two transfer functions is the following one

$$\hat{K}_{\mathrm{inc}}(\boldsymbol{\omega}) = \frac{1}{(2\pi)^2} \int \hat{K}_{\mathrm{coh}}(\boldsymbol{\omega} + \boldsymbol{\omega}') \, \hat{K}_{\mathrm{coh}}^*(\boldsymbol{\omega}') \, d\boldsymbol{\omega}' \ . \tag{2.79}$$

Very often, in Fourier optics, the following normalized transfer function is used

$$\hat{H}(\boldsymbol{\omega}) = \frac{\hat{K}_{\mathrm{inc}}(\boldsymbol{\omega})}{\hat{K}_{\mathrm{inc}}(0)} \tag{2.80}$$

which is called the *optical transfer function* (OTF). It is easy to show, thanks to the Schwarz inequality, that this function always satisfies the condition $|\hat{H}(\boldsymbol{\omega})| \le 1$.

We point out that the band of an optical system in the case of incoherent illumination is broader than the band of the same optical system in the case of coherent illumination. For instance, in the case of a square pupil, given by equation (2.71), with a cut-off frequency Ω, the cut-off frequency for the incoherent illumination is 2Ω. A similar result holds true for the circular pupil, given by equation (2.73). However, while the TF is constant over the band in the coherent case, it tends to zero at the boundary of the band in the incoherent case. Indeed, the incoherent PSF is integrable and therefore its TF is continuous according to the Riemann-Lebesgue theorem mentioned in Section 2.2. As a consequence, the gain in resolution with respect to the coherent case is smaller than a factor 2. In Figure 2.8, we compare the coherent and incoherent PSF and TF for a circular pupil in the absence of aberrations.

The effect of aberrations in the incoherent case is not a mere phase modulation as in the coherent case – see equation (2.75). Indeed, from equation (2.79), with $\hat{K}_{\mathrm{coh}}(\boldsymbol{\omega})$ given by equation (2.75), we see that aberrations can also produce variations of the modulus of the OTF.

In the case of a focusing error and of a circular pupil, using equation (2.76) one can find an expression of the OTF in terms of slowly convergent series [271]. The case of a square pupil is much simpler, because the OTF is then given by [137]

$$
\begin{aligned}
\hat{H}(\boldsymbol{\omega}) \ = \ & \left(1 - \frac{|\omega_1|}{2\Omega}\right)\left(1 - \frac{|\omega_2|}{2\Omega}\right) \\
& \times \ \mathrm{sinc}\left[\frac{8W}{\lambda}\left(\frac{\omega_1}{2\Omega}\right)\left(1 - \frac{|\omega_1|}{2\Omega}\right)\right] \mathrm{sinc}\left[\frac{8W}{\lambda}\left(\frac{\omega_2}{2\Omega}\right)\left(1 - \frac{|\omega_2|}{2\Omega}\right)\right]
\end{aligned}
\tag{2.81}
$$

where the sinc function is defined in equation (2.16), Ω is the cut-off frequency as given by equation (2.72) and $W = \alpha a^2/8$ (with α defined in equation (2.77)) is a number which indicates the severity of the focusing error. In Figure 2.9, we plot $\hat{H}(\omega_1, 0)$ for some values of W. When $W > \lambda/2$, the OTF can take negative values. In the limit $W \gg 1$, i.e. when

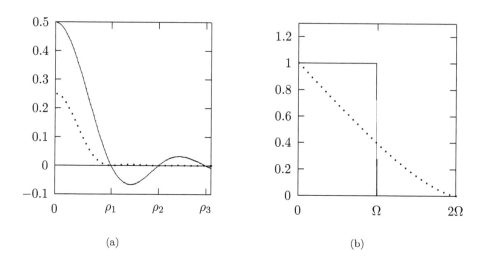

FIGURE 2.8 (a) PSF of a circular pupil both in the coherent (full line) and in the incoherent case (dotted line); (b) the corresponding transfer functions. ρ_1 denotes the Rayleigh distance associated with the bandwidth Ω.

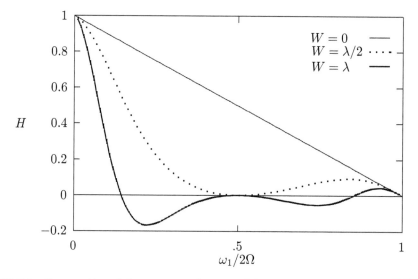

FIGURE 2.9 Cross section of the OTF, as a function of the normalized frequency $\omega_1/2\Omega$, for various focusing errors (case of a square pupil).

the focusing error is very severe, one can approximate the OTF (2.81) by means of its low frequency behavior

$$\hat{H}(\boldsymbol{\omega}) = \text{sinc}\left(\frac{8W}{\lambda}\frac{\omega_1}{2\Omega}\right)\text{sinc}\left(\frac{8W}{\lambda}\frac{\omega_2}{2\Omega}\right) \tag{2.82}$$

which, remarkably, yields the PSF predicted by geometrical optics, i.e. the characteristic function of a square. Indeed, the predictions of geometrical optics provide good approximations whenever aberrations of any sort are very severe.

2.6.1 Fluorescence microscopes

An optical microscope is a system that uses visible light and lenses to magnify images of small samples. Most microscopes used nowadays in biology are based on the physical phenomenon of fluorescence: the specimen (cells or living tissues) is stained by fluorophores, fluorescent chemical compounds that can re-emit light upon excitation by a light source, for instance a laser. The illuminating light has a specific wavelength and its absorption by the fluorophores causes the nearly simultaneous emission of light with a longer wavelength (Stokes shift). Therefore, in the case of fluorescence microscopy, the object $f^{(0)}(\mathbf{x})$ is the distribution of the fluorophores within the biological specimen (see, for instance, [33] for a very brief and synthetic account of fluorescence microscopy for users of image deconvolution).

A powerful instrument in fluorescence microscopy is the *confocal scanning laser microscope* (CSLM) [308] which is also able to produce 3D images thanks to the technique known as *optical sectioning*. The first CSLM images of a biological specimen, able to answer to biological questions, were published in 1985 by the group of Fred Brakenhoff [51]; they proved the importance of CSLM in biology. Since then CSLM has known continuous improvements and applications [228, 102], becoming an essential tool in the study of biological specimens.

The main features of a CSLM are: (i) the illumination of a small portion of the specimen in the focal plane by means of a focused laser beam; (ii) the use of a pinhole, in the image plane, in front of the detector. Then light originating from the illuminated portion of the specimen in the in-focus plane is imaged by the microscope objective and light coming from out-of-focus planes is blocked by the pinhole. Since only a small spot in the sample is imaged, a scanning procedure is required for illuminating the specimen point by point. Finally, the 3D image is obtained by moving the focal plane and repeating the scanning. An example of optical sectioning obtained with this procedure is shown in Figure 2.10.

It is important to point out that CSLM, besides optical sectioning, also provides an improvement of resolution in the lateral direction, i.e. the direction orthogonal to the optical axis of the microscope. Indeed, if we neglect the effect of the Stokes shift, it is easy to show that the relationship between image and object is given by

$$g^{(0)}(\mathbf{x}) = \int |K(\mathbf{x} - \mathbf{x}')|^4 f^{(0)}(\mathbf{x})d\mathbf{x} \tag{2.83}$$

where, in the case of aberration-free lenses and incoherent illumination (the light emitted by the fluorophores is incoherent), $K(\mathbf{x})$ is the diffraction pattern given in equation (2.73). Therefore the bandwidth of a confocal microscope is twice the bandwidth of a standard microscope with incoherent illumination. Moreover, the side lobes of the PSF are greatly reduced. In addition to such improvement, subsequent 2D image deconvolution can still be useful in view of refined image analysis [294].

In the case of optical sectioning, i.e. 3D imaging, a CSLM is also described by a convolution operator as in equation (2.9), where now $\mathbf{x}$ is a 3D vector. The 3D PSF $K(\mathbf{x})$ has a rather involved expression in terms of the properties of the optical instrument. Its main feature is that it is a bandlimited function with a band $\mathcal{B}$ having the following structure. If $\mathbf{x} =$

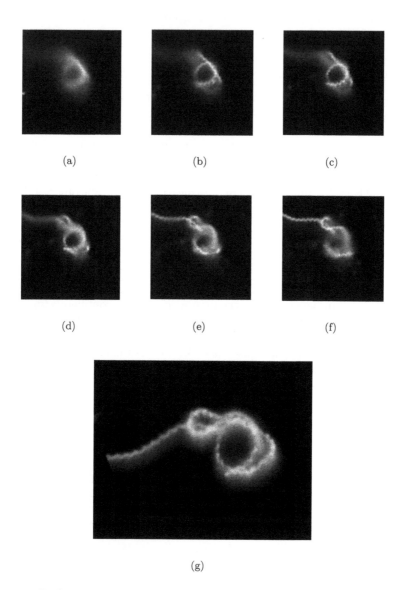

FIGURE 2.10   Confocal images of six slices (a–f) of a helical sperm head from an octopus Eledone Cirrhosa and (g) 3D representation of the same object obtained from 64 slices (images kindly provided by A. Diaspro).

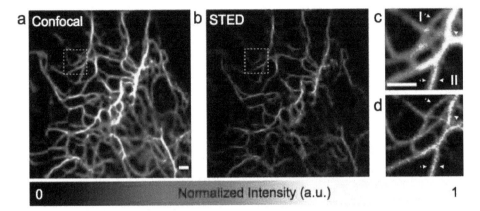

FIGURE 2.11   Bidimensional images of the microtubulin network of PtK2 cell as produced by CSLM (panel a) and STED microscopy (panel b) (512×512 pixels, dwell-time 10 μs, pixel size 32 nm). Panels c and d show a magnification of the small squares indicated in panels a and b, respectively (courtesy of G. Vicidomini).

$\{x_1, x_2, x_3\}$ are the space coordinates (with the x_3 coordinate in the direction of the optical axis of the instrument) and if $\boldsymbol{\omega} = \{\omega_1, \omega_2, \omega_3\}$ are the corresponding spatial frequencies, then $\mathcal{B}$ is rotationally invariant around the ω_3-axis but its size in the directions orthogonal to ω_3 is larger than its size in the direction of ω_3. This property implies that the axial resolution of a CSLM is poorer than the lateral resolution and thus the microscope provides orientation-dependent images. Image reconstruction is especially important for improving the quality of the images in the axial direction.

An extraordinary improvement of the resolution of CSLM is based on a technique proposed by Hell and Wichmann in 1994 [153] and experimentally demonstrated by Hell and Klar in 1999 [181]. For this basic invention and its development Stefan Hell was awarded the Nobel prize in Chemistry in 2014. This technique is now known as *STimulated Emission Depletion* (STED) microscopy and is able to bypass the diffraction limit of light microscopy (for a short review, see [296]). The most typical STED microscope uses a pair of synchronized laser pulses. The first excites the fluorophores of the specimen as in a conventional CSLM and produces an ordinary diffraction-limited focal point. The excitation pulse is followed by a red-shifted pulse which can desactivate the excited molecules before spontaneous emission. Therefore, in a suitable selected spatial region, the spontaneous decay can be anticipated by a decay stimulated by the second laser; this emitted light is suppressed by a suitable filter so that the image is formed by the light coming from the smaller region where spontaneous emission is occurring. It has been demonstrated that, by means of this technique, a resolution of few nanometers is possible. A comparison of CSLM image with the STED image of the same specimen is shown in Figure 2.11. The improvement in resolution is already evident by comparing the panels (a) and (b); it is even more evident by comparing the panels (c) and (d) which provide a zoom over the white squares indicated in the previous panels.

2.6.2   Optical and infrared telescopes

The images of ground-based optical and infrared telescopes are corrupted by the effect of atmospheric turbulence which, in practice, reduces the aperture of the telescope, providing images with a much smaller resolution. For this reason the Hubble space telescope (HST), an optical telescope, with a diameter of about 2.5 m, was launched in 1990 in a low Earth orbit outside the disturbing atmosphere. Soon after launch, in early 1990, a severe optical fault producing a remarkable blurring of the images was detected in the telescope. Spherical

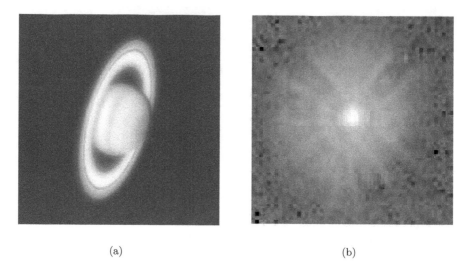

(a)   (b)

FIGURE 2.12   (a) 512×512 image of Saturn provided by the wide field planetary camera (WF/PC) of
HST; (b) 53×53 corresponding PSF (images produced by AURA/STScI).

aberration, caused by a manufacturing error, was taking light away from the central core of
the PSF. According to the original design, about 70% of the light had to be concentrated into
a disc with diameter of 0.1 arcseconds (diameter of the central lobe of the PSF). However,
in the configuration corresponding to the so-called Faint Object Camera (FOC) only 16%
of the light was concentrated into that disc, while the rest was diffused over a broader
disc with a diameter of about 2 arcseconds. In Figure 2.12, we show an example of these
aberrated images: it is a wide-field/planetary camera image of Saturn (512×512). In the
same figure, we also give the central part of the corresponding PSF (53×53), obtained by
imaging a distant star. For this reason, the correction of aberrations, by means of methods
of image deconvolution, was attempted before the correction of the manufacturing error [4].
In the four years between the launch and the mission for the aberration correction, several
efforts were made to improve the quality of the images and to this purpose both computed
and measured PSF were used. A few examples of aberrated HST images will be used in the
following chapters of this book to illustrate some of the deconvolution methods.

The originally designed capability of the telescope was restored by implementing the COSTAR
corrective optics during the shuttle servicing mission of December 1993. After the correction,
HST produced both fundamental scientific results and unprecedented beautiful images of
astronomical objects. In Figure 2.13, we show a comparison between the HST PSF before
and after the COSTAR correction; the images are obtained by pointing the telescope towards
the same star since a distant star is essentially a point object.

In the following, we first describe simple models of atmospheric perturbation and next
a new technology which allows to compensate for this perturbation. Turbulent velocity
fluctuations in the atmosphere generate a statistical temperature field and thus give rise to
random inhomogeneities in the index of refraction. As a consequence they cause distortions
in the propagation of electromagnetic and acoustic fields at all wavelengths and for this
reason, their effect is important in imaging through the atmosphere, as it occurs in optical
and radio astronomy, in remote sensing, in target identification, etc. [243].

We restrict our examples to the case of light propagation. Indeed, atmospheric turbulence
is a major problem in optical and infrared astronomy because it drastically reduces the
angular resolution of the ground-based telescopes. In the absence of turbulence (and also of

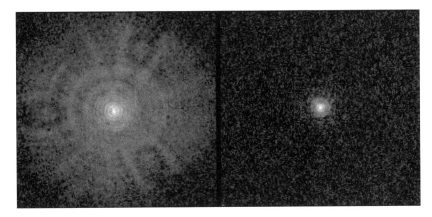

FIGURE 2.13 Comparison between the PSF of HST before (left panel) and after (right panel) the COSTAR correction.

aberrations in the optical components), the quality of the image is limited by the diffraction effects and the angular resolution is given by $\Delta\theta = 1.22\,(\lambda/D)$, where λ is the wavelength of the observed light and D is the diameter of the objective aperture, i.e. the diameter of the primary mirror in the case of a telescope. When turbulence is present, it produces a large and irregular blur. For instance, under good weather conditions at the best observatories, the angular resolution allowed by this blur varies approximately from 0.6 to 1.2 arcseconds. But according to the previous formula with $\lambda \simeq 5.6 \times 10^{-5}$ cm, these resolutions would correspond to telescope diameters ranging from about 30 to about 10 cm while, for instance, the diameter of the two telescopes of the Keck Observatory on the volcano Mauna Kea, Hawaii, is about 10 m. Therefore it is clear that the resolution limits due to atmospheric turbulence are much more severe than those due to diffraction effects.

Another feature of the atmospheric turbulence blur is that it changes rapidly with time. The characteristic time of these variations is about 1 millisecond. Therefore it is usual to distinguish between *long-exposure images* (the exposure time is large with respect to the characteristic time) and *short-exposure images* (the exposure time is comparable with the characteristic time of the fluctuations).

In a conventional astronomical image, the exposure time can be of the order of few seconds; therefore the recorded image is not random and is a typical example of a long-exposure image. The basic result in this case is that the OTF $\hat{K}(\boldsymbol{\omega})$ of the whole system, telescope and atmosphere, is the product of the OTF of the telescope, $\hat{H}(\boldsymbol{\omega})$, and of an atmospheric transfer function, $\hat{B}(\boldsymbol{\omega})$:

$$\hat{K}(\boldsymbol{\omega}) = \hat{H}(\boldsymbol{\omega})\hat{B}(\boldsymbol{\omega}) \tag{2.84}$$

(we notice that here the components of $\boldsymbol{\omega}$ are frequencies associated with angular variables). Then, it can be shown that [243]

$$\hat{B}(\boldsymbol{\omega}) = \exp\left[-3.44\left(\frac{\lambda|\boldsymbol{\omega}|}{r_0}\right)^{5/3}\right] \tag{2.85}$$

where λ is the wavelength of the observed radiation and r_0 is a parameter, called *critical diameter*, which depends on λ and is roughly proportional to $\lambda^{6/5}$. If the diameter of the telescope is small with respect to the critical diameter, then the effect of turbulence is negligible while it is dominant and determines the resolving power of the telescope when the critical diameter is smaller than the diameter of the telescope. As follows from equation (2.85), the atmospheric turbulence blur can be roughly approximated by a Gaussian

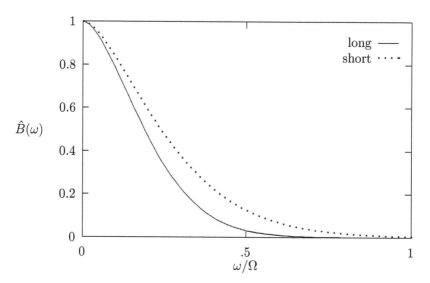

FIGURE 2.14 Comparison of the TF for long and short exposure time, as functions of ω/Ω, in the case $r_0/\lambda = \Omega/2$.

one

$$\hat{B}(\omega) = \exp(-s^2|\omega|^2) \ . \tag{2.86}$$

For short exposure times, i.e. of the order of 1 ms, the images looks random as well as the PSF and the OTF of atmospheric turbulence. It can be shown that the ensemble average of the OTF is given again by equation (2.84) where, under near-field conditions

$$\hat{B}(\omega) = \exp\left\{-3.44\left(\frac{\lambda|\omega|}{r_0}\right)^{5/3}\left[1-\left(\frac{|\omega|}{\Omega}\right)^{1/3}\right]\right\} \tag{2.87}$$

(Ω being the cut-off frequency of the telescope) while, under far-field conditions

$$\hat{B}(\omega) = \exp\left\{-3.44\left(\frac{\lambda|\omega|}{r_0}\right)^{5/3}\left[1-\frac{1}{2}\left(\frac{|\omega|}{\Omega}\right)^{1/3}\right]\right\} \ . \tag{2.88}$$

It is important to note that, at spatial frequencies close to the cut-off frequency of the optical system, the long-exposure OTF is considerably smaller than the short-exposure average. Therefore, short-exposure pictures are preferred because they can provide a better resolution than long-exposure pictures. In Figure 2.14, we compare the atmospheric turbulence TF for long-exposure images, equation (2.85), with that for short-exposure images in the case of far-field conditions, equation (2.88). It is assumed that $r_0/\lambda = \Omega/2$.

The previous models of the contribution of atmospheric turbulence to the OTF of a ground-based telescope could be useful for the deconvolution of images detected by means of telescopes with small primary mirrors, for instance with diameters of the order of 1 meter. However, turbulence correction is necessary for the large telescopes used in the scientific investigation of the deep sky such as the already mentioned Keck telescopes, the Very Large Telescopes (VLT) of the ESO Observatory, the Large Binocular Telescope (LBT) on Mount Graham Observatory, Arizona, or the Gran Telescopio Canarias, all with primary mirrors of the class $8-10$ m. The correction is even more important for the future extremely large telescopes such as the Extremely Large Telescope (ELT) of ESO, with a diameter of about 39 m or the proposed Thirty Meter Telescope (TMT) with a planned location on Mauna Kea.

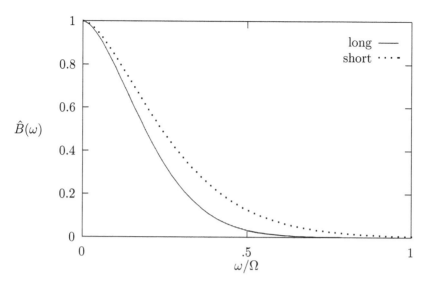

FIGURE 2.14   Comparison of the TF for long and short exposure time, as functions of ω/Ω, in the case $r_0/\lambda = \Omega/2$.

one

$$\hat{B}(\boldsymbol{\omega}) = \exp(-s^2|\boldsymbol{\omega}|^2) \; . \tag{2.86}$$

For short exposure times, i.e. of the order of 1 ms, the images looks random as well as the PSF and the OTF of atmospheric turbulence. It can be shown that the ensemble average of the OTF is given again by equation (2.84) where, under near-field conditions

$$\hat{B}(\boldsymbol{\omega}) = \exp\left\{-3.44\left(\frac{\lambda|\boldsymbol{\omega}|}{r_0}\right)^{5/3}\left[1 - \left(\frac{|\boldsymbol{\omega}|}{\Omega}\right)^{1/3}\right]\right\} \tag{2.87}$$

(Ω being the cut-off frequency of the telescope) while, under far-field conditions

$$\hat{B}(\boldsymbol{\omega}) = \exp\left\{-3.44\left(\frac{\lambda|\boldsymbol{\omega}|}{r_0}\right)^{5/3}\left[1 - \frac{1}{2}\left(\frac{|\boldsymbol{\omega}|}{\Omega}\right)^{1/3}\right]\right\} \; . \tag{2.88}$$

It is important to note that, at spatial frequencies close to the cut-off frequency of the optical system, the long-exposure OTF is considerably smaller than the short-exposure average. Therefore, short-exposure pictures are preferred because they can provide a better resolution than long-exposure pictures. In Figure 2.14, we compare the atmospheric turbulence TF for long-exposure images, equation (2.85), with that for short-exposure images in the case of far-field conditions, equation (2.88). It is assumed that $r_0/\lambda = \Omega/2$.

The previous models of the contribution of atmospheric turbulence to the OTF of a ground-based telescope could be useful for the deconvolution of images detected by means of telescopes with small primary mirrors, for instance with diameters of the order of 1 meter. However, turbulence correction is necessary for the large telescopes used in the scientific investigation of the deep sky such as the already mentioned Keck telescopes, the Very Large Telescopes (VLT) of the ESO Observatory, the Large Binocular Telescope (LBT) on Mount Graham Observatory, Arizona, or the Gran Telescopio Canarias, all with primary mirrors of the class $8-10$ m. The correction is even more important for the future extremely large telescopes such as the Extremely Large Telescope (ELT) of ESO, with a diameter of about 39 m or the proposed Thirty Meter Telescope (TMT) with a planned location on Mauna Kea.

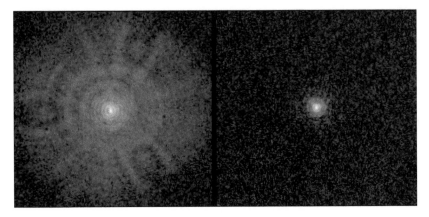

FIGURE 2.13 Comparison between the PSF of HST before (left panel) and after (right panel) the COSTAR correction.

aberrations in the optical components), the quality of the image is limited by the diffraction effects and the angular resolution is given by $\Delta\theta = 1.22 \, (\lambda/D)$, where λ is the wavelength of the observed light and D is the diameter of the objective aperture, i.e. the diameter of the primary mirror in the case of a telescope. When turbulence is present, it produces a large and irregular blur. For instance, under good weather conditions at the best observatories, the angular resolution allowed by this blur varies approximately from 0.6 to 1.2 arcseconds. But according to the previous formula with $\lambda \simeq 5.6 \times 10^{-5}$ cm, these resolutions would correspond to telescope diameters ranging from about 30 to about 10 cm while, for instance, the diameter of the two telescopes of the Keck Observatory on the volcano Mauna Kea, Hawaii, is about 10 m. Therefore it is clear that the resolution limits due to atmospheric turbulence are much more severe than those due to diffraction effects.

Another feature of the atmospheric turbulence blur is that it changes rapidly with time. The characteristic time of these variations is about 1 millisecond. Therefore it is usual to distinguish between *long-exposure images* (the exposure time is large with respect to the characteristic time) and *short-exposure images* (the exposure time is comparable with the characteristic time of the fluctuations).

In a conventional astronomical image, the exposure time can be of the order of few seconds; therefore the recorded image is not random and is a typical example of a long-exposure image. The basic result in this case is that the OTF $\hat{K}(\boldsymbol{\omega})$ of the whole system, telescope and atmosphere, is the product of the OTF of the telescope, $\hat{H}(\boldsymbol{\omega})$, and of an atmospheric transfer function, $\hat{B}(\boldsymbol{\omega})$:

$$\hat{K}(\boldsymbol{\omega}) = \hat{H}(\boldsymbol{\omega})\hat{B}(\boldsymbol{\omega}) \tag{2.84}$$

(we notice that here the components of $\boldsymbol{\omega}$ are frequencies associated with angular variables). Then, it can be shown that [243]

$$\hat{B}(\boldsymbol{\omega}) = \exp\left[-3.44\left(\frac{\lambda|\boldsymbol{\omega}|}{r_0}\right)^{5/3}\right] \tag{2.85}$$

where λ is the wavelength of the observed radiation and r_0 is a parameter, called *critical diameter*, which depends on λ and is roughly proportional to $\lambda^{6/5}$. If the diameter of the telescope is small with respect to the critical diameter, then the effect of turbulence is negligible while it is dominant and determines the resolving power of the telescope when the critical diameter is smaller than the diameter of the telescope. As follows from equation (2.85), the atmospheric turbulence blur can be roughly approximated by a Gaussian

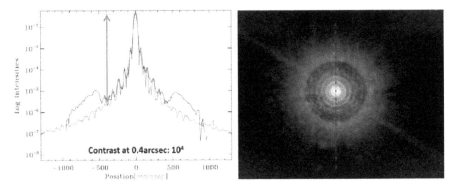

FIGURE 2.15 Left panel: comparison between PSF profiles: diffraction-limited PSF in red and AO corrected PSF in black. Right panel: image of the PSF of one mirror of LBT equipped with the FLAO system.

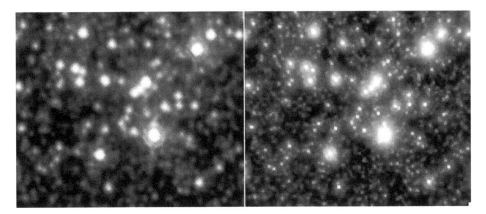

FIGURE 2.16 A central region of the globular cluster M92 at 1.6 μm as observed with the Hubble Space Telescope (left) and the LBT (Large Binocular Telescope) in adaptive mode (right). It is immediately clear that the resolution and depth achieved with LBT surpasses those of the Hubble image (courtesy of LBTO).

In the case of large telescopes, correction for turbulence can now be achieved by a technique known as Adaptive Optics (AO), first envisioned by Horace W. Babcock in 1953 [15], made practical during the 1990s thanks to advances in computer technology and continuously developed in recent years. AO technique compensates in real time the wavefront deformations introduced by the Earth's atmosphere [24]. Compensation is possible by inserting into the optical beam coming from the primary mirror of the telescope, a deformable mirror – in most recent AO systems, the secondary mirror of the telescope as in the FLAO (First Light Adaptive Optics) system of LBT [115] – capable of introducing a distortion of the wavefront equal and opposite to that generated by atmospheric turbulence. This correction must be performed in a very short time (of the order of millisecond) in order to track in real time the temporal evolution of the turbulence itself [242]. In Figure 2.15, we show the PSF of one mirror of LBT equipped with the FLAO system. The correction which produces a diffraction pattern is evident in the central part: up to four bright rings are visible; the external behavior is dominated by the *halo*, i.e. the residual part of the contribution due to the seeing produced by atmospheric turbulence, with an approximate Gaussian behavior. Moreover, in Figure 2.16, we compare the image of the same sky region as produced by HST (left panel) and by one mirror of LBT with FLAO system (right panel). It is evident that, thanks to AO correction, the much larger mirror of LBT (about 8.3 m against 2.5 m), is able to produce, as expected, a better resolution (thanks to the larger diameter) as well as a deeper image (thanks to the broader collecting area) so that much more distant and faint stars are detected.

2.7 Near-field acoustic holography

Near-Field Acoustic Holography (NFAH) was proposed in 1980 by Williams and Maynard [307] as a technique which can produce high-resolution images of sound sources. Since the resolution which can be achieved is much better than the wavelength of the used radiation, this method can provide a satisfactory resolution also in the case of long wavelengths.

The basic principle underlying NFAH is that the acoustic pressure is measured (for instance by means of an array of microphones) on a surface, containing the acoustic source inside, such that its distance from the surface of the source is smaller than the wavelength of the emitted radiation. Then the measured data are used for computing the acoustic pressure on the surface of the source. The good resolution achieved is due to the use of the information conveyed by the so-called *evanescent waves*, which are still present in the radiated field when the distance from the source is smaller than the wavelength. The same principle is used in so-called Scanning Near-field Optical Microscopy (SNOM) [39].

NFAH has gained popularity as an experimental tool for studying acoustical radiation from finite sources, including also industrial applications. An example is its use in vehicle design to optimize sound and vibration performance. From the mathematical point of view, the underlying problem is that of *inverse diffraction* which will be discussed in Section 6.5. Three configurations are frequently considered: planar, cylindrical and spherical sources [306].

A problem in the application is that Fourier-based NFAH does not produce satisfactory source reconstructions when the aperture size of the microphone array is smaller than the size of the source. An approach for overcoming this difficulty is proposed, for instance, in [250]. Most often the case of a stationary source (i.e. with a given wavelength) is considered while the case of nonstationary sources is investigated in [280].

Here we consider the simple case where the wavefield propagates from a plane, let us say $x_3 = 0$, to another parallel plane, let us say $x_3 = a$. The geometry of this problem is illustrated in Figure 2.17. More precisely, we assume that the sources are located in the

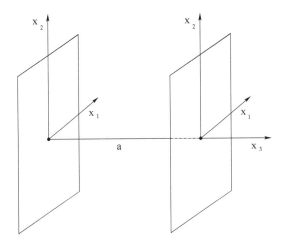

FIGURE 2.17 Schematic representation of the geometry of the problem of inverse diffraction.

half-space $x_3 < 0$, so that the plane $x_3 = 0$ is the boundary of the source region. We also assume that the sources emit a scalar monochromatic field, propagating in the half space $x_3 > 0$, where its complex amplitude $u(\mathbf{x}) = u(x_1, x_2, x_3)$ is a solution of the Helmholtz equation

$$\Delta u + k^2 u = 0 . \tag{2.89}$$

As usual, we denote by k the wave number, which is associated to the frequency ν of the monochromatic field by $k = 2\pi\nu/c$, c being the sound velocity. The wavelength λ is defined by $\lambda = 2\pi/k$ and it is the characteristic length of the problem.

The solution of equation (2.89) is uniquely determined by the following boundary conditions:
(1) the values $f^{(0)}(x_1, x_2)$ of the complex amplitude on the boundary plane $x_3 = 0$

$$u(x_1, x_2, 0) = f^{(0)}(x_1, x_2) ; \tag{2.90}$$

(2) the Sommerfeld radiation condition at infinity

$$\lim_{r \to \infty} \left[r \left(\frac{\partial u}{\partial r} - iku \right) \right] = 0 , \quad |\theta| < \pi/2 , \tag{2.91}$$

where r, θ are polar coordinates (the polar axis is the x_3-axis). In other words there exists a unique solution of equation (2.89), taking given values on the source place $x_3 = 0$ and behaving at infinity as an outgoing spherical wave. This is the forward problem, or also the direct problem in the sense of Section 1.1. It is natural to assume that the function $f^{(0)}$ is square integrable because its L^2-norm is proportional to the energy associated to the field. It is possible to give an integral representation of the solution of this problem in terms of the boundary values $f^{(0)}$, using the appropriate Green function [204]. Here we write the solution in a form suitable for the problem of inverse diffraction. Let us denote by $g_a^{(0)}(x_1, x_2)$ the values of the scalar field on the plane $x_3 = a$:

$$g_a^{(0)}(x_1, x_2) = u(x_1, x_2, a) . \tag{2.92}$$

Then, by solving the forward problem one gets [204] ($\mathbf{x} = \{x_1, x_2\}$)

$$g_a^{(0)}(\mathbf{x}) = \int K_a^{(+)}(\mathbf{x} - \mathbf{x}') f^{(0)}(\mathbf{x}') \, d\mathbf{x}' \tag{2.93}$$

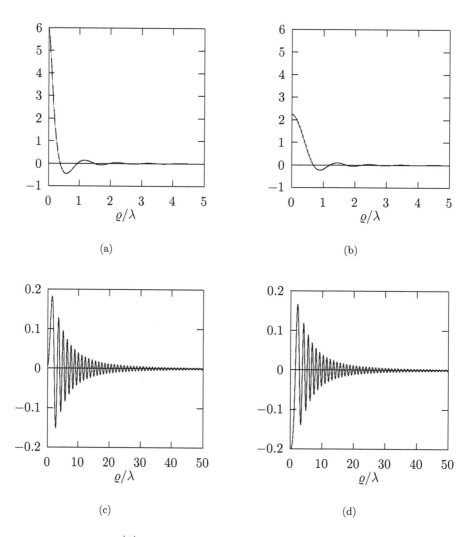

FIGURE 2.18   Plot of $\lambda K_a^{(+)}(\varrho)$ as a function of ϱ/λ. Top: real part (a) and imaginary part (b) in the case $a = \lambda/5$; bottom: real part (c) and imaginary part (d) in the case $a = 5\lambda$.

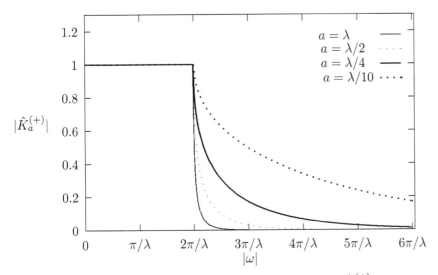

FIGURE 2.19   Behavior of the modulus of the transfer function $\hat{K}_a^{(+)}(\boldsymbol{\omega})$ for various values of a.

where

$$K_a^{(+)}(\mathbf{x}) = -\frac{1}{2\pi}\frac{\partial}{\partial a}\frac{e^{ikr_a}}{r_a} \ , \quad r_a = \sqrt{|\mathbf{x}|^2 + a^2} \ . \tag{2.94}$$

The FT of this function has a very simple expression [254, 255]

$$\hat{K}_a^{(+)}(\boldsymbol{\omega}) = e^{iam(\boldsymbol{\omega})} \tag{2.95}$$

where

$$m(\boldsymbol{\omega}) = \begin{cases} \sqrt{k^2 - |\boldsymbol{\omega}|^2}, & |\boldsymbol{\omega}| \le k \\ i\sqrt{|\boldsymbol{\omega}|^2 - k^2}, & |\boldsymbol{\omega}| > k \end{cases} \ . \tag{2.96}$$

The representation of $g_a^{(0)}(\mathbf{x})$, obtained from equations (2.93),(2.95) and (2.96), using the convolution theorem, is also called the *angular-spectrum representation* of the propagating field

$$g_a^{(0)}(\mathbf{x}) = \frac{1}{(2\pi)^2}\int e^{iam(\boldsymbol{\omega})}\,\hat{f}^{(0)}(\boldsymbol{\omega})\,e^{i\mathbf{x}\cdot\boldsymbol{\omega}}d\boldsymbol{\omega} \ . \tag{2.97}$$

In Figure 2.18, we plot the real and imaginary part of the PSF (2.94) for $a = \lambda/5$ and $a = 5\lambda$ while in Figure 2.19, we plot the modulus of the TF equation (2.95) for various values of a. The problem of *inverse diffraction from plane to plane* [18] is the problem of estimating $f^{(0)}$ from given and noisy values of $g_a^{(0)}$, i.e. $g_a = g_a^{(0)} + w$. One must know of course the amplitude and phase of the scalar field (for instance the acoustic pressure) and therefore this problem is quite similar to the problem of coherent imaging, considered in Section 2.6. In the present case, however, we have to deal with a PSF which is not rigorously bandlimited although it becomes practically bandlimited when a is sufficiently large. Indeed, the modulus of the TF has an exponential decay when $|\boldsymbol{\omega}| > k$ (see Figure 2.19). In practice, when $a > \lambda$, the TF rapidly drops to zero at $|\boldsymbol{\omega}| = k$ and therefore for such values of a the PSF $K_a^{(+)}(\mathbf{x})$ is essentially bandlimited, the band being the disc of radius k. The cut-off frequency is $\Omega = k$ and the corresponding Rayleigh distance is $\rho_1 = 1.22 \, (\lambda/2)$. Moreover, on the band, the exponential $\exp[iam(\boldsymbol{\omega})]$ is a pure phase factor. If we compare with equation (2.75) we see that the effect of propagation is analogous to that of an aberrated optical system with a phase distortion given by

$$\phi(\boldsymbol{\omega}) = a\sqrt{1 - |\boldsymbol{\omega}|^2/k^2} \ . \tag{2.98}$$

If such approximation holds true we can qualify the approach as *far-field acoustic holography* (FFAH). In the case where the object $f^{(0)}$ is bandlimited with a band whose size is much smaller than k, then, for small spatial frequencies one can approximate equation (2.98) by $\phi(\boldsymbol{\omega}) \simeq a(1 - |\boldsymbol{\omega}|^2/2k^2)$ and therefore the aberration due to propagation turns out to be equivalent to an aberration due to a focusing error (given by equation (2.76) with $\alpha d_i^2 = -a$).

3

The ill-posedness of image deconvolution

Ill-posedness characterizes linear inverse problems and, in particular, image deconvolution. It implies that the solution of the problem may not be unique, may not exist or else, when it exists as in the case of discrete problems, may be completely deprived of physical meaning as a consequence of error propagation and amplification from the data to the solution. This is essentially a consequence of the fact that the imaging system does not transmit complete information about the object. Therefore, even in the absence of noise, one could find difficulties in solving the problem.

A first consequence of ill-posedness is that one must reject the concept of exact solution of the problem and look for approximate solutions, i.e. objects which approximately reproduce the noisy images. This is the first basic point in the treatment of ill-posed problems. However, another consequence of ill-posedness is that the set of approximate solutions is too broad and contains not only physically acceptable objects but also objects which are too large and wildly oscillating. Then the second basic point is that one must use the knowledge of additional properties of the unknown object (constraints) for selecting from the set of approximate solutions those which are physically meaningful. This is the role played by the so-called *a priori information*.

3.1 Formulation of the problem

We consider a space-invariant imaging system, so that the model of image formation is given by equation (2.9). Then, the corresponding imaging operator is the *convolution operator* associated with the PSF $K(\mathbf{x})$; it is given by

$$(Af)(\mathbf{x}) = \int_{\mathbb{R}^d} K(\mathbf{x} - \mathbf{x}') \, f(\mathbf{x}') \, d\mathbf{x}' \, . \tag{3.1}$$

and defines the mapping from the object space into the image space, in agreement with the general lines discussed in Section 1.2. Indeed, let us consider a fixed function $K(\mathbf{x})$ whose FT

DOI: 10.1201/9781003032755-3

49

$\hat{K}(\boldsymbol{\omega})$ is bounded, and let us also consider the convolution product of this function with all square-integrable functions. The results are also square-integrable functions and therefore, in this way, we define an operator in $L^2(\mathbb{R}^d)$, the space of square-integrable functions, equipped with the usual scalar product (2.21), with $d = 1, 2, 3$ for 1D, 2D and 3D images, respectively. Such an operator is a particular case of more general integral operators which will be considered in the second part of the book.

From the convolution theorem, equation (2.27), and the inversion formula (2.14) of the FT, we obtain the following *spectral representation* of the convolution operator A

$$(Af)(\mathbf{x}) = \frac{1}{(2\pi)^d} \int_{\mathbb{R}^d} \hat{K}(\boldsymbol{\omega}) \, \hat{f}(\boldsymbol{\omega}) \, e^{i\mathbf{x}\cdot\boldsymbol{\omega}} \, d\boldsymbol{\omega} \ . \tag{3.2}$$

The operator A is *linear*, i.e. it satisfies the following condition

$$A(\alpha_1 f^{(1)} + \alpha_2 f^{(2)}) = \alpha_1 A f^{(1)} + \alpha_2 A f^{(2)} \tag{3.3}$$

where α_1, α_2 are arbitrary real or complex numbers.

Moreover it is *bounded* in the following sense. If we denote by $\hat{K}_{max}$ the maximum value of $\hat{K}(\boldsymbol{\omega})$

$$\hat{K}_{max} = \max_{\boldsymbol{\omega}} |\hat{K}(\boldsymbol{\omega})| \tag{3.4}$$

then, from equation (3.2), from the Parseval equality (2.20) and from the definition (2.22) of the norm, we find that

$$\|Af\| \le \hat{K}_{max}\|f\| \ . \tag{3.5}$$

From this inequality we see that the operator A is also *continuous*, in the sense that, if $\|f\|$ is small (tends to zero), then $\|Af\|$ is also small (tends to zero). In general, for a given operator A, the smallest constant C for which the inequality $\|Af\| \le C\|f\|$ holds true for all f, is called the *(spectral) norm* of the operator A and is denoted by $\|A\|$. Hence, we see that the norm of the convolution operator A is bounded by $\hat{K}_{max}$. This definition of norm generalizes to function spaces the notion of (spectral) norm of a matrix.

For future use, we also need to introduce the *adjoint operator* A^* of the bounded operator A. This is a generalization of the hermitian conjugate of a matrix, i.e. of the matrix whose elements are the complex conjugate of the elements of the transposed matrix. Therefore, in the following, the superscript $*$, when applied to an operator, will mean its adjoint while, when applied to a complex number, it will mean its complex conjugate. The adjoint operator A^* is the unique operator such that, for any pair of square-integrable functions f, g, we have

$$(Af, g) = (f, A^*g) \ , \tag{3.6}$$

the scalar product being defined in equation (2.21). Since from the Parseval equality (2.19) we have

$$\begin{aligned}(Af, g) &= \frac{1}{(2\pi)^d} \int_{\mathbb{R}^d} \hat{K}(\boldsymbol{\omega}) \, \hat{f}(\boldsymbol{\omega}) \, \hat{g}^*(\boldsymbol{\omega}) \, d\boldsymbol{\omega} \\ &= \frac{1}{(2\pi)^d} \int_{\mathbb{R}^d} \hat{f}(\boldsymbol{\omega}) \left[\hat{K}^*(\boldsymbol{\omega}) \, \hat{g}(\boldsymbol{\omega}) \right]^* d\boldsymbol{\omega} \\ &= (f, A^*g) \ , \end{aligned} \tag{3.7}$$

we see that A^* is given by

$$(A^*g)(\mathbf{x}) = \frac{1}{(2\pi)^d} \int_{\mathbb{R}^d} \hat{K}^*(\boldsymbol{\omega}) \, \hat{g}(\boldsymbol{\omega}) \, e^{i\mathbf{x}\cdot\boldsymbol{\omega}} \, d\boldsymbol{\omega} \ . \tag{3.8}$$

This is the spectral representation of the adjoint operator which is also a bounded convolution operator given by

$$(A^*g)(\mathbf{x}) = \int_{\mathbb{R}^d} K^*(\mathbf{x}' - \mathbf{x})\, g(\mathbf{x}')\, d\mathbf{x}' \ . \tag{3.9}$$

Here the convolution theorem has been used as well the fact that the inverse Fourier transform of $\hat{K}^*(\boldsymbol{\omega})$ is $K^*(-\mathbf{x})$.

We still introduce the notions of range and null space of an operator A. The subset of the object space constituted by the objects which are mapped by A on a zero image is called the *null space* of the operator A and will be denoted in the following by $\mathcal{N}(A)$. In the image space, the set of the noise-free images is the set of all square-integrable functions given by $g^{(0)} = Af^{(0)}$, where $f^{(0)}$ is an arbitrary square-integrable object. It is called the *range* of the operator A and will be denoted by $\mathcal{R}(A)$. Remark that, if the PSF is smooth, then the noise-free images are also smooth and therefore they form a subset of the space of square-integrable functions. On the other hand, the noisy images are modeled by equation (2.3) which can be rewritten as follows $g = Af^{(0)} + w$.

Similarly, in the case of 2D discrete objects and images, the discrete mapping $\mathbf{A}$, associated with the PSF $\mathbf{K}$, is defined in equation (2.48)

$$(\mathbf{Af})_{m,n} = \sum_{k,l=0}^{N-1} K_{m-k,n-l} f_{k,l} \ . \tag{3.10}$$

The object and image spaces are now Euclidean (Hermitian) spaces of 2D arrays, equipped with the standard scalar product. Again the subset of the noise-free images is the set of all 2D arrays given by $\mathbf{g}^{(0)} = \mathbf{Af}^{(0)}$, while the noisy images are modeled as $\mathbf{g} = \mathbf{Af}^{(0)} + \mathbf{w}$, after the addition of a noise term.

In rather broad terms, the problem of image deconvolution can now be formulated as the problem of estimating the unknown object $f^{(0)}(\mathbf{x})$, or its discrete version $\mathbf{f}^{(0)}$, being given the noisy image $g(\mathbf{x})$, or its discrete version $\mathbf{g}$. Moreover it is assumed that the PSF $K(\mathbf{x})$, or the discrete PSF $\mathbf{K}$, is also known.

REMARK 3.1   *In some cases one does not know the PSF or one only knows that it belongs to a class of functions depending on a certain number of parameters. For instance, one knows that the blur is due to uniform motion but the parameters of the motion are not known. Section 2.4 contains a short discussion of this problem and a way for its solution, by means of a processing of the image g. This is only a particular case of a more general problem, which is known as blind deconvolution (with equivalent names: blur identification, image-blur identification, a posteriori restoration) and consists in the attempt of estimating both $K(\mathbf{x})$ and $f^{(0)}(\mathbf{x})$ from the unique knowledge of $g(\mathbf{x})$. It is obvious that to solve such a problem one needs additional information about the two unknown functions $K(\mathbf{x})$ and $f^{(0)}(\mathbf{x})$. Even if nowadays there is a wide literature on this kind of problems, we will not consider blind deconvolution in this book and we will always assume that the PSF or a sufficiently good approximation of the PSF itself is known.*

In the models of noisy images with additive and signal-independent noise, both $f^{(0)}$ and w are unknown, but if the SNR is sufficiently large, it seems reasonable at first sight to neglect the noise term in these equations. If we do this approximation, then we can formulate the problem of image reconstruction as the problem of solving in $L^2(\mathbb{R}^d)$ the linear equation

$$Af = g \tag{3.11}$$

or, in the discrete case, as the problem of solving the linear algebraic system

$$\mathbf{Af} = \mathbf{g} \ . \tag{3.12}$$

As concerns equation (3.11), since we assume that $g(\mathbf{x})$ is a square-integrable function it is quite natural to search for a solution $f(\mathbf{x})$ which is also a square-integrable function. If we find such a function, we expect that it provides an approximation of the unknown object $f^{(0)}(\mathbf{x})$. Similarly we expect that a solution $\mathbf{f}$ of equation (3.12) provides an approximation of the discrete object $\mathbf{f}^{(0)}$.

If we use the Fourier transform, equation (3.11) becomes trivial, since we get by the convolution theorem

$$\hat{K}(\boldsymbol{\omega})\hat{f}(\boldsymbol{\omega}) = \hat{g}(\boldsymbol{\omega}) \ . \tag{3.13}$$

Analogously, by means of the DFT, from equation (3.12) we get

$$\hat{K}_{k,l} \, \hat{f}_{k,l} = \hat{g}_{k,l} \ . \tag{3.14}$$

These equations are the starting points of the analysis made in the next sections.

3.2   Well-posed and ill-posed problems

According to a definition which is now classical, a mathematical problem is said to be *well posed* in the sense of Hadamard if it satisfies the following conditions:

- the solution of the problem is unique;
- the solution exists for any data;
- the solution depends with continuity on the data.

The meaning of the first requirement is obvious. In the case of problem (3.11) the second requirement implies that the solution should exist for any square-integrable image $g(\mathbf{x})$. Finally the third requirement has the following precise formulation: let g_n be a sequence of square-integrable images such that $\|g_n\| \to 0$ when $n \to \infty$ (the norm being the usual norm of $L^2(\mathbb{R}^d)$), then the corresponding solutions f_n of the equation must have the same property, i.e. $\|f_n\| \to 0$ when $n \to \infty$.

The meaning of this property is clearly illustrated by the following statement of Courant [75]: *'the third requirement, particularly incisive, is necessary if the mathematical formulation is to describe observable natural phenomena. Data in nature cannot possibly be conceived as rigidly fixed; the mere process of measuring them involves small errors. [...] Therefore a mathematical problem cannot be considered as realistically corresponding to physical phenomena unless a variation of the given data in a sufficiently small range leads to an arbitrary small change in the solution. This requirement of 'stability' is not only essential for meaningful problems in mathematical physics, but also for approximation methods'.* This statement does not really apply to inverse problems because inverse problems do not describe natural phenomena; they rather describe the processing of data acquired by humans for understanding details of natural phenomena by means of suitable instrumentation. The loss of information implicit in the detection process induces a loss of stability in a naive approach to data inversion.

If one of the three conditions above is not satisfied, the problem is said to be *ill posed*. Therefore an ill-posed problem is a problem whose solution is not unique and/or does not exist for any data and/or does not depend continuously on the data. In this and the next section we will show that the problem (3.11) is always ill posed. The peculiar features of its discrete version, problem (3.12), which can be well posed in the sense specified above, are discussed in Section 3.4.

We first discuss the question of uniqueness. If the solution of equation (3.11) is not unique, then there exist at least two distinct objects, say f_1 and f_2, such that $Af_1 = g$ and $Af_2 = g$. Since the operator A is linear, we obtain $A(f_1 - f_2) = 0$ and therefore $f = f_1 - f_2$ is a non-trivial solution of the homogeneous equation $Af = 0$. Conversely, if this equation has a non-trivial solution f and if f_1 is a solution of equation (3.11), i.e. $Af_1 = g$, then also $f_2 = f_1 + f$ is a solution of the same equation because: $Af_2 = Af_1 + Af = Af_1 = g$. We see that *the solution of the equation $Af = g$ is unique if and only if the equation $Af = 0$ has only the solution $f = 0$*. The previous remark implies that the uniqueness or non-uniqueness of the solution can be established by investigating the solutions of the homogeneous equation. When there exist non-trivial solutions, these constitute the null space of A introduced above or, in our context, the space of the *invisible objects*, i.e. the objects whose noise-free image is exactly zero.

In order to establish the existence of invisible objects it is convenient to write the equation $Af = 0$ in terms of Fourier transforms, i. e.

$$\hat{K}(\boldsymbol{\omega})\hat{f}(\boldsymbol{\omega}) = 0 \ . \tag{3.15}$$

The existence of non-trivial solutions of this equation is related to the support of $\hat{K}(\boldsymbol{\omega})$, i.e. the closure of the set of values of ω where $\hat{K}(\boldsymbol{\omega}) \neq 0$. If the support of $\hat{K}(\boldsymbol{\omega})$ coincides with the whole frequency space, then $\hat{f}(\boldsymbol{\omega}) = 0$. Among the examples of Chapter 2, this is the case, for instance, of linear motion blur, of out-of-focus blur, of atmospheric turbulence blur and of near-field acoustic holography. In all these cases the solution of the reconstruction problem is unique. On the other hand, in the case of diffraction-limited imaging systems and in the case of far-field acoustic holography, the support of $\hat{K}(\boldsymbol{\omega})$ is the closure of its band $\mathcal{B}$ and is a bounded subset of the frequency space; therefore invisible objects do exist. An example in dimension one is given in Figure 3.1. We can conclude that, in the case of a bandlimited system, the solution of the deconvolution problem is not unique and the first condition required for well-posedness is not satisfied. The obvious reason is that the imaging system does not transmit any information about the object at the frequencies $\boldsymbol{\omega}$ which lie outside the band of the instrument.

As concerns the discretized problem given by equation (3.14), the equation $\mathbf{A}f = 0$ implies

$$\hat{K}_{k,l}\ \hat{f}_{k,l} = 0 \tag{3.16}$$

and therefore the solution of the problem (3.14) is unique if and only if $\hat{K}_{k,l} \neq 0$ for any k, l.

It is important to notice that, even if the problem (3.12) is a discrete version of the problem (3.11), the uniqueness of the solution of equation (3.12) is not directly related to the uniqueness of the solution of equation (3.11). Indeed, when the PSF $K(\mathbf{x})$ is discretized and the DFT of the discrete PSF is computed, as a consequence of the discretization errors all values of the discrete TF $\hat{K}_{k,l}$ can be different from zero even if $K(\mathbf{x})$ is bandlimited: the values of $\hat{K}_{k,l}$ outside the band can be extremely small but not exactly zero. On the other hand, in the case of linear motion blur where uniqueness holds true, some values of $\hat{K}_m$ can be zero for the discrete problem – see equation (2.61) – and therefore in such a case uniqueness does not hold.

3.3   Existence of the solution and inverse filtering

In this section we investigate the existence of the solution in the case where uniqueness holds true. Indeed, as shown in the previous section, the uniqueness can be investigated independently of the existence. We consider first the case of equation (3.11) and then the case of its discrete version, equation (3.12).

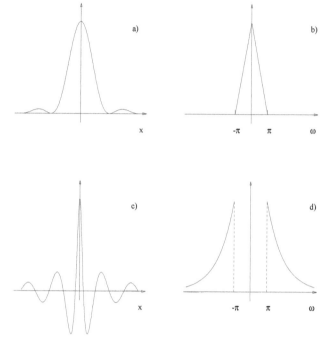

FIGURE 3.1 An example of invisible object: (a) the function $K(x)$ defining the convolution operator; (b) the Fourier transform of $K(x)$; (c) a function $f(x)$ such that $(K * f)(x) = 0$; (d) the Fourier transform of $f(x)$.

If the support of $\hat{K}(\omega)$ is the whole ω-space, equation (3.13) implies that the FT of the solution of equation (3.11) is given by

$$\hat{f}(\omega) = \frac{\hat{g}(\omega)}{\hat{K}(\omega)} \ . \tag{3.17}$$

However this equation defines a solution in $L^2(\mathbb{R}^d)$ if and only if the r.h.s. is the FT of a square-integrable function. In order to investigate this point, let us recall the model of image formation discussed in Section 2.1. According to this model, the FT of the noisy image $\hat{g}(\omega)$ is given by equation (2.29), where $\hat{f}^{(0)}(\omega)$ is the FT of the true object and $\hat{w}(\omega)$ is the FT of the noise contribution. If we substitute this equation into equation (3.17) we get

$$\hat{f}(\omega) = \hat{f}^{(0)}(\omega) + \frac{\hat{w}(\omega)}{\hat{K}(\omega)} \ . \tag{3.18}$$

We conclude that the function $\hat{f}(\omega)$ is the sum of the FT of the true object $\hat{f}^{(0)}(\omega)$ plus a term which comes from the inversion of the noise contribution. This second term may be responsible for the non-existence of the solution, i.e. the non-existence of a square-integrable inverse FT of $\hat{f}(\omega)$.

We first consider the case where $\hat{K}(\omega)$ is zero for some values of ω (this situation applies, for instance, to the linear-motion blur and to the out-of-focus blur). In this case $\hat{w}(\omega)$, in general, is not zero at the frequencies where the TF is zero because the noise is a process which is independent of the imaging process. As a consequence, in equation (3.18) we have division by zero and $\hat{f}(\omega)$ has singularities at the zeroes of $\hat{K}(\omega)$. This fact implies that the inverse FT of $\hat{f}(\omega)$ may not exist in the space of square-integrable functions and therefore equation (3.11) may not have a solution.

Moreover, even if $\hat{K}(\boldsymbol{\omega})$ does not vanish for some values of $\boldsymbol{\omega}$, it tends to zero when $|\boldsymbol{\omega}| \to \infty$. This property holds true for all TF considered in Chapter 2 and for all TF associated with integrable PSF (Riemann-Lebesgue theorem). Since the behavior of $\hat{w}(\boldsymbol{\omega})$ for large values of $|\boldsymbol{\omega}|$ is not related to the behavior of $\hat{K}(\boldsymbol{\omega})$, the ratio $\hat{w}(\boldsymbol{\omega})/\hat{K}(\boldsymbol{\omega})$ may not tend to zero. Depending on the relationship between the high-frequency behavior of the noise and that of the TF, this ratio may tend to infinity, or to a constant, or to zero. Possibly it may have no limit. The typical situation is that the inverse FT of this ratio does not exist in $L^2(\mathbb{R}^d)$ and therefore no solution of the problem exists.

The previous discussion can be synthesized in a precise mathematical form as follows: the solution of equation (3.11) exists and is square integrable if and only if the image $g(\mathbf{x})$ satisfies the following condition

$$\int_{\mathbb{R}^d} \left| \frac{\hat{g}(\boldsymbol{\omega})}{\hat{K}(\boldsymbol{\omega})} \right|^2 d\boldsymbol{\omega} < \infty , \tag{3.19}$$

in which case the solution (denoted by $f^\dagger$ to have the same notation as in Section 3.5 for the generalized solution) is given by

$$f^\dagger(\mathbf{x}) = \frac{1}{(2\pi)^d} \int_{\mathbb{R}^d} \frac{\hat{g}(\boldsymbol{\omega})}{\hat{K}(\boldsymbol{\omega})} e^{i\mathbf{x}\cdot\boldsymbol{\omega}} d\boldsymbol{\omega} . \tag{3.20}$$

Since condition (3.19) is not satisfied by arbitrary square-integrable functions $\hat{g}(\boldsymbol{\omega})$ and since, in general, noisy images do not satisfy this condition, the solution of equation (3.11), in general, does not exist. Obviously it does exist, however, in the case of a noise-free image. As concerns the continuous dependence of the solution on the data, it is sufficient to observe that, if we consider, for instance, the object $f^{(0)}(\mathbf{x}) = A_0 \exp(i\mathbf{x} \cdot \boldsymbol{\omega}_0)$, with a sufficiently large frequency $\boldsymbol{\omega}_0$, then its image is given by $g^{(0)}(\mathbf{x}) = A_0 \hat{K}(\boldsymbol{\omega}_0) \exp(i\mathbf{x} \cdot \boldsymbol{\omega}_0)$, whose amplitude can be very small even when the amplitude A_0 of the object is very large. For instance, if one takes $A_0 = |\hat{K}(\boldsymbol{\omega}_0)|^{-1/2}$, then, for $|\boldsymbol{\omega}_0| \to \infty$, the amplitude of the image tends to zero while the amplitude of the object tends to infinity.

The previous discussion makes clear that both the non-existence of the solution and the lack of uniqueness (discussed in the previous section), are due to the fact that the imaging system does not transmit complete information about the Fourier transform of the object at certain frequencies. This lack of information is a point which is very important and which must never be forgotten. To this purpose one must always remember a very incisive statement of Lanczos [188]: '*a lack of information cannot be remedied by any mathematical trickery*'. As we will see the methods developed for solving ill-posed problems are not based on mathematical trickeries but rather on a reformulation of the problem based on the use of additional information on the object. This additional information is compensating, at least partially, the lack of information due to the imaging system. This point will be further discussed in Section 3.6.

We conclude that, when uniqueness holds true, the problem of image deconvolution is ill posed because the second and third condition for well-posedness are not satisfied.

Consider now the discrete equation (3.12). In this case the situation is quite different. We know that the solution is unique if and only if $\hat{K}_{k,l} \neq 0$ for all values of k and l. If this condition is satisfied, from equation (3.14) we get

$$\hat{f}_{k,l} = \frac{\hat{g}_{k,l}}{\hat{K}_{k,l}} \tag{3.21}$$

and, by taking the inverse DFT, we obtain the following solution, denoted again as the generalized solution discussed in Section 3.5,

$$(\mathbf{f}^\dagger)_{m,n} = \frac{1}{N^2} \sum_{k,l=0}^{N-1} \frac{\hat{g}_{k,l}}{\hat{K}_{k,l}} \exp\left[i\frac{2\pi}{N}(km+ln)\right]. \qquad (3.22)$$

This solution exists for any noisy image $\mathbf{g}$ and also depends continuously on the data. Therefore we conclude that, when uniqueness holds true, the discrete problem (3.12) is well posed in the sense of Hadamard.

The procedure outlined above is usually called *inverse filtering* in the literature on image deconvolution. We also observe that equation (3.22) can be written in terms of the inverse of the matrix $\mathbf{A}$ associated with the PSF $\mathbf{K}$

$$\mathbf{f}^\dagger = \mathbf{A}^{-1}\mathbf{g} \qquad (3.23)$$

the matrix $\mathbf{A}^{-1}$ being the 2D matrix associated with the 2D *inverse* PSF or *inverse filter*

$$(\mathbf{K}^{-1})_{m,n} = \frac{1}{N^2} \sum_{k,l=0}^{N-1} \frac{1}{\hat{K}_{k,l}} \exp\left[i\frac{2\pi}{N}(km+ln)\right]. \qquad (3.24)$$

It follows that equation (3.23) can also be written as

$$(\mathbf{f}^\dagger)_{m,n} = \sum_{k,l=0}^{N-1} (\mathbf{K}^{-1})_{m-k,n-l}\, g_{k,l} \, . \qquad (3.25)$$

We have given the equations for the 2D case. Their modification in order to cover the 1D and 3D case is easy, since, in these cases, the 2D DFT has simply to be replaced by the 1D and 3D DFT respectively.

3.4 Discretization: from ill-posedness to ill-conditioning

A puzzling implication of the previous analysis is that, even if equation (3.11) is ill posed because, in general, its solution does not exist, its discrete version (3.12) can be well posed because it has a unique solution which depends continuously on the data. Nevertheless, in this section we show that this solution is, in general, unacceptable from the physical point of view because it is completely corrupted by noise. The usual analysis of error propagation from the data to the solution proceeds as follows.

Let us consider a (small) variation $\delta\mathbf{g}$ of the discrete image $\mathbf{g}$. The corresponding variation of the solution $\delta\mathbf{f}^\dagger$ is given by

$$\delta\mathbf{f}^\dagger = \mathbf{A}^{-1}\delta\mathbf{g} \qquad (3.26)$$

thanks to the linearity of equation (3.23). Then one easily obtains the bound

$$\|\delta\mathbf{f}^\dagger\| \leq \frac{1}{\hat{K}_{min}}\, \|\delta\mathbf{g}\| \qquad (3.27)$$

where $\|\cdot\|$ is the usual Euclidean norm and $\hat{K}_{min}$ is the minimum value of $|\hat{K}_{k,l}|$. Analogously, from equation (3.12) we get

$$\|\mathbf{g}\| \leq \hat{K}_{max}\, \|\mathbf{f}^\dagger\| \qquad (3.28)$$

where $\hat{K}_{max}$ is now the maximum value of $|\hat{K}_{k,l}|$. By combining the two bounds we obtain

$$\frac{\|\delta\mathbf{f}^\dagger\|}{\|\mathbf{f}^\dagger\|} \leq \frac{\hat{K}_{max}}{\hat{K}_{min}}\, \frac{\|\delta\mathbf{g}\|}{\|\mathbf{g}\|} \, . \qquad (3.29)$$

The quantity

$$\alpha = \frac{\hat{K}_{max}}{\hat{K}_{min}} \tag{3.30}$$

is called the *condition number* of the problem; it is the quantity which controls error propagation from the data to the solution.

Let us first remark that *the inequality* (3.29) *can not be improved because, in some cases, equality holds true*. To show this let us consider, for simplicity, a 1D case and let us represent vectors as in (2.40), i.e. as expansions in terms of the orthonormal vectors $\mathbf{v}_0, \mathbf{v}_1, \ldots, \mathbf{v}_{N-1}$. Let us also assume that $\hat{K}_{max} = \hat{K}_0$ (i.e. the maximum value of the TF corresponds to the zero frequency component) while $\hat{K}_{min} = |\hat{K}_{N/2}|$ (i.e. the minimum value of the TF corresponds to the highest frequency component). This assumption is not really restrictive because very often in practical problems the maximum corresponds to a low-frequency component while the minimum corresponds to a high-frequency component. We consider now the following particular case: a data vector $\mathbf{g}$ such that only its component associated with $\mathbf{v}_0$ is different from zero and an error vector $\delta\mathbf{g}$ such that only its component associated with $\mathbf{v}_{N/2}$ is different from zero:

$$\mathbf{g} = \frac{1}{\sqrt{N}} \, \hat{g}_0 \mathbf{v}_0 \, , \ \delta\mathbf{g} = \frac{1}{\sqrt{N}} \, \hat{\delta g}_{N/2} \mathbf{v}_{N/2} \, . \tag{3.31}$$

The corresponding solution $\mathbf{f}^\dagger$ and error $\delta\mathbf{f}^\dagger$ are given by

$$\mathbf{f}^\dagger = \frac{1}{\sqrt{N}} \frac{\hat{g}_0}{\hat{K}_0} \mathbf{v}_0 \, , \quad \delta\mathbf{f}^\dagger = \frac{1}{\sqrt{N}} \frac{\hat{\delta g}_{N/2}}{\hat{K}_{N/2}} \mathbf{v}_{N/2} \tag{3.32}$$

so that

$$\|\delta\mathbf{f}^\dagger\| = \frac{\|\delta\mathbf{g}\|}{\hat{K}_{min}} \, , \ \|\mathbf{f}^\dagger\| = \frac{\|\mathbf{g}\|}{\hat{K}_{max}} \, . \tag{3.33}$$

These equations imply that, for this particular example, equality holds true in equation (3.29). Hence it is not possible in general to improve the inequality (3.29), even if it may be pessimistic in practical situations.

A second remark is that, *when discretizing an ill-posed problem, the condition number of the corresponding discrete problem can be very large and even extremely large*. Indeed, the values of the discrete TF are approximations of sampled values of $\hat{K}(\omega)$, which can be zero at certain frequencies and, in any case, tends to zero when $|\omega| \to \infty$. Moreover, we expect that, if we discretize more finely the problem, we will obtain even smaller values of $\hat{K}_{min}$. Consider, for example, a PSF with a Gaussian shape, i.e. $K(x) = (2\pi)^{-1/2} \exp(-x^2/2)$. If we sample this function in 32 points in the interval $[-3, 3]$ and we compute the DFT of these samples, we find that the value of the condition number (3.30) is $\alpha = 4.35 \times 10^3$. However, if we sample the same function in 128 points in the interval $[-6, 6]$ and we compute again the DFT of these samples, we find that the condition number is now $\alpha = 3.42 \times 10^7$. From such example we can infer that in general *the finer is the discretization of the ill-posed problem, the larger is the condition number of the corresponding discrete problem*.

A problem with a large condition number is called *ill conditioned* while a problem whose condition number is close to one is called *well conditioned*. An example of a well-conditioned problem is image deconvolution when the TF is a phase factor, i.e. $\hat{K}_m = \exp(i\phi_m)$ with ϕ_m real.

The previous analysis implies that when discretizing an ill-posed problem we usually get an ill-conditioned problem and, in fact, a very ill-conditioned one if the discretization is very fine. If the discretization is coarse, however, the discrete problem can be moderately ill conditioned and even nearly well conditioned if the discretization is very coarse. It is obvious that a large condition number implies *numerical instability*: for instance, if $\alpha = 10^6$, a relative error on the data of the order of 10^{-6} may imply an error of 100% on the solution. Therefore we see that *continuous dependence of the solution on the data is necessary but not sufficient to guarantee numerical stability.*

An example showing the numerical instability of the inverse filter is provided in Figure 3.2. In this figure, we give two reconstructions of the object of Figure 2.5. In (a) we reproduce the object while in (b) we give its image corrupted by linear motion blur. If no noise is added to this image, the result obtained by applying the inverse filtering is shown in (c). It is clear that the reconstruction is quite good. If white Gaussian noise is added to the image in (b) and inverse filtering is applied again, the result is shown in (d). It is obvious that the reconstruction is quite bad, even if it is difficult to represent graphically its main feature: large positive and negative values in adjacent pixels. In fact, instability appears, in general, in the form of wild oscillations superimposed to the image since it is mainly generated by the noise corrupting the high-frequency components of the image.

In the case of synthetic data, when the undistorted object $\mathbf{f}^{(0)}$ is available (see the discussion in Section 14.2) a measure of the improvement introduced by the process of deconvolution is given by the so-called *Mean Square Error Improvement Factor* (MSEIF) defined by

$$\text{MSEIF} = 20 \, \log_{10} \frac{\|\mathbf{g} - \mathbf{f}^{(0)}\|}{\|\mathbf{f} - \mathbf{f}^{(0)}\|} (\text{dB}) \tag{3.34}$$

where $\mathbf{f}^{(0)}$ is the object, $\mathbf{g}$ is the noisy image and $\mathbf{f}$ is the deconvolved image. In the case of Figure 3.2, MSEIF= 88 dB for the reconstruction obtained from the noise-free image and MSEIF= -19 dB for the reconstruction from the noisy image. In the noisy case the inverse filtering has severely degraded the quality of the image.

REMARK 3.2   *The condition number (3.30) quantifies error propagation not only in the solution of the inverse problem but also in the solution of the direct one, which consists in computing the image $\mathbf{g}^{(0)} = \mathbf{A}\mathbf{f}^{(0)}$. Indeed inequality (3.27) holds true when $\delta\mathbf{f}^\dagger$ and $\delta\mathbf{g}$ are replaced respectively by $\mathbf{f}^{(0)}$ and $\mathbf{g}^{(0)}$, while inequality (3.28) holds true when $\mathbf{g}$ and $\mathbf{f}^\dagger$ are replaced by $\delta\mathbf{g}^{(0)}$ and $\delta\mathbf{f}^{(0)}$. As a consequence we obtain*

$$\frac{\|\delta\mathbf{g}^{(0)}\|}{\|\mathbf{g}^{(0)}\|} \leq \frac{\hat{K}_{max}}{\hat{K}_{min}} \frac{\|\delta\mathbf{f}^{(0)}\|}{\|\mathbf{f}^{(0)}\|} . \tag{3.35}$$

We should conclude that, in the case of a large condition number, the direct problem is affected by uncontrolled error propagation from the data $\mathbf{f}^{(0)}$ to the solution $\mathbf{g}^{(0)}$, even if this result is in conflict with the everyday experience that the image $\mathbf{g}^{(0)}$ can be accurately computed also when the object $\mathbf{f}^{(0)}$ is affected by errors.

The inequality (3.35) can be understood by looking at the case where equality holds true. Under the same assumptions as for equation (3.29), we have equality in equation (3.35) when $\mathbf{f}^{(0)}$ is proportional to $\mathbf{v}_{N/2}$ (and therefore is a rapidly oscillating vector) while $\delta\mathbf{f}^{(0)}$ is proportional to $\mathbf{v}_0$ (and therefore is a constant vector representing, for instance, a systematic error); indeed, we then have $\|\mathbf{g}^{(0)}\| = \hat{K}_{min}\|\mathbf{f}^{(0)}\|$ and $\|\delta\mathbf{g}^{(0)}\| = \hat{K}_{max}\|\delta\mathbf{f}^{(0)}\|$.

In conclusion, uncontrolled error propagation appears in the solution of the inverse problem when the noise-free image is smooth and the error is rapidly varying (and this is precisely the usual situation in practice) whereas it appears in the solution of the direct problem when the object is rapidly varying and the error is smooth (and this is a very unusual situation). When we compute the solution of the direct problem in the case of a reasonably smooth object, the small round-off errors (which can be rapidly oscillating) not only do not produce numerical instability, but are in fact reduced by the small values of $\hat{\mathbf{K}}$ at high frequencies.

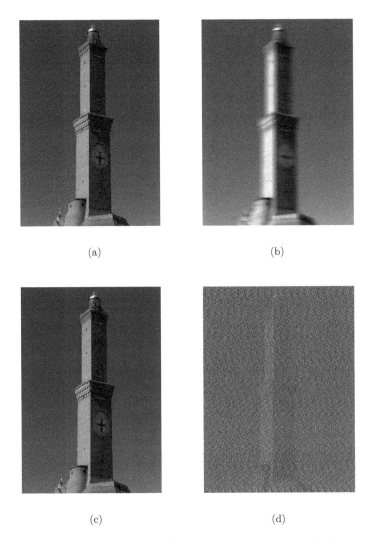

(a)   (b)

(c)   (d)

FIGURE 3.2 (a) The object of Figure 2.9; (b) noise-free image corrupted by linear motion blur; (c) solution obtained by inverse filtering in the noise-free case; (d) solution obtained by inverse filtering in the noisy case (white Gaussian noise added to the image in (b)).

The remark emphasizes the fact that the condition number can be a pessimistic estimate of error propagation even if, in the case of the inverse problems, the practical situation is close to the situation corresponding to equality in equation (3.29). A better estimate of error propagation is the so-called *Average Relative Error Magnification Factor* (AREMF) introduced by Twomey [290]. For image reconstruction (1D case) it is given by

$$\beta = \frac{1}{N} \left(\sum_{m=0}^{N-1} |\hat{K}_m|^2 \right)^{1/2} \left(\sum_{m=0}^{N-1} |\hat{K}_m|^{-2} \right)^{1/2} . \tag{3.36}$$

For example in the case of the Gaussian PSF considered above when $\alpha = 4.35 \times 10^3$, we have $\beta = 6.25 \times 10^2$, and when $\alpha = 3.42 \times 10^7$, we have $\beta = 1.83 \times 10^6$.

3.5 Bandlimited systems: least-squares solutions and generalized solution

In Section 3.2 we proved that, if the imaging system is bandlimited, then the solution of the image deconvolution problem is not unique as a consequence of the existence of invisible objects. We now discuss the existence of the solution and we introduce a modification of the concept of solution which is very useful in many circumstances.

Let us first consider equation (3.11) and its version in terms of Fourier transforms, equation (3.13). When ω does not belong to the band $\mathcal{B}$ of the instrument, the first member of this latter equation is zero because $\hat{K}(\omega)$ is zero. On the other hand, the second member $\hat{g}(\omega)$ is given by equation (2.29) and therefore, in general, is not zero because the noise term $\hat{w}(\omega)$, due to the recording process, may be different from zero. Therefore, equation (3.13) becomes $0 = \hat{w}(\omega)$ at these frequencies and it is inconsistent, i.e. the solution of equation (3.11) does not exist. The function $\hat{w}_{out}(\omega)$, which is zero when ω is in $\mathcal{B}$ and coincides with $\hat{w}(\omega)$ when ω is not in $\mathcal{B}$, is called the *out-of-band noise*. We notice that a realization of the out-of-band noise is also an invisible object, in the sense that, if it is imaged by the bandlimited imaging system, its image is zero. Therefore, the out-of-band noise is always orthogonal to the noise-free image, i.e. the image which is in the range of the operator A and, more generally, to all bandlimited functions with band $\mathcal{B}$.

If the non-existence of the solution is only due to out-of-band noise, then it is easy to remove this obstacle. We introduce the projection operator over the subspace of the bandlimited functions whose band coincides with or is contained in the band $\mathcal{B}$ of the imaging system

$$(P^{(\mathcal{B})}g)(\mathbf{x}) = \frac{1}{(2\pi)^d} \int \chi_{\mathcal{B}}(\omega)\,\hat{g}(\omega)\,e^{i\mathbf{x}\cdot\omega}\,d\omega \tag{3.37}$$

where $\chi_{\mathcal{B}}(\omega)$ is the characteristic function of the set $\mathcal{B}$. The operator $P^{(\mathcal{B})}$ can be called *bandlimiting operator*. If the function $g(\mathbf{x})$ is not bandlimited, it is obvious that $P^{(\mathcal{B})}g$ is bandlimited and orthogonal to the out-of-band noise. Then, a quite natural idea is to modify equation (3.11) by suppressing the out-of-band noise with the operator $P^{(\mathcal{B})}$, i.e. to replace equation (3.11) by the following one

$$Af = P^{(\mathcal{B})}g . \tag{3.38}$$

The FT's of the two members of this equation are zero outside the band while on the band we reobtain equation (3.13). At this point one can repeat the discussion of Section 3.3 and conclude that a solution of equation (3.38) exists if and only if the following condition is satisfied

$$\int_{\mathcal{B}} \left| \frac{\hat{g}(\omega)}{\hat{K}(\omega)} \right|^2 d\omega < \infty \tag{3.39}$$

which is analogous to condition (3.19), the integration being now restricted to the band $\mathcal{B}$ of the imaging system.

If condition (3.39) is satisfied for any g, then equation (3.38) has a solution for any g. If we look at the examples of Chapter 2, we find that condition (3.39) is satisfied for any g in the case of a diffraction-limited system with coherent illumination (with or without aberrations) as well as in the case of far-field acoustic holography. In all these cases we have $|\hat{K}(\boldsymbol{\omega})| = 1$. On the other hand, if the bandlimited PSF is integrable, then, as we already remarked, the TF $\hat{K}(\boldsymbol{\omega})$ is continuous and therefore, since it is zero outside the band it must be zero also on the boundary of the band. Examples are provided by diffraction-limited imaging systems in the case of incoherent illumination. In these cases, equation (3.39) is not satisfied by a noisy image because the noise, in general, is not zero on the boundary of the band. As a consequence equation (3.38) has, in general, no solution.

It is easy to prove that equation (3.38) is equivalent to the following one

$$A^*Af = A^*g \tag{3.40}$$

in the sense that any solution of this equation is also a solution of equation (3.38) and conversely. Indeed, by taking the Fourier transform of both sides of equation (3.40), we obtain

$$|\hat{K}(\boldsymbol{\omega})|^2 \hat{f}(\boldsymbol{\omega}) = \hat{K}^*(\boldsymbol{\omega})\, \hat{g}(\boldsymbol{\omega}) \; ; \tag{3.41}$$

when $\hat{K}(\boldsymbol{\omega}) \neq 0$, $\hat{f}(\boldsymbol{\omega})$ is given by equation (3.17) whereas, when $\hat{K}(\boldsymbol{\omega}) = 0$, the equation (3.41) is identically satisfied and $\hat{f}(\boldsymbol{\omega})$ is arbitrary. These are precisely the properties of the solutions of equation (3.38)

It is also easy to prove that equation (3.40) is the Euler-Lagrange equation of the variational problem

$$\|Af - g\| = \text{minimum} \tag{3.42}$$

and therefore its solutions are the objects which provide the best approximation of the given image g in the sense of the L^2-norm. Any function f which solves this minimization problem is also called a *least-squares solution*. The name comes from the fact that, in the discrete case, such a solution minimizes the sum of the squares of the differences between the components of $\mathbf{Af}$ and the components of $\mathbf{g}$.

Since any least-squares solution is also a solution of equation (3.38), it follows from the previous analysis that *least-squares solutions exist for a given g if and only if g satisfies condition* (3.39). The solution of the least-squares problem (3.42), however, is not unique because to any given least-squares solution we can add an arbitrary invisible object and obtain in this way another least-squares solution. If we do not have additional information on the object, it can be reasonable to consider the unique least-squares solution which is zero outside the band of the imaging system. When condition (3.39) is satisfied, this particular least-squares solution, which is called the *generalized solution* and denoted by $f^{\dagger}(\mathbf{x})$, is given by

$$f^{\dagger}(\mathbf{x}) = \frac{1}{(2\pi)^d} \int_{\mathcal{B}} \frac{\hat{g}(\boldsymbol{\omega})}{\hat{K}(\boldsymbol{\omega})}\, e^{i\mathbf{x}\cdot\boldsymbol{\omega}}\, d\boldsymbol{\omega} \; . \tag{3.43}$$

One can show that *the generalized solution is the unique least-squares solution of minimal norm. It is also the unique least-squares solution orthogonal to the subspace of all invisible objects.*

If the function $1/\hat{K}(\boldsymbol{\omega})$ is not bounded as a consequence of zeroes of $\hat{K}(\boldsymbol{\omega})$ inside or at the boundary of the band $\mathcal{B}$, the generalized solution (3.43) does not exist for arbitrary $\hat{g}(\boldsymbol{\omega})$ – and, in general, does not exist for noisy $\hat{g}(\boldsymbol{\omega})$ – so that *the problem of determining the generalized solution is ill-posed.* Nevertheless, equation (3.43) defines a linear mapping from the set of all noise-free images into the set of all bandlimited objects. This linear mapping is called the *generalized inverse operator* of the operator A and is denoted by $A^\dagger$, so that $f^\dagger = A^\dagger g$. If we consider a noise-free image $g = Af$, then, from equation (3.43) with $\hat{g}(\boldsymbol{\omega}) = \hat{K}(\boldsymbol{\omega})\hat{f}(\boldsymbol{\omega})$ we find that

$$A^\dagger g = P^{(\mathcal{B})} f \tag{3.44}$$

where $P^{(\mathcal{B})}$ is the bandlimiting operator defined in equation (3.37).

The problem of determining the generalized solution is well posed in the important cases, discussed in Chapter 2, where $|\hat{K}(\boldsymbol{\omega})| = 1$ on the band $\mathcal{B}$. In these cases, if we introduce the *generalized inverse* PSF defined by

$$K^\dagger(\mathbf{x}) = \frac{1}{(2\pi)^d} \int_{\mathcal{B}} \frac{1}{\hat{K}(\boldsymbol{\omega})} \, e^{i\mathbf{x}\cdot\boldsymbol{\omega}} \, d\boldsymbol{\omega} \;, \tag{3.45}$$

we can rewrite equation (3.43) as follows

$$f^\dagger = K^\dagger * g \tag{3.46}$$

and we see that the generalized inverse is a linear and continuous operator in the space of the square-integrable functions, given by

$$A^\dagger g = K^\dagger * g \;. \tag{3.47}$$

A similar analysis of the discrete case is quite easy. In the 2D case the band is now the set B of the pairs of indices $\{k, l\}$ such that $\hat{K}_{k,l} \neq 0$. Then by a trivial repetition of the various steps discussed above one concludes that there exists a unique generalized solution given by

$$(\mathbf{f}^\dagger)_{m,n} = \frac{1}{N^2} \sum_{k,l \in B} \frac{\hat{g}_{k,l}}{\hat{K}_{k,l}} \exp\left[i\frac{2\pi}{N}(km + ln)\right], \tag{3.48}$$

the difference with respect to equation (3.22) being that the summation is only extended to the pairs of indices in the set B. This generalized solution can also be written in terms of a *generalized inverse matrix* $\mathbf{A}^\dagger$ (also called *Moore-Penrose inverse matrix*)

$$\mathbf{f}^\dagger = \mathbf{A}^\dagger \mathbf{g} \;. \tag{3.49}$$

The matrix $\mathbf{A}^\dagger$ is the matrix associated with the 2D array defined as in equation (3.24) but with the summation restricted to the set B. This array will be denoted by $\mathbf{K}^\dagger$ and will be called the *generalized inverse filter* or the *generalized inverse* PSF. Error propagation from $\mathbf{g}$ to $\mathbf{f}^\dagger$ can be analyzed along the lines indicated in Section 3.4 and the result is that numerical stability is still controlled by the condition number as given in equation (3.30), but with $\hat{K}_{min}$ being now the minimum of $|\hat{K}_{k,l}|$ over the set B.

In conclusion, let us consider a few examples where the generalized solution can be used. The first one is that of an ideal diffraction-limited system in the case of coherent illumination. The TF of the imaging system is simply the characteristic function of the band $\mathcal{B}$ of the instrument and therefore, from equation (3.45) with $\hat{K}(\boldsymbol{\omega}) = \chi_\mathcal{B}(\boldsymbol{\omega})$, we see that $K^\dagger(\mathbf{x})$ coincides with the PSF of the imaging system. In such a case the generalized solution is simply the noisy image after suppression of the out-of-band noise.

More interesting results can be obtained in the case of aberrations or in the case of far-field acoustic holography. In the latter case, from equation (2.97) and the far-field approximation we deduce that the generalized inverse PSF is given by

$$K^\dagger(\mathbf{x}) = \frac{1}{(2\pi)^2} \int_{|\boldsymbol{\omega}| \leq k} \exp\left(-ia(k^2 - |\boldsymbol{\omega}|^2)^{1/2}\right) \, e^{i\mathbf{x}\cdot\boldsymbol{\omega}} \, d\boldsymbol{\omega} \;. \tag{3.50}$$

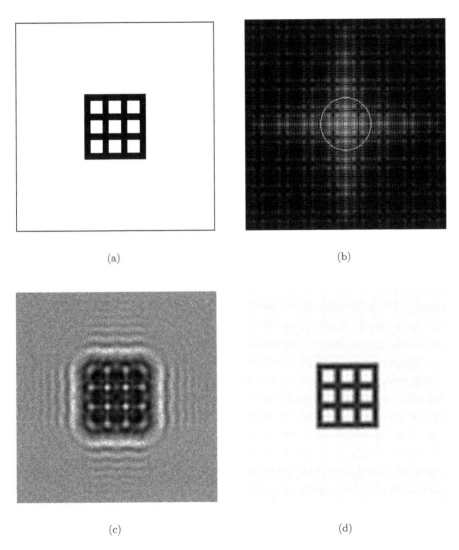

(a)

(b)

(c)

(d)

FIGURE 3.3 Example of the use of the generalized solution in far-field acoustic holography: (a) the object, $10\lambda \times 10\lambda$ wide; (b) modulus of the FT of the object (the white circle indicates the band of the far-field data); (c) modulus of the noisy image at a distance 5λ from the source plane; (d) the reconstruction obtained by means of the generalized inverse.

The operator defined by the convolution of the image g with the generalized inverse PSF is usually called the *back-propagation operator* because it transforms the field amplitude on the plane $x_3 = a$ into the field amplitude on the plane $x_3 = 0$.

In Figure 3.3, we present the result of a numerical simulation. The original object is a binary object consisting of a square grid, with overall dimensions $10\lambda \times 10\lambda$. The size of the horizontal and vertical bars is λ and therefore is greater than the Rayleigh resolution distance (see Section 2.6). The receiving plane is located at a distance 5λ from the source plane. At this distance the effect of evanescent waves is negligible. The image $g^{(0)}(\mathbf{x})$, which is the complex amplitude of the acoustic field in the receiving plane, has been corrupted by white Gaussian noise (SNR $\simeq 50$ dB). The picture of $|g(\mathbf{x})|$, given in Figure 3.3(c), is a picture of the shadow of the object. The generalized solution corresponding to the noisy image $g(\mathbf{x})$ is given in Figure 3.3(d). The result is quite satisfactory and the reason is that the object considered does not contain details smaller than the Rayleigh resolution distance which is of the order of $\lambda/2$.

3.6 Approximate solutions and the use of 'a priori' information

The analysis made in the previous sections has shown that the solution of the problem of image deconvolution may not exist while the solution (or generalized solution) of the corresponding discrete problem always exists but may be completely corrupted by noise and therefore completely deprived of physical meaning. It should be clear, however, that the image g certainly contains information about the object $f^{(0)}$ and therefore that it should be possible to find some estimate of $f^{(0)}$ better than g itself. It should also be clear that, for doing some progress in this direction, one must modify the concept of *solution* of the problem. A first step in this direction has been already performed in the case of bandlimited imaging systems where the concept of solution has been replaced by that of least-squares solution and of generalized solution. However this step is not yet sufficient, because also the generalized solution may not exist or may be completely corrupted by noise in the discrete case.

Since the solution, when it exists, is the object which reproduces exactly the noisy data, i.e. the noisy image g (the generalized solution is the object which reproduces the data in the best possible way), and since this solution is unacceptable, the only possible way is to look for solutions which reproduce only approximately the noisy image, i.e. for *approximate solutions*. We point out that the true object $f^{(0)}$ is such an approximate solution because, as a consequence of the noise, it reproduces the noisy image g not exactly but only up to the noise term w.

If f is an approximate solution, then the image Af associated with f does not coincide in general with g; the difference between the computed image Af and the measured image g is called *residual* and will be denoted by r

$$r = g - Af . \tag{3.51}$$

The square norm of the residual will be called the *discrepancy* and will be denoted by $\varepsilon^2(f;g)$. In the discrete case the norm of the residual $\mathbf{r}$ is the absolute root-mean-square (rms) error (RMSE) we commit when estimating $\mathbf{g}$ by means of $\mathbf{Af}$.

The point now is to understand the structure of the set of all approximate solutions compatible with the noisy image within a given noise level. To simplify the analysis we only consider the discrete case. Next, let us assume that we know an estimate ε of the absolute rms-error affecting the noisy image $\mathbf{g}$. Under these conditions, the set of all approximate solutions compatible with the noisy image is the set of all objects $\mathbf{f}$ whose discrepancy does

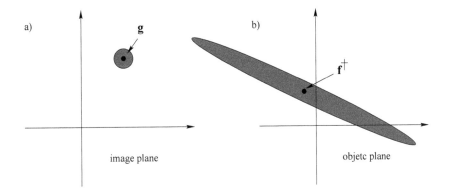

FIGURE 3.4 Two-dimensional representation of the situation described by equation (3.52). The set in (b) is the set of all objects which correspond to images in the small circle around **g** in (a).

not exceed ε^2, i.e.

$$\|\mathbf{Af} - \mathbf{g}\|^2 \leq \varepsilon^2 \ . \tag{3.52}$$

If ε^2 is a reliable estimate then it is obvious that both the true object $\mathbf{f}^{(0)}$ and the unphysical solution, discussed in the previous sections, belong to this set, which therefore contains completely different elements. As a consequence, we expect that this set is extremely large and this can be quantified by a more detailed analysis.

For the sake of simplicity we consider the 1D case, assuming that all components of the TF, $\hat{K}_m$, are different from zero. In such a case the solution of the discrete problem is given by

$$(\mathbf{f}^{\dagger})_n = \frac{1}{N} \sum_{m=0}^{N-1} \frac{\hat{g}_m}{\hat{K}_m} \exp\left(i\frac{2\pi}{N}nm\right) \ . \tag{3.53}$$

Then, using the Parseval equality for the DFT, we can write condition (3.52) as follows

$$\frac{1}{N} \sum_{m=0}^{N-1} |\hat{K}_m \hat{f}_m - \hat{g}_m|^2 \leq \varepsilon^2 \tag{3.54}$$

or else

$$\sum_{m=0}^{N-1} \left(\frac{|\hat{K}_m|^2}{N\varepsilon^2}\right) \left|\hat{f}_m - \frac{\hat{g}_m}{\hat{K}_m}\right|^2 \leq 1 \ . \tag{3.55}$$

This condition defines, in the space of the DFT of the objects, the set of the interior points of an N-dimensional ellipsoid whose center is the DFT of the solution (3.53) and whose half axes have lengths

$$a_m = \frac{\sqrt{N}\,\varepsilon}{|\hat{K}_m|} \ . \tag{3.56}$$

Since the DFT defines a unitary transformation, in the space of the objects we still have an N-dimensional ellipsoid with center $\mathbf{f}^{\dagger}$ and half axes in the directions of the inverse DFT of the vectors $\mathbf{v}_m$, equation (2.38), whose lengths are given again by equation (3.56), up to the normalization factor $\sqrt{N}$. This ellipsoid is the inverse image of a small ball of radius ε, centered at the noisy image **g** (see Figure 3.4). The ratio between the maximum half axis and the minimum half axis of the ellipsoid is precisely the condition number introduced in Section 3.4.

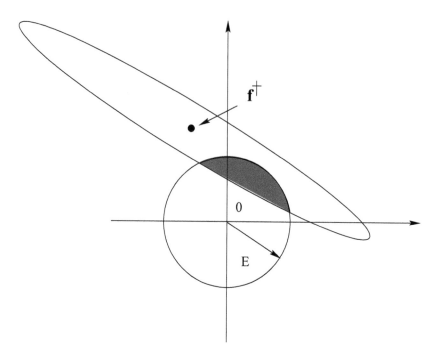

FIGURE 3.5 Two-dimensional representation of the effect of a bound on the norm of the unknown object. The estimates of the object compatible with the data and the bound E on their norm are the points in the intersection of the ellipse with the disc of radius E.

As we see this set is too broad and it is hopeless to extract from it a useful estimate of $\mathbf{f}^{(0)}$ without using additional information. Indeed when we say that the solution or generalized solution $\mathbf{f}^{\dagger}$ is unacceptable from the physical point of view, some additional information on the object $\mathbf{f}^{(0)}$ is implicit in the statement. For instance we may know that the components of $\mathbf{f}^{\dagger}$ are too large and/or too rapidly oscillating to correspond to a physical object. This remark suggests a first remedy to the pathology of the problem. If we know an upper bound on the norm of the unknown object $\mathbf{f}^{(0)}$, then we can look only for the objects $\mathbf{f}$ which are compatible both with the data and with this bound (see Figure 3.5). More generally, any kind of property of the solution which can be expressed in a suitable mathematical form should be used. Before providing a list of such possible properties, we will discuss further the basic idea. The picture of Figure 3.4 visualizes the fact that the image does not contain sufficient information about the object. A very large set of objects (side (b) of the figure) is compressed by the imaging system into a very small set of images (side (a) of the same figure). This loss of information must be compensated in some way by additional information, coming from other sources or from previous information we have about the object. Mathematically, such information will be expressed in the form of constraints on the solution. This is usually called *a priori information* (or prior information) and, in a more or less explicit form, this is the basic principle on which all inversion methods rely. In Figure 3.5, we see that this additional information can considerably reduce the uncertainty about the unknown object. We assume for example that the norm of the object is bounded by a given constant E. The objects compatible with the noisy image are represented by the points interior to the ellipse while the objects compatible with the constraint are represented by the points interior to the circle of radius E. Any point of the dashed region may be considered as a satisfactory estimate of the unknown object because it is compatible both with the data and with the additional information. The picture makes also clear that a further reduction of the size of the compatible objects is obtained if we know that they are nonnegative.

Let us remark that information about the noise, in particular, about its statistical properties, is also useful. This is the starting point of the maximum-likelihood methods which will be discussed in Chapter 11. However, it must be clear that information about noise can somehow modify the structure of the set of objects compatible with the data but does not reduce in a significant way the lack of information about the object. Indeed, this is an intrinsic property of the imaging system, i.e. of its PSF, which cannot be compensated by information about the noise. A priori information about the object is then also required and this will distinguish the Bayesian methods from the maximum-likelihood methods, as discussed in Chapter 11.

The most convenient way of expressing *a priori* information about the solution of the image reconstruction problem is to state that the object must belong to some given subset of the space of all possible objects. We give here some examples of sets which are used very frequently in image deconvolution and, more generally, in the solution of inverse problems. We consider mainly the case of functions but the analogous sets for discrete images can be derived in a straightforward way.

- The set of all functions whose 'energy', defined as the square of their L^2-norm, does not exceed a prescribed value

$$\|f\|^2 = \int |f(\mathbf{x})|^2 d\mathbf{x} \le E^2 \ . \tag{3.57}$$

  This is the ball, with radius E and the zero function as center, in the space of all square-integrable functions.

- The set of all functions whose derivatives up to a certain order satisfy some prescribed bound. In general, if D is a differential operator (for instance, the Laplace operator, $D = \Delta$), then the set is defined by

$$\|Df\|^2 = \int |(Df)(\mathbf{x})|^2 d\mathbf{x} \le E^2 \ . \tag{3.58}$$

  Examples are considered in the next chapter.

- The set of all functions which are zero outside a bounded region $\mathcal{D}$. This constraint can be used in image deconvolution when we know that the object is localized in one or more small portions of its domain of definition.

- The set of all functions whose Fourier transform is zero outside a bounded domain $\mathcal{B}$. In signal processing this constraint is used for the extrapolation of bandlimited signals.

- The set of all functions which are nonnegative. This is a constraint which is, in general, very important in image deconvolution.

- The set of all functions which coincide with a prescribed function on some bounded region $\mathcal{D}$ or which take some prescribed value in some points of the domain where they are defined.

- A set defined as in the previous case but with the Fourier transform in place of the function itself.

- Any intersection of some of the previous sets.

A common feature of the above sets is that they are *closed* and *convex*. If we denote by $\mathcal{C}$ such a set, then closure means that, if we consider a sequence of functions f_n which belong

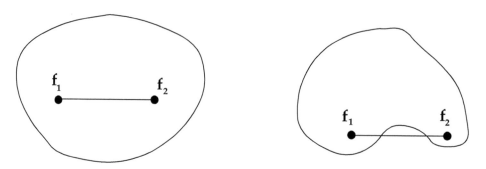

FIGURE 3.6   Two-dimensional example of a convex set (left) and of a non-convex set (right).

to $\mathcal{C}$ and if this sequence is convergent to a function f, then also f belongs to $\mathcal{C}$. On the other hand, convexity is a geometrical property of $\mathcal{C}$ which can be formulated as follows: if $f^{(1)}$ and $f^{(2)}$ are two functions in $\mathcal{C}$, then all functions f given by

$$f = tf^{(1)} + (1-t)f^{(2)} , \quad 0 \le t \le 1 \tag{3.59}$$

also belong to $\mathcal{C}$. This means that the segment of straight line joining $f^{(1)}$ to $f^{(2)}$ belongs also to $\mathcal{C}$ (see Figure 3.6).

A basic property of a closed and convex set $\mathcal{C}$ in a Hilbert space is that, given an arbitrary element f of the space, which does not belong to $\mathcal{C}$, there exists a unique element $f_\mathcal{C}$ of $\mathcal{C}$ which has minimal distance from f [17], i.e.

$$\|f_\mathcal{C} - f\| = \min_{h \in \mathcal{C}} \|h - f\| . \tag{3.60}$$

This element is called the *convex projection* (or metric projection) of f onto $\mathcal{C}$. The operator which associates to any object f its convex projection $f_\mathcal{C}$ is called *projection operator onto the set $\mathcal{C}$* and denoted by $P_\mathcal{C}$

$$P_\mathcal{C} f = f_\mathcal{C} . \tag{3.61}$$

This operator, which is used in Section 5.2, is in general nonlinear. For example, in the case of the set $\mathcal{C}$ defined by condition (3.57), the operator $P_\mathcal{C}$ is given by

$$P_\mathcal{C} f = \frac{E}{\|f\|} f \tag{3.62}$$

if $\|f\| > E$, while $P_\mathcal{C} f$ coincides with f otherwise. In the case of the set of nonnegative functions, the operator $P_\mathcal{C}$ is given by

$$(P_\mathcal{C} f)(\mathbf{x}) = \begin{cases} f(\mathbf{x}) & \text{if } f(\mathbf{x}) > 0 \\ 0 & \text{if } f(\mathbf{x}) \le 0 . \end{cases} \tag{3.63}$$

The effect of this projection operator is illustrated in Figure 3.7.

A basic property of the convex projections is the following one [268]: *if f_1, f_2 is any pair of elements of $\mathcal{X}$ and $f_{1,c}$, $f_{2,c}$ the corresponding convex projections onto the closed and convex set $\mathcal{C}$, then*

$$\|f_{1,c} - f_{2,c}\| \le \|f_1 - f_2\| . \tag{3.64}$$

This result is rather intuitive, as suggested by Figure 3.8. The equality in equation (3.64) holds true, for instance, if both f_1 and f_2 are elements of $\mathcal{C}$. This property of the convex projections means that the operator $P_\mathcal{C}$, defined in equation (3.60) is a so-called nonexpansive mapping (see Section 5.3).

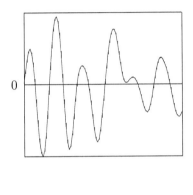 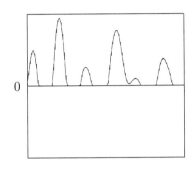

(a) (b)

FIGURE 3.7 Illustrating the effect of the projection onto the convex set of nonnegative functions: (a) the original function; (b) the projected function.

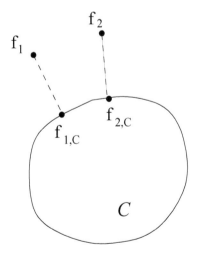

FIGURE 3.8 Illustrating the inequality (3.64) for convex projections.

If we know that the solution of the image reconstruction problem belongs to some closed and convex set $\mathcal{C}$, it is quite natural to look, among all possible objects satisfying this condition, for that or those which minimize the discrepancy, i.e. for the solutions of the following constrained problem

$$\|Af - g\| = \text{minimum} , \quad f \in \mathcal{C} . \tag{3.65}$$

Any solution of this problem is called a *constrained least-squares solution*.

As we will discuss in the following, this problem can be ill posed or well posed, according to properties of the set $\mathcal{C}$ and of the operator A. If it is well posed, this strategy provides a way for obtaining a reasonable solution to the problem of image deconvolution.

If we know that the solution of the image reconstruction problem belongs to some closed and convex set $\mathcal{C}$, it is quite natural to look, among all possible objects satisfying this condition, for that or those which minimize the discrepancy, i.e. for the solutions of the following constrained problem

$$\|Af - g\| = \text{minimum} \;, \quad f \in \mathcal{C} \;. \tag{3.65}$$

Any solution of this problem is called a *constrained least-squares solution.*

As we will discuss in the following, this problem can be ill posed or well posed, according to properties of the set $\mathcal{C}$ and of the operator A. If it is well posed, this strategy provides a way for obtaining a reasonable solution to the problem of image deconvolution.

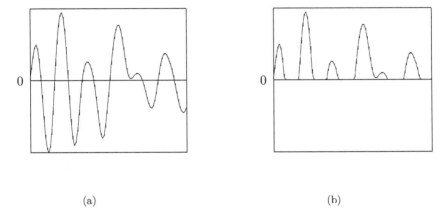

(a) (b)

FIGURE 3.7 Illustrating the effect of the projection onto the convex set of nonnegative functions: (a) the original function; (b) the projected function.

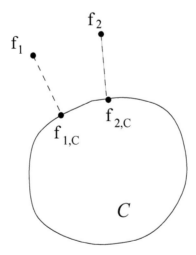

FIGURE 3.8 Illustrating the inequality (3.64) for convex projections.

4

Quadratic Tikhonov regularization and filtering

The first approach to linear ill-posed problems based on constrained least-squares solutions was proposed by V. K. Ivanov in 1962 [169]. In the same year D. L. Phillips [230] introduced a method based on the determination of the smoothest approximate solution compatible with the data within a given noise level. One year later the Russian mathematician A. N. Tikhonov proposed, independently, a general approach named *regularization* or *regularization method* of ill-posed problems. This approach also provided a unification of the methods of Ivanov and of Phillips.

The basic idea of regularization consists in considering a family of approximate solutions depending on a positive parameter called *regularization parameter*. The main property is that, in the case of noise-free data, the functions of the family converge to the exact solution of the problem when the regularization parameter tends to zero. In the case of noisy data one can obtain an optimal approximation of the exact solution for a non-zero value of the regularization parameter. Moreover for suitable values of the regularization parameter one recovers both the solutions of Ivanov and of Phillips.

In this chapter we present these regularization methods in the case of image deconvolution. For this particular problem it is easy to show that the simplest form of Tikhonov regularization is equivalent to a linear filtering of the solution or generalized solution discussed in the previous chapter. This remark suggests the introduction of a wider class of regularization methods called *spectral filtering*.

The description of regularization in terms of a global PSF and the problem of the choice of the regularization parameter are also discussed.

4.1 Least-squares solutions with prescribed energy

We call *energy* of a given object $f(\mathbf{x})$ the square of its L^2-norm

$$E^2(f) = \|f\|^2 = \int |f(\mathbf{x})|^2 d\mathbf{x} . \tag{4.1}$$

DOI: 10.1201/9781003032755-4

In some cases (for instance, in the case of coherent imaging) this quantity is proportional to the physical energy of the signal $f(\mathbf{x})$.

Recalling that a *functional* is a mapping which associates a real (or complex) number to each object, then the *energy functional* is the mapping which associates to each object the value of its energy. Analogously the *discrepancy functional* is the mapping which associates to each object the value of its discrepancy, which was defined in Section 3.6 as the squared norm of the residual $r = g - Af$. This functional will be denoted by $\varepsilon^2(f;g)$.

Now a *least-squares solution with prescribed energy* is an object $\tilde{f}_1(\mathbf{x})$ which solves the following constrained least-squares problem: *find the minimum (or the minima) of the discrepancy functional*

$$\varepsilon^2(f;g) = \|Af - g\|^2 \tag{4.2}$$

in the set of all objects whose energy does not exceed a prescribed bound E^2

$$E^2(f) = \|f\|^2 \le E^2 \ . \tag{4.3}$$

This problem is a particular example of the problem (3.65) and is precisely the problem considered by Ivanov in the paper mentioned before, although in a more general setting.

In the terminology used in optimization theory, a functional to be minimized (possibly under some constraints) is also called a *cost function* and its minimizers or set of minimizers are called the *arguments of the minimum*, in short 'arg min'.

The bound E^2 must be acceptable from the point of view of physics, i.e. it must be a reasonable value of the energy of the unknown object $f^{(0)}$. As discussed in Chapter 3, the solution or generalized solution of equation (3.11) does not necessarily satisfy this constraint because of noise propagation, i.e. we can have

$$E^2(f^\dagger) = \frac{1}{(2\pi)^d} \int_B \left| \frac{\hat{g}(\boldsymbol{\omega})}{\hat{K}(\boldsymbol{\omega})} \right|^2 d\boldsymbol{\omega} > E^2 \tag{4.4}$$

(possibly $E^2(f^\dagger) = +\infty$). This fact implies that the unconstrained minimum of the discrepancy functional (4.2) is not interior to the ball defined by the inequality (4.3). If $E^2(f^\dagger) = +\infty$ the unconstrained minimum does not exist because, in such a case, no solution or least-squares solution exists, as explained in Sections 3.3 and 3.5. It follows that the constrained minimum must stay on the surface of this ball (a pictorial justification of this point is given in Figure 4.1) and therefore the inequality (4.3) can be replaced by the following equality

$$E^2(f) = E^2 \ . \tag{4.5}$$

The problem of minimizing the functional (4.2) over the set of functions satisfying condition (4.5) can be solved by means of the *method of Lagrange multipliers*. According to this method, we introduce a functional which is a linear combination of the functional to be minimized and of the functional expressing the constraint,

$$\Phi_\mu(f;g) = \varepsilon^2(f;g) + \mu\, E^2(f) = \|Af - g\|^2 + \mu\, \|f\|^2 \tag{4.6}$$

where μ is an arbitrary real and positive number which is called the *Lagrange multiplier*. Then the proof of existence and uniqueness of the solution of the problem formulated above follows from the proof of the following points:

- for any $\mu > 0$, there exists a unique function $f_\mu(\mathbf{x})$ which minimizes the functional (4.6);

- there exists a unique value of μ, let us say $\mu_1 = \mu_1(E)$, such that

$$E(f_{\mu_1}) = E \ . \tag{4.7}$$

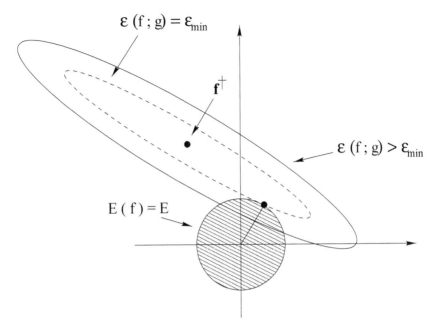

FIGURE 4.1 Two-dimensional picture showing that the minimum of the discrepancy is reached on the surface of the ball of radius E. The ellipses are the level curves of the discrepancy functional. The center of these ellipses is the solution provided by the inverse filtering method.

If these results are true, then the object $\tilde{f}_1 = f_{\mu_1}$ is the unique object which minimizes the discrepancy functional (4.2) under the condition (4.5). Once the two points above are proved, we can conclude that: *for any image g there exists a unique object $\tilde{f}_1$ which is the unique least-squares solution with prescribed energy E^2.*

For the problem of image deconvolution the previous points can be proved in a simple way. We assume that the imaging system is bandlimited and that $\mathcal{B}$ is its band while its complement $\mathcal{B}^C$ is the set of the out-of-band frequencies. This is not restrictive since a system which is not bandlimited can be viewed as the particular case where $\mathcal{B}$ coincides with the whole frequency space and $\mathcal{B}^C$ is empty. Therefore, using the Parseval equality we can write the functional (4.6) in the following form (for d-dimensional images):

$$
\begin{aligned}
\Phi_\mu(f;g) = \; & \frac{1}{(2\pi)^d} \int_{\mathcal{B}} |\hat{K}(\boldsymbol{\omega})\hat{f}(\boldsymbol{\omega}) - \hat{g}(\boldsymbol{\omega})|^2 d\boldsymbol{\omega} \qquad (4.8) \\
& + \frac{1}{(2\pi)^d} \int_{\mathcal{B}^C} |\hat{g}(\boldsymbol{\omega})|^2 d\boldsymbol{\omega} + \frac{\mu}{(2\pi)^d} \int_{\mathcal{B}} |\hat{f}(\boldsymbol{\omega})|^2 d\boldsymbol{\omega} \\
& + \frac{\mu}{(2\pi)^d} \int_{\mathcal{B}^C} |\hat{f}(\boldsymbol{\omega})|^2 d\boldsymbol{\omega} \; .
\end{aligned}
$$

The first two terms correspond to the discrepancy functional (the second one is obtained by observing that $\hat{K}(\boldsymbol{\omega}) = 0$ when $\boldsymbol{\omega}$ is not in the band) while the two other terms correspond to the decomposition of the total energy of $f(\mathbf{x})$ into the sum of the in-band and out-of-band energy. The integrands of the first and of the third term of equation (4.8) can be rearranged as follows

$$|\hat{K}(\boldsymbol{\omega})\hat{f}(\boldsymbol{\omega}) - \hat{g}(\boldsymbol{\omega})|^2 + \mu|\hat{f}(\boldsymbol{\omega})|^2 \tag{4.9}$$
$$= (|\hat{K}(\boldsymbol{\omega})|^2 + \mu)|\hat{f}(\boldsymbol{\omega})|^2 - 2Re\{\hat{K}(\boldsymbol{\omega})\hat{f}(\boldsymbol{\omega})\hat{g}^*(\boldsymbol{\omega})\} + |\hat{g}(\boldsymbol{\omega})|^2 =$$
$$= (|\hat{K}(\boldsymbol{\omega})|^2 + \mu)\left|\hat{f}(\boldsymbol{\omega}) - \frac{\hat{K}^*(\boldsymbol{\omega})\hat{g}(\boldsymbol{\omega})}{|\hat{K}(\boldsymbol{\omega})|^2 + \mu}\right|^2 + \frac{\mu}{|\hat{K}(\boldsymbol{\omega})|^2 + \mu}|\hat{g}(\boldsymbol{\omega})|^2$$

so that the functional (4.8) takes the following form

$$\Phi_\mu(f;g) = \frac{1}{(2\pi)^d}\int_{\mathcal{B}}(|\hat{K}(\boldsymbol{\omega})|^2 + \mu)\left|\hat{f}(\boldsymbol{\omega}) - \frac{\hat{K}^*(\boldsymbol{\omega})\hat{g}(\boldsymbol{\omega})}{|\hat{K}(\boldsymbol{\omega})|^2 + \mu}\right|^2 d\boldsymbol{\omega} \tag{4.10}$$
$$+ \frac{1}{(2\pi)^d}\int_{\mathcal{B}^C}|\hat{f}(\boldsymbol{\omega})|^2 d\boldsymbol{\omega} + \frac{\mu}{(2\pi)^d}\int_{\mathcal{B}}\frac{|\hat{g}(\boldsymbol{\omega})|^2}{|\hat{K}(\boldsymbol{\omega})|^2 + \mu}d\boldsymbol{\omega}$$
$$+ \frac{1}{(2\pi)^d}\int_{\mathcal{B}^C}|\hat{g}(\boldsymbol{\omega})|^2 d\boldsymbol{\omega} \ .$$

From this expression we deduce that, for any $\mu > 0$, there exists a unique function $f_\mu(\mathbf{x})$ which minimizes the functional. Its FT can be obtained by making the first two terms equal to zero and therefore $\hat{f}_\mu(\boldsymbol{\omega}) = 0$ when $\boldsymbol{\omega}$ is in $\mathcal{B}^C$ and

$$\hat{f}_\mu(\boldsymbol{\omega}) = \frac{\hat{K}^*(\boldsymbol{\omega})}{|\hat{K}(\boldsymbol{\omega})|^2 + \mu}\hat{g}(\boldsymbol{\omega}) \tag{4.11}$$

when $\boldsymbol{\omega}$ is in $\mathcal{B}$. Since $\hat{K}(\boldsymbol{\omega})$ is zero in $\mathcal{B}^C$, the last expression can be used for any $\boldsymbol{\omega}$ and therefore

$$f_\mu(\mathbf{x}) = \frac{1}{(2\pi)^d}\int_{\mathbb{R}^d}\frac{\hat{K}^*(\boldsymbol{\omega})}{|\hat{K}(\boldsymbol{\omega})|^2 + \mu}\,\hat{g}(\boldsymbol{\omega})\,e^{i\mathbf{x}\cdot\boldsymbol{\omega}}\,d\boldsymbol{\omega} \ . \tag{4.12}$$

We point out that *if the imaging system is bandlimited then $f_\mu(\mathbf{x})$ is also bandlimited and has the same band as the system.* We also observe that the minimum value of the functional $\Phi_\mu(f;g)$, which is the sum of the last two terms in equation (4.10), can be written in the following form

$$\Phi_\mu(f_\mu;g) = \frac{\mu}{(2\pi)^d}\int_{\mathbb{R}^d}\frac{|\hat{g}(\boldsymbol{\omega})|^2}{|\hat{K}(\boldsymbol{\omega})|^2 + \mu}d\boldsymbol{\omega} \tag{4.13}$$

where we have used again the fact that $\hat{K}(\boldsymbol{\omega}) = 0$ when $\boldsymbol{\omega}$ is in $\mathcal{B}^C$.
The function $f_\mu(\mathbf{x})$ has the following properties:

- it is square integrable for any g and any $\mu > 0$;
- it depends continuously on g;
- it is orthogonal to the subspace of all invisible objects.

The first two properties are a simple consequence of the following elementary inequality

$$\frac{\xi^2}{(\xi^2 + \mu)^2} \leq \frac{1}{4\mu} \ , \tag{4.14}$$

the r.h.s. being the maximum value of the function. From this inequality with $\xi = |\hat{K}(\boldsymbol{\omega})|$ and from the Parseval equality, we get

$$E^2(f_\mu) = \int |f_\mu(\mathbf{x})|^2 dx = \frac{1}{(2\pi)^d}\int_{\mathcal{B}}\frac{|\hat{K}(\boldsymbol{\omega})|^2}{(|\hat{K}(\boldsymbol{\omega})|^2 + \mu)^2}|\hat{g}(\boldsymbol{\omega})|^2 d\boldsymbol{\omega}$$
$$\leq \frac{1}{4\mu}\frac{1}{(2\pi)^d}\int_{\mathcal{B}}|\hat{g}(\boldsymbol{\omega})|^2 d\boldsymbol{\omega} \leq \frac{1}{4\mu}\|g\|^2 \ . \tag{4.15}$$

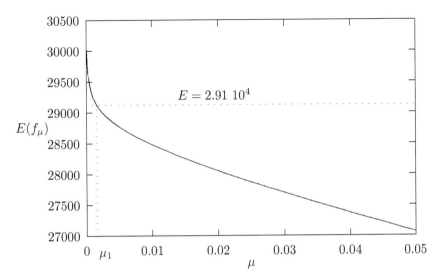

FIGURE 4.2 Behavior of $E(f_\mu)$ as a function of μ in the case of the example of Section 4.6. The point μ_1 where $E(f_{\mu_1}) = E$ is also indicated. In such a case we have $\mu_1 = 1.6 \; 10^{-3}$.

This inequality proves the first two statements above. As concerns the third one, it is obvious because, as we already remarked, $f_\mu(\mathbf{x})$ is bandlimited with the same band as the imaging system and therefore it is orthogonal to all invisible objects.

Having proved that the functional (4.6) has a unique minimum for any $\mu > 0$, in order to complete the program of the method of Lagrange multipliers, we still have to show that there exists a unique value of μ satisfying equation (4.7). Now, from the expression of $E(f_\mu)$ provided by the Parseval equality – see equation (4.15) – it follows that $E(f_\mu)$ is a decreasing function of μ. Moreover it tends to $E(f^\dagger) > E$ (possibly, $E(f^\dagger) = \infty$) when $\mu \to 0$ and tends to zero when $\mu \to \infty$. It follows that there exists a unique value of μ, $\mu_1 = \mu_1(E)$, such that $E(f_{\mu_1}) = E$. In Figure 4.2, we give the plot of the function $E(f_\mu)$ in the case of the numerical example discussed in Section 4.6. We conclude that, for any given g, there exists a unique least-squares solution with prescribed energy.

REMARK 4.1   *The continuous dependence of $\tilde{f}_1 = f_{\mu_1}$ on the data and its convergence to the exact generalized solution $(f^{(0)})^\dagger = P^{(\mathcal{B})} f^{(0)}$, equation (3.44), when the error on the data tends to zero are rather subtle questions and go beyond the scope of this book (see e.g. [169] and [31]). One can prove that $\|\tilde{f}_1 - P^{(\mathcal{B})} f^{(0)}\|$ tends to zero when $g \to g^{(0)}$ if and only if E is a precise estimate of the energy of $P^{(\mathcal{B})} f^{(0)}$, i.e. $\|P^{(\mathcal{B})} f^{(0)}\| = E$ [31]. If this energy has been overestimated, i.e. $\|P^{(\mathcal{B})} f^{(0)}\| < E$, then it is only possible to prove convergence in a weaker sense, namely that the scalar products $(\tilde{f}_1 - P^{(\mathcal{B})} f^{(0)}, h)$ tend to zero, when $g \to g^{(0)}$, for any arbitrary square-integrable function h. These results about convergence are important because they ensure that, when the noise is small, the least-squares solution of minimal energy is close to the in-band component of the true object $f^{(0)}$.*

The method considered above can be easily applied to discrete images if the L^2-norm is replaced by the canonical Euclidean norm and the FT is replaced by the DFT. The function (4.12) is now replaced by the following array (2D image)

$$(\mathbf{f}_\mu)_{m,n} = \frac{1}{N^2} \sum_{k,l=0}^{N-1} \frac{\hat{K}^*_{k,l}}{|\hat{K}_{k,l}|^2 + \mu} \; \hat{g}_{k,l} \; \exp\left[i\frac{2\pi}{N}(km + ln)\right]. \qquad (4.16)$$

Since the Euclidean norm of $\mathbf{f}_\mu$ is a decreasing function of μ, one can easily prove again that there exists a unique least-squares solution with prescribed energy. If E is an upper bound of the Euclidean norm of $\mathbf{f}^{(0)}$, this least-squares solution converges to $(\mathbf{f}^{(0)})^\dagger$ when the error on the data tends to zero.

4.2 Approximate solutions with minimal energy

An alternative to the least-squares solution with prescribed energy is provided by a method investigated by Ivanov [169] which is a simplified version of the method proposed by Phillips [230]. The starting point is the assumption that an estimate ε^2 of the energy of the noise is known (see equation (2.4) and the related discussion). Then it is quite natural to consider the set of the approximate solutions such that their discrepancy does not exceed ε^2. This set has been already investigated in the case of discrete images – see Section 3.6 – and we know that it is very broad and that it contains both physical and unphysical approximate solutions as well as the true object $f^{(0)}(\mathbf{x})$. Since the unphysical solutions are those which are too large and oscillatory, it is quite natural to look for the approximate solution which minimizes the energy functional (4.1).

An *approximate solution of minimal energy* is an object $\tilde{f}_2(\mathbf{x})$ which solves the following problem: *find the minimum (or the minima) of the energy functional*

$$E^2(f) = \|f\|^2 \tag{4.17}$$

in the set of all objects whose discrepancy does not exceed a prescribed bound ε^2

$$\varepsilon^2(f;g) = \|Af - g\|^2 \leq \varepsilon^2 . \tag{4.18}$$

It is obvious that, if the set (4.18) contains $f = 0$, then $f = 0$ is the solution of minimal energy. But this is a trivial case. Indeed, if $f = 0$ belongs to the set (4.18), this means that $\|g\| \leq \varepsilon$, i.e. that the image g consists only of noise and does not contain any significant and detectable signal. If we exclude this situation, then it is rather intuitive, as shown in Figure 4.3, that the solution of minimal energy stays on the boundary of the set defined by equation (4.18) so that the inequality (4.18) can be replaced by the following equality

$$\varepsilon^2(f;g) = \|Af - g\|^2 = \varepsilon^2 . \tag{4.19}$$

This result can be proved rigorously (see, for instance, [169]) and is related to the well-known fact that a closed and convex set of a Hilbert space, not containing $f = 0$, has always a point with minimum distance from the origin which lies on the boundary of the set.

The problem we are considering is, in a sense, the dual of the problem considered in Section 4.1 because the role of the two functionals is exchanged: *in the previous setting the discrepancy is minimized, given the value of the energy, whereas now the energy is minimized, given the value of the discrepancy.* This remark makes obvious that we can use again the method of Lagrange multipliers for solving the present problem.

We consider the functional (4.6), with μ real and positive, and the function $f_\mu(\mathbf{x})$, given by equation (4.12), which is the unique minimizer of this functional. Then the second point of the method of Lagrange multipliers, as formulated in Section 4.1, must be modified as follows:

- there exists a unique value of μ, let us say $\mu_2 = \mu_2(\varepsilon)$, such that

$$\varepsilon(f_{\mu_2};g) = \varepsilon . \tag{4.20}$$

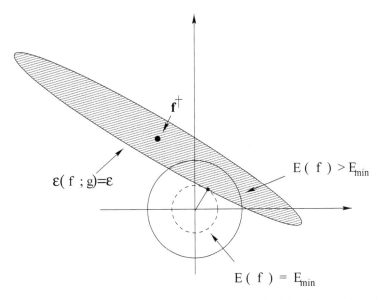

FIGURE 4.3 Two-dimensional picture showing that the point of minimal energy (= minimal distance from the origin) is on the boundary of the ellipse representing the set of approximate solutions whose discrepancy does not exceed ε^2. The center of the ellipse is the solution provided by the inverse filtering method. The circles are the level lines of the energy functional.

If such a value exists, then $\tilde{f}_2 = f_{\mu_2}$ is the unique object of minimal energy that is compatible with the image g within the error ε. Once this result is proved, we can conclude that: *for any image g there exists a unique object $\tilde{f}_2$ which is the approximate solution of minimal energy for a prescribed value of the discrepancy.*

In order to prove the existence of a unique value μ_2 such that equation (4.20) holds true, we compute $\varepsilon(f_\mu; g)$ using again the Parseval equality; if we take into account that $\hat{K}(\boldsymbol{\omega}) = 0$ when $\boldsymbol{\omega}$ is in $\mathcal{B}^C$, we get

$$\varepsilon^2(f_\mu; g) = \frac{1}{(2\pi)^d} \int_{\mathbb{R}^d} \left| \frac{|\hat{K}(\boldsymbol{\omega})|^2}{|\hat{K}(\boldsymbol{\omega})|^2 + \mu} \hat{g}(\boldsymbol{\omega}) - \hat{g}(\boldsymbol{\omega}) \right|^2 d\boldsymbol{\omega} \qquad (4.21)$$

$$= \frac{1}{(2\pi)^d} \int_{\mathcal{B}} \frac{\mu^2 |\hat{g}(\boldsymbol{\omega})|^2}{(|\hat{K}(\boldsymbol{\omega})|^2 + \mu)^2} d\boldsymbol{\omega} + \frac{1}{(2\pi)^d} \int_{\mathcal{B}^C} |\hat{g}(\boldsymbol{\omega})|^2 d\boldsymbol{\omega} .$$

It follows that $\varepsilon(f_\mu; g)$ is an increasing function of μ which takes, for $\mu = 0$, the value $\|g_{out}\|$, i.e. the norm of the out-of-band noise (if the system is not band-limited, then $\|g_{out}\| = 0$), and tends to $\|g\|$ when $\mu \to \infty$. In Figure 4.4, we plot the behavior of the function $\varepsilon(f_\mu; g)$, again in the case of the numerical example of Section 4.6.

Therefore, if ε satisfies the conditions

$$\|g_{out}\| < \varepsilon < \|g\| \qquad (4.22)$$

there exists a unique value of μ, $\mu_2 = \mu_2(\varepsilon)$, such that $\varepsilon(f_{\mu_2}; g) = \varepsilon$. This value is also indicated in Figure 4.4. We point out that the conditions (4.22) make sense from the physical point of view. Indeed the condition $\varepsilon < \|g\|$ must be satisfied if the image contains not only noise but also a detectable signal (this point has been already discussed). Moreover the condition $\varepsilon > \|g_{out}\|$ means that the energy of the total noise must be greater than the energy of the out-of-band noise.

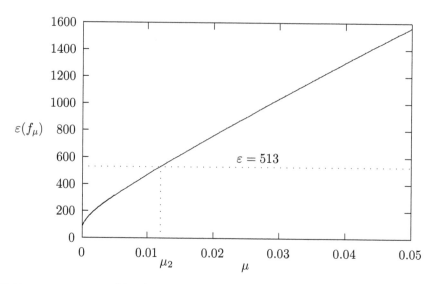

FIGURE 4.4   Behavior of $\varepsilon(f_\mu; g)$ as a function of μ in the case of the numerical example of Section 4.6. The point μ_2 where $\varepsilon(f_{\mu_2}; g) = \varepsilon$ is also indicated. In such a case, we have $\mu_2 = 1.2 \times 10^{-2}$.

REMARK 4.2   *As concerns the continuous dependence of $\tilde{f}_2$ on the data and its convergence to the exact generalized solution $(f^{(0)})^\dagger = P^{(\mathcal{B})} f^{(0)}$, one can prove that $\|\tilde{f}_2 - P^{(\mathcal{B})} f^{(0)}\|$ tends to zero when the error on the data tends to zero (see for instance [31] or [169]). Therefore the convergence properties of $\tilde{f}_2$ are stronger than those of $\tilde{f}_1$.*

All results can be easily extended to the case of discrete images $\mathbf{f}_\mu$ using equation (4.16).

4.3   Regularization methods in the sense of Tikhonov

The method of Lagrange multipliers used for the solution of the constrained minimization problems of the previous sections introduces, as an intermediate step, the family of functions defined in equation (4.12) (in equation (4.16) for discrete images). This family of approximate solutions has very interesting properties which have been elucidated and stressed by Tikhonov [283], who based on these properties the general definition of a *regularization algorithm* [284]. In order to introduce the basic idea of this method it is convenient to write the functions $f_\mu(\mathbf{x})$ in a slightly different form.

It is easy to show that the function $f_\mu(\mathbf{x})$ is the solution of the Euler-Lagrange equation associated with the quadratic functional (4.6)

$$(A^* A + \mu I) f_\mu = A^* g \tag{4.23}$$

where I denotes the identity operator. This result can be easily verified in the case of image deconvolution because, by taking the FT of both members of this equation, we get

$$(|\hat{K}(\boldsymbol{\omega})|^2 + \mu) \hat{f}_\mu(\boldsymbol{\omega}) = \hat{K}^*(\boldsymbol{\omega}) \, \hat{g}(\boldsymbol{\omega}) \tag{4.24}$$

and therefore the expressions (4.11) and (4.12) are re-obtained. It is also evident that, by putting $\mu = 0$ in equation (4.23), we get back the equation of the least-squares solutions, equation (3.40).

Equation (4.23) implies that f_μ can be written in the following form

$$f_\mu = R_\mu \, g \tag{4.25}$$

where
$$R_\mu = (A^*A + \mu I)^{-1} A^* .\tag{4.26}$$

This is a useful way of writing equation (4.12) because the operator R_μ is an approximation, in a sense which will be specified in a moment, of the generalized inverse $A^\dagger$ defined in equation (3.47). In order to investigate this point we observe that the operator R_μ is a convolution operator
$$R_\mu g = K_\mu^\dagger * g\tag{4.27}$$

where
$$K_\mu^\dagger(\mathbf{x}) = \frac{1}{(2\pi)^d} \int_{\mathbb{R}^d} \frac{\hat{K}^*(\boldsymbol{\omega})}{|\hat{K}(\boldsymbol{\omega})|^2 + \mu} e^{i\mathbf{x}\cdot\boldsymbol{\omega}} d\boldsymbol{\omega} .\tag{4.28}$$

Thanks to the factor $\hat{K}^*(\boldsymbol{\omega})$ the integral extends only to the band $\mathcal{B}$ of the imaging system. Then, in the limit $\mu = 0$, we re-obtain equation (3.45) if the integral in this equation is convergent. Otherwise the limit does not exist or, more precisely, in the case of an ill-posed problem, the limit is not a continuous operator. Indeed, for $\mu > 0$, $K_\mu^\dagger(\mathbf{x})$ is a well-defined function and R_μ is a bounded operator but, from equations (4.25) and (4.15), we get
$$\|R_\mu g\| \leq \frac{1}{2\sqrt{\mu}} \|g\| ,\tag{4.29}$$

and the bound tends to infinity when $\mu \to 0$, coherently with the fact that the generalized inverse is not a bounded operator.

However, we know that the generalized solution exists if $g = g^{(0)} = Af^{(0)}$. By looking at equation (4.12) we can argue intuitively that f_μ should converge to $f^\dagger$. This property of f_μ can be easily confirmed as follows. By Parseval's equality we get
$$\|R_\mu g^{(0)} - A^\dagger g^{(0)}\|^2 = \frac{1}{(2\pi)^d} \int_{\mathcal{B}} \frac{\mu^2}{(|\hat{K}(\boldsymbol{\omega})|^2 + \mu)^2} \frac{|\hat{g}^{(0)}(\boldsymbol{\omega})|^2}{|\hat{K}(\boldsymbol{\omega})|^2} d\boldsymbol{\omega}\tag{4.30}$$

and, since $|\hat{g}^{(0)}(\boldsymbol{\omega})|^2 / |\hat{K}(\boldsymbol{\omega})|^2$ is integrable, by assumption, while the other factor is bounded by one and tends to zero everywhere for $\mu \to 0$, the well-known *dominated convergence theorem* implies that the integral tends to zero and $R_\mu g^{(0)}$ converges to $A^\dagger g^{(0)}$, strongly i.e. in norm.

REMARK 4.3   *We point out that the rate of convergence to zero can be arbitrarily slow and depends on the properties of the object $f^{(0)}$. Indeed, from the elementary inequality*
$$\frac{x}{x+a} \leq (1-\alpha)^{1-\alpha} \left(\frac{\alpha}{a}\right)^\alpha x^\alpha , \quad 0 < \alpha \leq 1 ,\tag{4.31}$$

with $x = \mu$ and $a = |\hat{K}(\boldsymbol{\omega})|^2$, we obtain
$$\|R_\mu g^{(0)} - A^\dagger g^{(0)}\| \leq (1-\alpha)^{(1-\alpha)} \alpha^\alpha \frac{\mu^\alpha}{(2\pi)^{d/2}} \left(\int_{\mathcal{B}} \frac{|\hat{f}^{(0)}(\boldsymbol{\omega})|^2}{|\hat{K}(\boldsymbol{\omega})|^{4\alpha}} d\boldsymbol{\omega}\right)^{1/2} ;\tag{4.32}$$

therefore, the approximation error provided by R_μ tends to zero as μ^α if the object satisfies the condition
$$\int_{\mathcal{B}} \frac{|\hat{f}^{(0)}(\boldsymbol{\omega})|^2}{|\hat{K}(\boldsymbol{\omega})|^{4\alpha}} d\boldsymbol{\omega} < \infty .\tag{4.33}$$

These conditions are sometimes called source conditions (see [114] for an abstract formulation), where source is synonymous with our term object. The higher the value of α, the *smoother the object has to be. Indeed, they are essentially conditions on the behavior of the Fourier transform of $f^{(0)}$ for large values of ω and, as it is known, this behavior is related to the differentiability of the function. In a sense, they imply that, if the object is very smooth, then it is possible to obtain quite good reconstructions. Obviously the point is that one may need to reconstruct functions which do not satisfy these source conditions. Examples are provided in the second part of the book.*

From the previous results it follows that the family of linear operators $\{R_\mu\}_{\mu>0}$ has the following properties:

- for any $\mu > 0$, R_μ is a linear and continuous operator
- for any noise-free image $g^{(0)} = Af^{(0)}$

$$\lim_{\mu \to 0} R_\mu g^{(0)} = P^{(\mathcal{B})} f^{(0)} = A^\dagger g^{(0)} \qquad (4.34)$$

where the limit is in the L^2-norm and $P^{(\mathcal{B})}$ is the bandlimiting operator defined in equation (3.37).

The second property clarifies in what sense R_μ is an approximation of $A^\dagger$: when R_μ is applied to a noise-free image $g^{(0)}$ then one gets an approximation of the generalized solution $f^\dagger = A^\dagger g^{(0)}$ and the L^2-norm of the approximation error tends to zero when μ tends to zero.

Any family $\{R_\mu\}_{\mu>0}$ of linear and bounded operators, depending on a parameter $\mu > 0$, is called a family of *linear regularization operators for the approximation of the generalized inverse of the ill-posed problem* $Af = g$ if it satisfies the two conditions above. The parameter μ is called *regularization parameter* and the functions $f_\mu = R_\mu g$ are called *regularized solutions* . In the particular examples we have used for introducing this definition, the regularization parameter is the Lagrange multiplier introduced in Section 4.1. Other examples of regularization algorithms with regularization parameters not related to Lagrange multipliers are presented and investigated in the next sections and chapters. We also observe that the condition of linearity can be dropped in the definition of regularization operators because there are examples of nonlinear regularization operators for the solution of linear inverse problems.

The important question is: what happens in the case of a noisy image g? In such a case we need the definition of *regularization method*, introduced by Tikhonov [283, 114], which is essentially based on the noise model discussed in Section 2.1 and more precisely in equations (2.3) and (2.4).

A linear regularization method consists of the following ingredients:

- let $\{R_\mu\}_{\mu>0}$ be a family of regularization operators of the generalized inverse $A^\dagger$;
- for any given $g^{(0)}$, let g_ε be a sequence of noisy images approaching $g^{(0)}$ for $\varepsilon \to 0$ as follows

$$\|g_\varepsilon - g^{(0)}\| \leq \varepsilon \ ; \qquad (4.35)$$

- there exists a choice of the regularization parameter $\mu = \mu(\varepsilon)$ such that, if $\varepsilon \to 0$, then

$$\mu(\varepsilon) \to 0 \ , \quad \text{and} \quad \|R_{\mu(\varepsilon)} g_\varepsilon - A^\dagger g^{(0)}\| \to 0 \ . \qquad (4.36)$$

Therefore a regularization method consists of a family of regularization operators and of a selection of the regularization parameter as a function of the noise level allowing to approach the exact object. In general, for a given family of regularization operators it is not easy to prove that they can lead to a regularization method. We give the proof only in a couple of cases.

Let us now come back to the particular example of regularization algorithm provided by equations (4.26)-(4.28); this is usually called the *Tikhonov regularization algorithm*, which provides the most simple case of a regularization method, as we show in the subsequent remark.

We already know that, in the case of noise-free images $g^{(0)}$, the regularized solutions $f_\mu = R_\mu g^{(0)}$ provide better and better approximations of the true solution (or generalized solution) when the regularization parameter μ tends to zero. If we use the model for image formation, i.e. $g = Af^{(0)} + w$, we have

$$R_\mu g = R_\mu Af^{(0)} + R_\mu w \ . \tag{4.37}$$

From equation (4.34) we know that we can restore at most $P^{(\mathcal{B})}f^{(0)}$ when the noise is zero and μ tends to zero. When the image is noisy and μ is not zero, the error we commit in the reconstruction of $P^{(\mathcal{B})}f^{(0)}$ is given by

$$R_\mu g - P^{(\mathcal{B})}f^{(0)} = (R_\mu Af^{(0)} - P^{(\mathcal{B})}f^{(0)}) + R_\mu w \ . \tag{4.38}$$

The first term in the second member of this equation can be called the *approximation error* because this is the error due to the fact that we approximate the operator $A^\dagger$ by means of R_μ. This error depends only on the value of μ and does not depend on the noise. The second term $R_\mu w$ is the error due to the noise corrupting the image and it can be called the *noise-propagation error* since it quantifies the propagation of the noise from the image g to the regularized solution f_μ. It depends both on μ and on w.

If we use the triangular inequality for the norm we have

$$\|R_\mu g - P^{(\mathcal{B})}f^{(0)}\| \leq \|R_\mu Af^{(0)} - P^{(\mathcal{B})}f^{(0)}\| + \|R_\mu w\| \tag{4.39}$$

i.e. *the norm of the reconstruction error is bounded by the sum of the norm of the approximation error and of the norm of the noise-propagation error.*

REMARK 4.4 *Before investigating the behavior of the reconstruction error as a function of μ, we show that the Tikhonov regularization algorithm provides a regularization method as it easily follows from the previous inequality and from inequality (4.29). Indeed, let g_ε be the family of noisy images considered in the definition, satisfying the condition $\|g_\varepsilon - g^{(0)}\| \leq \varepsilon$ and therefore converging to the noise-free image $g^{(0)}$ when $\varepsilon \to 0$. Since $w_\varepsilon = g_\varepsilon - g^{(0)}$, thanks to inequality (4.29), the second term in the r.h.s. of equation (4.39) is bounded by $\varepsilon/2\sqrt{\mu}$. Therefore, if we select $\mu(\varepsilon) = \varepsilon^\alpha$ with $\alpha < 2$, the second term tends to zero when ε tends to zero. On the other hand, also the first term tends to zero, thanks to equation (4.34) and the fact that $\mu(\varepsilon)$ tends to zero. In conclusion, the basic properties of a regularization method are satisfied.*

To understand the behavior of the reconstruction error as a function of μ, we investigate the behavior of the two terms in the r.h.s. of equation (4.39). If we use again Parseval's equality, we find that the norm of the approximation error is given by

$$\|R_\mu Af^{(0)} - P^{(\mathcal{B})}f^{(0)}\|^2 = \frac{1}{(2\pi)^d} \int_\mathcal{B} \left| \frac{|\hat{K}(\boldsymbol{\omega})|^2}{|\hat{K}(\boldsymbol{\omega})|^2 + \mu} \hat{f}^{(0)}(\boldsymbol{\omega}) - \hat{f}^{(0)}(\boldsymbol{\omega}) \right|^2 d\boldsymbol{\omega}$$

$$= \frac{1}{(2\pi)^d} \int_\mathcal{B} \frac{\mu^2}{(|\hat{K}(\boldsymbol{\omega})|^2 + \mu)^2} |\hat{f}^{(0)}(\boldsymbol{\omega})|^2 d\boldsymbol{\omega} \ . \tag{4.40}$$

Therefore the norm of the approximation error is a continuous and increasing function of μ, equal to zero for $\mu = 0$ and tending to $\|P^{(\mathcal{B})}f^{(0)}\|$ when $\mu \to \infty$. On the other hand, the

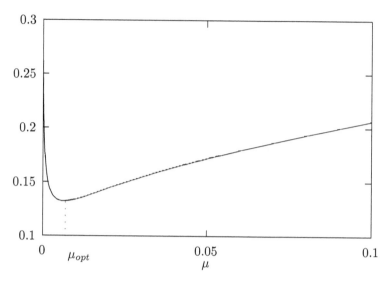

FIGURE 4.5   Plot of the norm of the relative reconstruction error, as a function of μ, in the case of the example of Section 4.6. The optimal value of the regularization parameter, corresponding to the minimum of the function is also indicated. In such a case, we have $\mu_{opt} = 7 \times 10^{-3}$.

norm of the noise-propagation error is given by

$$\|R_\mu w\|^2 = \frac{1}{(2\pi)^d} \int_{\mathcal{B}} \frac{|\hat{K}(\boldsymbol{\omega})|^2}{(|\hat{K}(\boldsymbol{\omega})|^2 + \mu)^2} \, |\hat{w}(\boldsymbol{\omega})|^2 \, d\boldsymbol{\omega} \qquad (4.41)$$

and therefore it is a continuous and decreasing function of μ, which tends to a very large value (possibly $+\infty$) when $\mu \to 0$ and tends to zero when $\mu \to +\infty$.

We find that the norm of the approximation error and the norm of the noise-propagation error have opposite behavior as functions of μ. If we wish to reduce the approximation error, then we must choose a small value of the regularization parameter but the noise-propagation error can be very large. On the other hand, if we wish to reduce the noise-propagation error, then we must choose a large value of μ but the approximation error can be too large. The r.h.s. of the inequality (4.39) can be expected to typically present a U-shape and to pass through a minimum since it is the sum of a decreasing function of μ and of an increasing one. The same behavior is expected to hold true for the norm of the reconstruction error because this norm is approximately equal to the norm of the noise-propagation error (decreasing) for small μ and to the norm of the approximation error (increasing) for large μ. The behavior of the reconstruction error in the case of the numerical example of Section 4.6 is given in Figure 4.5. Here we plot the so-called *relative reconstruction error* defined by

$$\rho(\mu) = \frac{\|f_\mu - f^{(0)}\|}{\|f^{(0)}\|} \ . \qquad (4.42)$$

In simulations like this one, the value of μ minimizing the reconstruction error corresponds to the regularized solution which provides the best approximation of $\mathcal{P}^{(\mathcal{B})} f^{(0)}$, hence the best compromise between approximation and noise-propagation. This value of μ is called the *optimal value* of the regularization parameter and it is denoted by μ_{opt}. The corresponding *optimal regularized solution* is denoted by $\tilde{f}_{opt}$. It is clear that, here, optimality is defined in the sense of the L^2-norm. Other measures of optimality could of course be introduced, depending on the particular problem.

The existence of a minimum of the reconstruction error and the related existence of an optimal value of the regularization parameter for any given noisy image is a typical property of the Tikhonov regularization algorithm. This result implies that for decreasing values of μ the regularized solutions first approach the exact object $f^{(0)}(\mathbf{x})$ and then go away. This property is usually called *semi-convergence* [114, 218] so that one has convergence in the case of noise-free images and semi-convergence in the case of noisy images. *Semi-convergence is an additional property one should require for any regularization algorithm to be used in practice.*

It is obvious that the optimal value of the regularization parameter cannot be determined in the case of a real image because one does not know $f^{(0)}$. It can only be determined in the case of numerical simulations which are very useful for the understanding of any specific inverse problems. In the case of real images we need methods for estimating sensible values of the regularization parameter and this point is discussed in Section 4.6.

REMARK 4.5 *Although the semi-convergence behavior described above appears as pretty much typical of regularizing methods for ill-posed problems and although it will constitute a recurrent theme throughout this book (in particular, for iterative methods), let us stress the fact that the reasoning made is purely heuristic and does by no means constitute a proof of the property. We are not aware of any rigorous mathematical result in this sense and we believe that such results would be very difficult to establish. We believe, however, that the understanding of the semi-convergence phenomenon is of uttermost importance both for numerical practice and for gaining theoretical insight into the concepts of ill-posedness and regularization.*

It is interesting to describe the behavior of the regularized solutions $f_\mu(\mathbf{x})$ in terms of trajectories in the space of all possible square-integrable functions or also the behavior of the regularized 2D arrays, $\mathbf{f}_\mu$, given by equation (4.16), in terms of trajectories in the space of all possible $N \times N$ arrays. Indeed, to each $f_\mu(\mathbf{x})$ or $\mathbf{f}_\mu$ there corresponds a point in the appropriate object space and, as μ varies, this point moves describing a trajectory in this space.

Consider for simplicity the discrete case and let $\mathbf{f}^\dagger$ be the solution of equation (3.12) as given by equation (3.22). As we know, this solution provides, in general, an unacceptable reconstruction of the true object $\mathbf{f}^{(0)}$. It is the starting point of the trajectory at $\mu = 0$. As μ grows, the distance of the point from the origin, i.e. $\|\mathbf{f}_\mu\|$, decreases (this is precisely the behavior which is observed in equation (4.15), in the continuous case); first the point approaches $\mathbf{f}^{(0)}$, the true object, and then it goes away, arriving at the origin only in the limit $\mu \to \infty$. This is the geometric description of the semi-convergence property. A two-dimensional picture of this behavior is given in Figure 4.6, where we also indicate the ellipse that is the boundary of the set of approximate solutions defined by equation (4.18), as well as the circle that is the boundary of the set of images with prescribed energy defined by the equation (4.3). We assume that the true object $\mathbf{f}^{(0)}$ belongs to both sets so that their intersection is not empty (and is indicated by the dashed region in Figure 4.6).

The picture of Figure 4.6 suggests that, if the intersection of the two sets, equation (4.3) and equation (4.18), is not too broad, then the approximate solutions $\tilde{f}_1$ and $\tilde{f}_2$ considered in the previous sections can provide satisfactory approximations of the optimal regularized solution and that, in general, $\|\tilde{f}_1\|$ will be greater than $\|\tilde{f}_2\|$

$$\|\tilde{f}_2\| \leq \|\tilde{f}_1\| \tag{4.43}$$

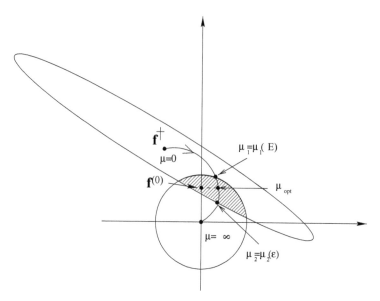

FIGURE 4.6   Two-dimensional picture of the trajectory described by the regularized solutions. The ellipse is the boundary of the set (4.18) while the circle is the boundary of the set (4.3). We indicate on the trajectory the three points corresponding to μ_1, μ_{opt} and μ_2.

with the following inequality between the corresponding values of the regularization parameter

$$\mu_1(E) \leq \mu_2(\varepsilon) \ . \tag{4.44}$$

It can be proved [32] that these inequalities hold true whenever the conditions indicated above are satisfied – see also Section 4.6. Moreover Figure 4.6 suggests that the optimal value of the regularization parameter should be intermediate between $\mu_1(E)$ and $\mu_2(\varepsilon)$. This relationship can be investigated by means of numerical simulations and it turns out that it is not always true (even if it is verified in the case of the numerical example of Section 4.6). The sequence of pictures of Figure 4.7 is another way for visualizing the variations of the regularized solutions f_μ when μ moves from 0 to ∞. Also in this case the existence of an optimal value of the regularization parameter is evident. The reconstructions corresponding to small values of the regularization parameter are very noisy. The noise propagation decreases when μ increases and, for a suitable value of μ, a satisfactory reconstruction of the image appears. If we further increase the value of μ, then the reconstructed image becomes more and more blurred as an effect of the increase of the approximation error.

4.4   Regularization and spectral filtering

The methods of Phillips [230] and Tikhonov [283] are formulated for linear integral equations of the first kind in one variable and are based on the use of differential operators for the smoothness condition. In this section we apply these methods to the problem of image deconvolution.

For 1D deconvolution problems, the method of Phillips consists in looking for the approximate solution (in the sense of equation (4.19)) which minimizes the L^2-norm of the second derivative

$$\|f''\|^2 = \int_{-\infty}^{+\infty} |f''(x)|^2 dx \ . \tag{4.45}$$

FIGURE 4.7 Sequence of regularized solutions obtained from the image of Saturn given in Figure 2.12, (a) $\mu = 0$, (b) $\mu = 10^{-5}$, (c) $\mu = 10^{-3}$, (d) $\mu = 10^{-2}$, (e) $\mu = 1$, (f) $\mu = 1000$.

This requirement replaces the minimization of the energy considered in Section 4.2. Roughly speaking, by minimizing the functional (4.45) with the condition (4.19), one looks for the approximate solution having minimal curvature in the sense of the quadratic mean.

In order to solve the problem we can use again the method of Lagrange multipliers and, accordingly, minimize the following functional

$$\begin{aligned}\Phi_\mu(f;g) &= \|Af - g\|^2 + \mu\|f''\|^2 \\ &= \frac{1}{2\pi}\int_{-\infty}^{+\infty}|\hat{K}(\omega)\hat{f}(\omega) - \hat{g}(\omega)|^2 d\omega + \frac{\mu}{2\pi}\int_{-\infty}^{+\infty}\omega^4|\,\hat{f}(\omega)|^2 d\omega\;.\end{aligned}\tag{4.46}$$

where the Parseval equality is used again and also the fact that the FT of $f''(x)$ is $-\omega^2\hat{f}(\omega)$. By repeating computations analogous to those of Section 4.1, we easily find that, if $\hat{K}(0)\neq 0$, then, for any $\mu > 0$, there exists a unique minimum of the functional (4.46) given by

$$f_\mu(x) = \frac{1}{2\pi}\int_{-\infty}^{+\infty}\frac{\hat{K}^*(\omega)}{|\hat{K}(\omega)|^2 + \mu\,\omega^4}\,\hat{g}(\omega)\,e^{ix\omega}d\omega\;.\tag{4.47}$$

If the imaging system is bandlimited, this function is also bandlimited and has the same band as the system. Moreover we find again that the discrepancy function $\varepsilon^2(f_\mu;g)$ is an increasing function of μ and therefore that there exists a unique value of μ, let us say $\mu_3 = \mu_3(\varepsilon)$, such that $\varepsilon(f_{\mu_3};g) = \varepsilon$. The family of approximate solutions defined by equation (4.47) provides another example of a regularization algorithm.

The first example given by Tikhonov [283] is somehow different and based on the minimization of a functional containing the L^2-norm of the function and of its first derivative. In the case of 1D deconvolution problems the corresponding family of regularized solutions is the family of the minimizers, for different values of μ, of the following functional

$$\begin{aligned}\Phi_\mu(f;g) &= \|Af - g\|^2 + \mu(\|f\|^2 + a\|f'\|^2) \\ &= \frac{1}{2\pi}\int_{-\infty}^{+\infty}|\hat{K}(\omega)\hat{f}(\omega) - \hat{g}(\omega)|^2 d\omega \\ &\quad + \frac{\mu}{2\pi}\int_{-\infty}^{+\infty}(1 + a\omega^2)|\hat{f}(\omega)|^2 d\omega\end{aligned}\tag{4.48}$$

where a is a given positive constant defining the relative weight of $\|f\|^2$ and $\|f'\|^2$. Also in this case it is easy to show that the minimizer is given by

$$f_\mu(x) = \int_{-\infty}^{+\infty}\frac{\hat{K}^*(\omega)}{|\hat{K}(\omega)|^2 + \mu(1 + a\omega^2)}\,\hat{g}(\omega)\,e^{ix\omega}\,d\omega\tag{4.49}$$

and that this family of functions defines a regularization algorithm in the sense of Section 4.3. Again, if the imaging system is bandlimited, the functions (4.49) have the same band as the system.

The extension to 2D and 3D problems of the previous regularization algorithms is rather straightforward. If we consider derivatives up to the second order, then we can introduce the following functional

$$\Sigma^2(f) = a_0\|f\|^2 + \sum_{i=1}^{d}a_i\left\|\frac{\partial f}{\partial x_i}\right\|^2 + \sum_{i,j=1}^{d}a_{i,j}\left\|\frac{\partial f}{\partial x_i\partial x_j}\right\|^2\tag{4.50}$$

FIGURE 4.7 Sequence of regularized solutions obtained from the image of Saturn given in Figure 2.12, (a) $\mu = 0$, (b) $\mu = 10^{-5}$, (c) $\mu = 10^{-3}$, (d) $\mu = 10^{-2}$, (e) $\mu = 1$, (f) $\mu = 1000$.

where a_0, a_i and $a_{i,j}$ are given nonnegative numbers and the $a_{i,j}$'s define a positive semi-definite matrix. In the case $a_0 = 1, a_i = 0, a_{i,j} = 0$ we re-obtain the energy functional considered in the previous sections. It is also evident that, in the 1D case, the functional (4.50) includes, as particular cases, the functionals of Phillips and Tikhonov.

The regularized solutions are defined as the minimizers of the functional

$$\Phi_\mu(f; g) = \varepsilon^2(f; g) + \mu \, \Sigma^2(f) \qquad (4.51)$$

and are given by

$$f_\mu(\mathbf{x}) = \frac{1}{(2\pi)^d} \int \frac{\hat{K}^*(\boldsymbol\omega)}{|\hat{K}(\boldsymbol\omega)|^2 + \mu \, \hat{P}(\boldsymbol\omega)} \, \hat{g}(\boldsymbol\omega) \, e^{i\mathbf{x}\cdot\boldsymbol\omega} d\boldsymbol\omega \qquad (4.52)$$

where $\hat{P}(\boldsymbol\omega)$ is the polynomial

$$\hat{P}(\boldsymbol\omega) = a_0 + \sum_{i=1}^{d} a_i \omega_i^2 + \sum_{i,j=1}^{d} a_{i,j} (\omega_i \omega_j)^2 \, . \qquad (4.53)$$

The derivation of equation (4.52) is left to the reader.

The expression (4.52) contains as particular cases all regularized solutions considered previously. This equation can also be written in the same form as equation (4.25) where R_μ is a convolution operator $R_\mu g = K_\mu^\dagger * g$, the convolution kernel $K_\mu^\dagger$ now being given by

$$K_\mu^\dagger(\mathbf{x}) = \frac{1}{(2\pi)^d} \int \frac{\hat{K}^*(\boldsymbol\omega)}{|\hat{K}(\boldsymbol\omega)|^2 + \mu \, \hat{P}(\boldsymbol\omega)} \, e^{i\mathbf{x}\cdot\boldsymbol\omega} \, d\boldsymbol\omega \, . \qquad (4.54)$$

If we compute the norm of the approximation error and the norm of the noise-propagation error, we find that they have behaviors, as a function of μ, similar to those derived in the case of Tikhonov regularization algorithm. As a consequence, the norm of the reconstruction error is supposed to have a minimum, as shown in Figure 4.5, so that an optimal value of the regularization parameter should exist for these regularization algorithms. In other words all regularization methods considered so far have the semi-convergence property.

An alternative way of writing equation (4.52), which suggests further generalizations, is the following one

$$f_\mu(\mathbf{x}) = \frac{1}{(2\pi)^d} \int_{\mathcal{B}} \hat{W}_\mu(\boldsymbol\omega) \, \frac{\hat{g}(\boldsymbol\omega)}{\hat{K}(\boldsymbol\omega)} \, e^{i\mathbf{x}\cdot\boldsymbol\omega} \, d\boldsymbol\omega \qquad (4.55)$$

where

$$\hat{W}_\mu(\boldsymbol\omega) = \frac{|\hat{K}(\boldsymbol\omega)|^2}{|\hat{K}(\boldsymbol\omega)|^2 + \mu \, \hat{P}(\boldsymbol\omega)} \, . \qquad (4.56)$$

This form makes clear that a family of regularized solutions is a family of filtered versions of the solution or generalized solution of equation (3.11) (see equation (3.20) or equation (3.43)), which can be obtained by means of a family of *window functions* satisfying the following conditions

- $|\hat{W}_\mu(\boldsymbol\omega)| \leq 1$, for any $\mu > 0$
- $\lim_{\mu\to 0}\hat{W}_\mu(\boldsymbol\omega) = 1$, for any $\boldsymbol\omega$ such that $\hat{K}(\boldsymbol\omega) \neq 0$
- $\hat{W}_\mu(\boldsymbol\omega)/\hat{K}(\boldsymbol\omega)$ is bounded for any $\mu > 0$.

For the family of window functions defined by equation (4.56) the first two conditions are obviously satisfied. The third one is satisfied if the denominator does not vanish.

Conversely, one can prove [31] that any family of window functions satisfying the three conditions above defines a linear regularization algorithm since the family of linear operators

$$(R_\mu g)(\mathbf{x}) = \frac{1}{(2\pi)^d} \int_\mathcal{B} \hat{W}_\mu(\boldsymbol{\omega}) \, \frac{\hat{g}(\boldsymbol{\omega})}{\hat{K}(\boldsymbol{\omega})} \, e^{i\mathbf{x}\cdot\boldsymbol{\omega}} d\boldsymbol{\omega} \qquad (4.57)$$

satisfies the conditions stated in Section 4.3.

This remark suggests many other regularization methods. A very simple one is obtained by chopping off the spatial frequencies where the TF $\hat{K}(\boldsymbol{\omega})$ is smaller than some threshold value. This threshold plays the role of the regularization parameter. The corresponding regularization algorithm, investigated by Miller in a more general setting [210], is defined by the following family of window functions which are called the *truncated window functions*

$$\hat{W}_\mu(\boldsymbol{\omega}) = \begin{cases} 1 & \text{if } |\hat{K}(\boldsymbol{\omega})| > \sqrt{\mu} \\ 0 & \text{if } |\hat{K}(\boldsymbol{\omega})| \le \sqrt{\mu} \end{cases} . \qquad (4.58)$$

These window functions, as well as those defined by equation (4.56), depend on the TF $\hat{K}(\boldsymbol{\omega})$. If $\hat{K}(\boldsymbol{\omega})$ does not vanish, as in the case, for instance, of Gaussian blur or of near-field acoustic holography, then it is also possible to use window functions which are independent of $\hat{K}(\boldsymbol{\omega})$. In particular, it is possible to use window functions which have a bounded support so that they provide bandlimited regularized solutions in a case where the imaging system is not bandlimited.

For simplicity we give a list of such windows in the 1D case. They are different from zero on a bounded interval $[-\Omega, \Omega]$ and Ω can take any value from 0 to $+\infty$. For any value of Ω, the window functions satisfy the first and third conditions stated above. The second one is satisfied when $\Omega \to \infty$, so that the relationship between Ω and the regularization parameter μ, as it is usually defined, is given by

$$\mu = \frac{1}{\Omega} . \qquad (4.59)$$

Herebelow we indicate the window functions as depending on Ω, which is the bandwidth of the corresponding regularized solutions.

- Rectangular window

$$\hat{W}_\Omega(\omega) = \begin{cases} 1 & \text{if } |\omega| < \Omega \\ 0 & \text{if } |\omega| \ge \Omega \end{cases} \qquad (4.60)$$

- Triangular window

$$\hat{W}_\Omega(\omega) = \begin{cases} 1 - \frac{|\omega|}{\Omega} & \text{if } |\omega| < \Omega \\ 0 & \text{if } |\omega| \ge \Omega \end{cases} \qquad (4.61)$$

- Generalized Hamming window

$$\hat{W}_\Omega(\omega) = \begin{cases} \alpha + (1-\alpha)\cos(\frac{\pi\omega}{\Omega}) & \text{if } |\omega| < \Omega \\ 0 & \text{if } |\omega| \ge \Omega \end{cases} \qquad (4.62)$$

For $\alpha = 0.5$ we get the Hanning window, while for $\alpha = 0.54$ we get the Hamming window [183].

The window functions which have a bounded support can be used for any TF (without zeroes) independently of its behavior for $|\omega| \to \infty$. It is also possible to use window functions which do not have a bounded support but depend on a parameter Ω defining an interval such that they are negligible outside this set. Their behavior at infinity must be such that, for any $\Omega > 0$, they tend to zero more rapidly than $\hat{K}(\omega)$ when $|\omega| \to \infty$. We give a few examples.

- Gaussian window

$$\hat{W}_\Omega(\omega) = \exp\left[-\frac{1}{2}\left(\frac{\omega}{\Omega}\right)^2\right] \tag{4.63}$$

This window can be used whenever $|\hat{K}(\omega)|$ tends to zero less rapidly than any Gaussian at infinity. An example is provided by near-field acoustic holography.

- Butterworth window

$$\hat{W}_\Omega(\omega) = \left[1 + \left(\frac{|\omega|}{\Omega}\right)^n\right]^{-1/2} \tag{4.64}$$

This window function, which is frequently used in tomography (see Chapter 8), can be used whenever $|\hat{K}(\omega)|$ tends to zero less rapidly than $|\omega|^{-n/2+1}$.

4.5 The global point spread function: resolution and Gibbs oscillations

The families of regularized solutions introduced in the previous sections are defined in terms of linear and bounded operators R_μ which are convolution operators, $R_\mu g = K_\mu^\dagger * g$, with kernels $K_\mu^\dagger(\mathbf{x})$ given by

$$K_\mu^\dagger(\mathbf{x}) = \frac{1}{(2\pi)^d}\int_B \frac{\hat{W}_\mu(\boldsymbol{\omega})}{\hat{K}(\boldsymbol{\omega})}\, e^{i\mathbf{x}\cdot\boldsymbol{\omega}}\, d\boldsymbol{\omega}\ , \tag{4.65}$$

$\hat{W}_\mu(\boldsymbol{\omega})$ being a family of window functions satisfying the conditions stated in the previous section. Analogously, in the discrete case, we have

$$\mathbf{f}_\mu = \mathbf{R}_\mu\, \mathbf{g} = \mathbf{K}_\mu^\dagger * \mathbf{g} \tag{4.66}$$

where

$$(\mathbf{K}_\mu^\dagger)_{m,n} = \frac{1}{N^2}\sum_B \frac{(\hat{\mathbf{W}}_\mu)_{k,l}}{\hat{K}_{k,l}}\exp\left[i\frac{2\pi}{N}(mk+nl)\right]. \tag{4.67}$$

If we use the model discussed in Chapter 2 for the noisy image g we obtain

$$f_\mu = R_\mu A f^{(0)} + R_\mu w\ . \tag{4.68}$$

A similar equation holds true in the discrete case. Equation (4.68) has the same structure as the model of the noisy image: the first term corresponds to the blurring while the second one corresponds to the noise. Therefore the operator $R_\mu A$ is a new blurring operator which describes the effect of two successive operations: the blurring due to the imaging operator and the partial deblurring due to the use of a regularized inversion algorithm. This operator is again a convolution operator whose kernel is the PSF of a linear system consisting of two linear systems in cascade: the first one is the imaging system (without the recording system, whose effect is the addition of noise) while the second one is the computer where the linear inversion algorithm R_μ has been implemented. It follows that it is quite natural to say that the PSF associated with the operator $R_\mu A$ is the *global* PSF since it is the PSF of the global system described above. The scheme of this global system is represented in Figure 4.8.
In order to compute the global PSF we take the FT of both sides of equation (4.68). Using equation (4.65) we obtain

$$\hat{f}_\mu(\boldsymbol{\omega}) = \hat{W}_\mu(\boldsymbol{\omega})\hat{f}^{(0)}(\boldsymbol{\omega}) + \hat{W}_\mu(\boldsymbol{\omega})\frac{\hat{w}(\boldsymbol{\omega})}{\hat{K}(\boldsymbol{\omega})} \tag{4.69}$$

object   imaging system   noise-free image

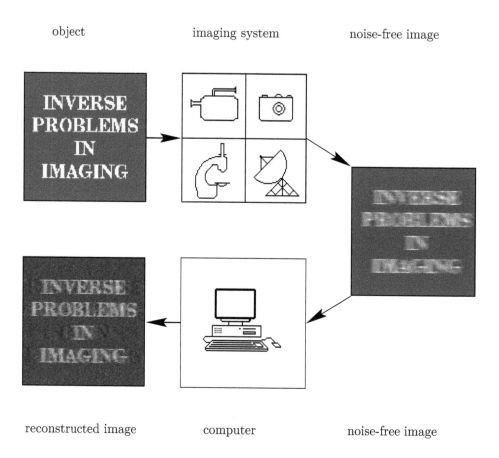

reconstructed image   computer   noise-free image

FIGURE 4.8   Scheme of the global system consisting of the imaging system and of the computer (where the inversion algorithm is implemented) in cascade.

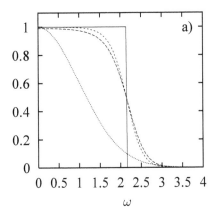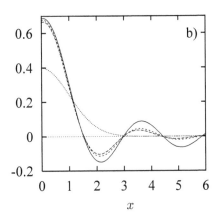

FIGURE 4.9 (a) Plot of the Gaussian TF (dotted line) compared with the Tikhonov window (dashed line), the Phillips window (thin dashed line) and the truncated window (solid line). (b) Plot of the Gaussian blur compared with the global PSF corresponding to the window functions in (a). The values of the regularization parameters are indicated in the text.

and this equation implies that the global TF is the window function $\hat{W}_\mu(\boldsymbol{\omega})$ so that *the global PSF is the inverse FT of the window function*:

$$W_\mu(\mathbf{x}) = \frac{1}{(2\pi)^d} \int_{\mathcal{B}} \hat{W}_\mu(\boldsymbol{\omega}) \, e^{i\mathbf{x}\cdot\boldsymbol{\omega}} d\boldsymbol{\omega} \ . \tag{4.70}$$

In conclusion we have

$$R_\mu A f^{(0)} = W_\mu * f^{(0)} \ . \tag{4.71}$$

Similar equations hold true in the discrete case where the discrete global PSF is the inverse DFT of the discrete window function

$$(\mathbf{W}_\mu)_{m,n} = \frac{1}{N^2} \sum_B (\hat{\mathbf{W}}_\mu)_{k,l} \, \exp\left[i\frac{2\pi}{N}(mk+nl)\right]. \tag{4.72}$$

The knowledge of the global PSF provides information about the approximation errors so that one can gain insight into their effect on image reconstruction. To this purpose we analyze two examples which can be viewed as paradigms of the PSF's most frequently encountered in practice: Gaussian blur and linear motion blur.

(A) *Gaussian blur*

For simplicity we consider 1D Gaussian blur with unit variance so that the PSF is given by $K(x) = (2\pi)^{-1/2} \exp(-x^2/2)$, and the corresponding TF by $\hat{K}(\omega) = \exp(-\omega^2/2)$. The TF is plotted in the panel (a) of Figure 4.9 and compared with the *Tikhonov window function*, as given by equation (4.56) with $\hat{P}(\omega) = 1$ and $\mu = 10^{-2}$; the *Phillips window function*, as given by equation (4.56) with $\hat{P}(\omega) = \omega^2$ and $\mu = 10^{-2}/2\ln(10)$; the *truncated window function*, defined in equation (4.58), also in the case $\mu = 10^{-2}$. The values of the regularization parameters are chosen in such a way that the values of both the Tikhonov and the Phillips window functions are equal to $1/2$ at the cut-off frequency of the truncated window function, which is given by $\Omega = [\ln(1/\mu)]^{1/2} = [2\ln(10)]^{1/2}$. In panel (b) of the same figure we give the PSF and the inverse FT of the window functions, i.e. the global PSF's.

These window functions define low-pass filters. While the truncated window function defines a perfect low-pass filter, the Tikhonov window has a smoother behavior and the Phillips window is intermediate between the two others even if closer to the Tikhonov one. The corresponding global PSF's have a central peak which is narrower than that of the Gaussian PSF. This means that in the reconstructed image details are sharper than in the original blurred image. However all global PSF's have rather important side lobes (i.e. the local negative minima and positive maxima to the right of the central peak) which can produce important artifacts.

A glance to panel (a) of Figure 4.9 makes clear that, once the value of the regularization parameter has been chosen, the regularized solution is an essentially bandlimited function with an effective bandwidth Ω defined, for instance, by the condition $\hat{W}_\mu(\Omega) = 1/2$. Since $\hat{W}_\mu(\omega) \cong 1$ for $|\omega| < \Omega$, it follows that $\hat{f}(\omega)$ approximately coincides with $\hat{f}^{(0)}(\omega)$ for these values of ω (if we neglect noise contribution), i.e. $f_\mu(x)$ provides an Ω-bandlimited approximation of $f^{(0)}(x)$.

The bandlimiting intrinsic to the regularization methods considered in this section has two main effects:

- the object $f^{(0)}(x)$ can only be estimated with a limited resolution;
- the reconstructed object may be affected by Gibbs oscillations.

Both effects are related to the behavior of the global PSF as indicated above.

As concerns resolution we only observe that details of the object $f^{(0)}(x)$ smaller than the width of the central peak of the global PSF are usually lost in the reconstructed image. We also observe that this width is approximately related to the effective bandwidth Ω by the sampling theorem, i.e. it is of the order of π/Ω.

As concerns the second effect, it is well known that Gibbs oscillations are produced by the truncation of Fourier series or Fourier integrals in the case of discontinuous functions. Therefore they are also produced by the bandlimiting implied by regularization and are related to the side lobes of the global PSF. To this purpose it is important to note that a reduction of the value of μ (which is possible if the noise on the image is reduced) does not imply a reduction of the side lobes. A smaller value of μ, indeed, implies a narrower and higher central peak but also higher side lobes. In the particular case of the truncated window function, the global PSF is a 'sinc' function, i.e. $W_\mu(x) = \sin(\Omega x)/\pi x$ with $\Omega = [\ln(1/\mu)]^{1/2}$. Hence the modulus of the ratio between the value at the first side lobe and the central value is $|\cos\xi_1|$, where ξ_1 is the root, between $\pi/2$ and $3\pi/2$, of the equation $\tan(\xi) = \xi$. This ratio is independent of Ω and it does not change when Ω is increased as an effect of noise reduction. Similar considerations apply to the Tikhonov and Phillips window functions.

Gibbs oscillations appear, for instance, in the 2D reconstruction of a point-like object (a bright star on a black background): the side lobes of the global PSF originate a series of rings around the central spot; hence this feature of the reconstructed image is also referred to as *ringing*. Ringing also appears in the reconstruction of sharp intensity variations in the object. In Figure 4.10, we give the reconstruction of a discontinuous function whose image is provided by the Gaussian blur considered above. The Gibbs oscillations clearly appear at the edges of the object. The definition of the position of the discontinuity is more precise than in the blurred image but the reconstruction is distorted by this effect. If the value of μ is decreased, the width of the maxima and minima around the discontinuity decreases but their height does not decrease. In order to make evident this effect, which is due to the approximation error, we also provide in Figure 4.10, some reconstructions obtained from noise-free data in a neighborhood of the right discontinuity and corresponding to various values of the regularization parameter.

The previous analysis applies to all cases where the PSF is essentially bandlimited with a TF which does not vanish at low frequencies. In these cases, the global PSF does not strongly

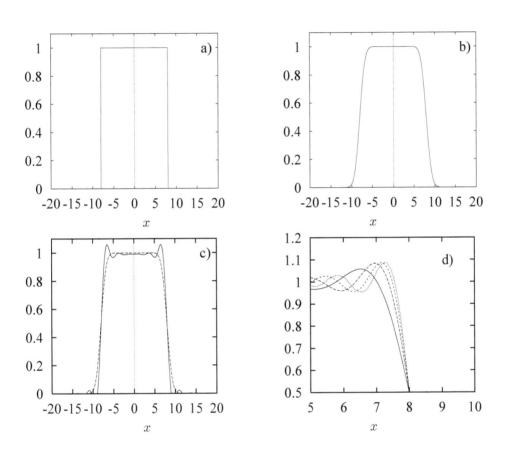

FIGURE 4.10 (a) The object; (b) the image provided by the Gaussian blur considered in the text; (c) Tikhonov's reconstruction, with $\mu = 10^{-2}$, of the object in (a) (full line) compared with the image in (b) (dashed line); (d) magnification of the reconstructions obtained with $\mu = 10^{-2}, 10^{-4}, 10^{-6}, 10^{-8}$ in a neighborhood of the discontinuity.

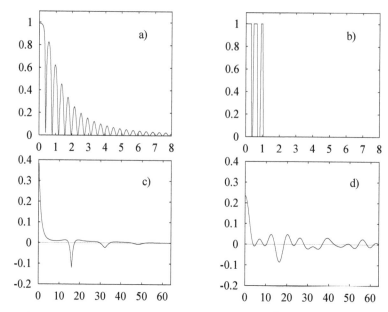

FIGURE 4.11 (a) The Tikhonov window function with $\mu = 10^{-2}$ for the linear motion blur in the case $s = 16$; (b) the truncated window function with the same value of μ and for the same problem; (c) the global PSF corresponding to the Tikhonov window; (d) the global PSF corresponding to the truncated window.

depend on the PSF and, for some filtering methods, it is even independent of it.
Removal of ringing and Gibbs oscillations can be obtained by some of the nonlinear methods described in the next chapters. In particular, Gibbs oscillations are completely removed by the methods discussed in Chapter 9.

(B) *Linear motion blur*

A quite different situation occurs for the case of linear motion blur. If the PSF is given by equation (2.59), then from equation (2.56) we get that the modulus of the TF is given by

$$|\hat{K}(\omega)| = \left|\text{sinc}\left(\frac{s\,\omega}{2\pi}\right)\right| . \tag{4.73}$$

In Figure 4.11, we plot the Tikhonov window function for this problem (with $s = 16$, arbitrary units) and the truncated window function, both with $\mu = 10^{-2}$. We also give the corresponding global PSF. We point out that the window functions vanish at the zeroes of the TF.
As a first remark we observe that, in such a case, the truncated window is certainly less convenient than the Tikhonov window with the same value of the regularization parameter. The reason is that the modulus of the TF (4.73) at high frequencies is bounded by a factor $|\omega|^{-1}$ which tends to zero rather slowly. As a consequence, the use of a truncated window removes a large part of the frequency domain where the TF is still important. This is not true for the Gaussian blur. The reader can compare Figure 4.9, (a) with Figure 4.11, (a) and (b).
If we consider the global PSF corresponding to the truncated window we find that it is much more irregular than that corresponding to the Tikhonov window. The behavior of the latter is rather interesting. It has a rather narrow central peak, whose FWHM (*Full Width at Half Maximum*) is about 1.6 while the translation parameter of the motion blur is $s = 16$. This

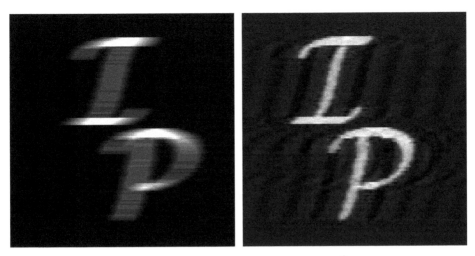

FIGURE 4.12   Reconstruction by the Tikhonov method, with $\mu = 10^{-2}$, of an image blurred by motion ($S = 16$ pixels). In order to make the effect of the approximation errors evident, the image, shown to the left, was not corrupted by noise. In the reconstruction, shown to the right, the ghosts discussed in the text clearly appear. In the case of a noisy image these ghosts are perturbed by the noise contribution.

means that in the reconstructed image it is possible to identify many more details than in the blurred one. Moreover, a reduction of the value of μ (which is possible if the noise is reduced), produces a reduction of the width of the central peak, hence an improvement in resolution. Indeed the width of this central peak is related to the high-frequency cut-off introduced by the Tikhonov window.

The central peak is flanked by a secondary and negative peak at a distance $d_0 = 16$, equal to the translation distance. The position of this peak does not depend on the value of μ but depends on the value of s because it depends on the zeroes of the TF given in equation (4.73). If we reduce μ, the peak becomes narrower but its position does not change. Its effect on the reconstructed image is the appearance of two ghosts (one on the right and one on the left) reproducing translated and negative versions of the original object. Panel (c) of Figure 4.11 also suggests the appearance of other weaker ghosts at distances that are multiples of d_0.

This effect can be understood if we observe that the Tikhonov window function of Figure 4.11, panel (a), approximates the profile of a modulated grating with period equal to the distance between the zeroes of the TF given in equation (4.73), i.e. $\omega_0 = 2\pi/s$. Then the inverse FT of the profile of the grating has peaks spaced by the reciprocal period $d_0 = 2\pi/\omega_0 = s$.

The appearance of ghosts is shown in Figure 4.12. A picture 128×128 of the calligraphic initials of Inverse Problems was blurred by uniform motion corresponding to a horizontal translation of 16 pixels. In the reconstructed image at least four ghosts are visible on both sides of the two letters.

We point out that this effect cannot be removed by means of the linear filtering methods considered in this chapter because it is due to the lack of information in the neighborhoods of the zeroes of the TF. In some particular cases, however, it can be removed by means of some of the nonlinear methods which will be discussed in the next chapters, in particular, methods implementing the constraint of non-negativity when the image is nonnegative.

4.6   Choice of the regularization parameter

The choice of the value of the regularization parameter is a crucial and difficult problem in the theory of regularization. This point has been widely discussed in the mathematical literature. No precise recipe has been discovered which could be used for any problem. In this section we consider the case of the Tikhonov regularization algorithm, i.e. of a regularized solution $f_\mu(\mathbf{x})$ given by equation (4.12), and we summarize the main methods which are used in practice.

As we know from the discussion of Section 4.3, for any image g there exists in general an optimal value, μ_{opt}, of the regularization parameter. For that value of μ, the corresponding regularized solution $f_\mu(\mathbf{x})$ has minimal distance from the true object $f^{(0)}(\mathbf{x})$. The problem is that the determination of this optimal value implies the knowledge of $f^{(0)}(\mathbf{x})$. On the other hand, the methods described in Sections 4.1 and 4.2 can also be considered as methods for estimating μ_{opt} when $f^{(0)}(\mathbf{x})$ is not known and an intuitive relation between μ_1, μ_2 and μ_{opt} is discussed at the end of Section 4.3. We describe again these methods herebelow, as well as other methods which can be used in the analysis of real images.

In order to compare the various methods we also report the results obtained in the following numerical experiment. A 256×256 picture of the bell tower of the San Donato church in Genoa, given in Figure 4.13, is blurred by linear motion corresponding to a horizontal translation of $S = 8$ pixels. In such a case the discrete TF, given by equation (2.61), is zero for $m = Nk/S = 32k$ ($k = 1, 2, \ldots, 8$). However, as a consequence of round-off errors, in our experiment the condition number (given by equation (3.30)) is not infinite but of the order of the inverse of the machine precision, i.e. $\alpha = 9.68 \ 10^7$. Finally the blurred image is perturbed by white Gaussian noise with a variance whose square root is equal to 0.8% of the maximum value of the noise-free image. The values of the square root of the energy and of the discrepancy of the object are the following

$$E = 2.91 \ 10^4 \ , \quad \varepsilon = 513 \ . \tag{4.74}$$

Since we have $\|g\| = 2.78 \ 10^4$, the relative error due to noise is about 1.84 %.

The behavior of the relative reconstruction error as a function of μ is given in Figure 4.5 and the optimal value of μ, corresponding to the minimum, is $\mu_{opt} = 7.0 \ 10^{-3}$. The reconstructed image is given in Figure 4.13 together with a detail showing the appearance of the ghosts discussed in the previous section. The reconstruction error is about 13.2%, corresponding to a MSEIF equal to 8.7 dB (the MSEIF is defined in equation (3.34)). The reconstructed image contains some negative values which are not very important because if we replace these values by zero (projection onto the set of positive images), the reconstruction error does not decrease significantly (it becomes 12.9%).

(A) *Regularized solution with prescribed energy*

If we do not know $f^{(0)}(\mathbf{x})$ but do know its energy E^2, then the constrained least-squares problem discussed in Section 4.1 can be considered as a method for estimating the optimal value of the regularization parameter. The estimate is the value $\mu_1 = \mu_1(E)$ such that the corresponding regularized solution has the same energy as the true object, i.e.

$$\|f_{\mu_1}\| = E \ . \tag{4.75}$$

As one can easily deduce from Figure 4.2, the value of $\mu_1(E)$ is a decreasing function of E. Therefore, if we overestimate the energy of the object, we obtain a value of the regularization parameter smaller than the one corresponding to the exact energy of the object. In such a case the reconstructed image shows a higher noise contamination. Conversely, if we underestimate the energy value, we obtain a reconstruction with a higher blur contribution.

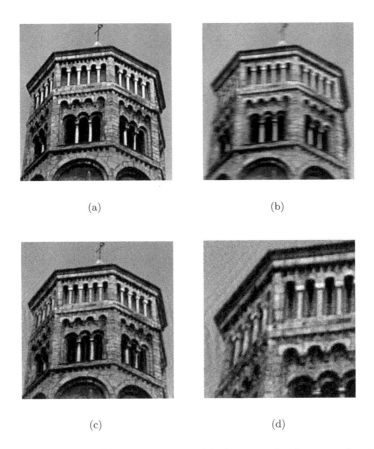

(a)

(b)

(c)

(d)

FIGURE 4.13   (a) The object; (b) the image blurred by linear motion (corresponding to a translation of 8 pixels) and perturbed by white Gaussian noise with SNR=35 (dB); (c) the optimal reconstruction corresponding to the minimum of the reconstruction error; (d) detail of the reconstruction in (c) showing the appearance of ghosts both in the sky and in the space between the columns.

In the numerical experiment described above, using the value of E given in equation (4.74), we find $\mu_1 = 1.6 \ 10^{-3}$. This value is smaller than the optimal value by a factor of 4. However the increase in the reconstruction error is not dramatic, since we find a relative error of about 15.1%. The visualization (not reproduced here) of the reconstructed image confirms that the result is still satisfactory.

(B) *Regularized solution with prescribed discrepancy*

If we know a precise estimate ε of the energy of the noise, then the method discussed in Section 4.2 can be considered as another method for estimating μ_{opt}. The estimate is the value $\mu_2 = \mu_2(\varepsilon)$ such that the discrepancy of the corresponding regularized solution is equal to ε^2, i.e.

$$\|Af_{\mu_2} - g\| = \varepsilon \ . \tag{4.76}$$

In the literature on inverse problems this method is known as *Morozov's discrepancy principle* [213] . From Figure 4.4 we deduce that $\mu_2(\varepsilon)$ is an increasing function of ε. Therefore, if we overestimate the energy of the noise, we get a value of the regularization parameter which is larger than the one corresponding to the exact energy of the noise. In such a case the reconstructed image will show a smaller deblurring effect.

The value found in our numerical experiment, using the value of ε given in equation (4.74), is $\mu_2 = 1.2 \ 10^{-2}$. This value is greater than the optimal value and slightly closer to μ_{opt} than μ_1. The increase in the reconstruction error is quite small since we find a relative error of about 13.5% (13.2% in the case of μ_{opt}). We point out that the range of the values of μ between μ_1 and μ_2 is rather broad and that all reconstructions corresponding to this interval are satisfactory. Therefore, in this example, the choice of the value of the regularization parameter is not very critical.

In the many numerical experiments we have performed, using the exact value of ε, we have found values of μ_2 sometimes greater and sometimes smaller than μ_{opt}. In all cases, however, the value of μ_2 was closer to μ_{opt} than the value of μ_1. Therefore the choice of μ based on the knowledge of the discrepancy seems more satisfactory than the one based on the knowledge of the energy. This conclusion is coherent with an important theoretical result proved for this particular choice of the regularization parameter.

Indeed, let us consider a family of images g_ε satisfying, for any ε, the conditions (4.22) and strongly converging, for $\varepsilon \to 0$, to a noise-free image $g^{(0)}$ as assumed in Section 4.3. Moreover, for any g_ε, let us denote by $f_{\mu(\varepsilon)}$ the unique regularized solution solving equation (4.76). Then the following result has been proved [212, 213, 140, 31]: *if $\varepsilon \to 0$, then $f_{\mu(\varepsilon)}$ strongly converges to the unique generalized solution associated with $g^{(0)}$, i.e.*

$$\lim_{\varepsilon \to 0} \|f_{\mu(\varepsilon)} - A^\dagger g^{(0)}\| = 0 \ . \tag{4.77}$$

The proof of this result is not difficult but requires some background in functional analysis. Its relevance is obvious: if in an experiment we can reduce the additive noise affecting the image, then Morozov's discrepancy principle ensures that we can obtain a regularized solution approaching the true object.

(C) *The Miller method*

An approach to ill-posed problems proposed by Miller [210] can also be considered as a method for estimating the value of the regularization parameter. In this approach it is assumed that one knows both a bound on the energy and a bound on the discrepancy of the unknown object $f^{(0)}(\mathbf{x})$. Then the set of all objects $f(\mathbf{x})$ satisfying the two conditions

$$\|Af - g\|^2 \le \varepsilon^2 \ , \quad \|f\|^2 \le E^2 \tag{4.78}$$

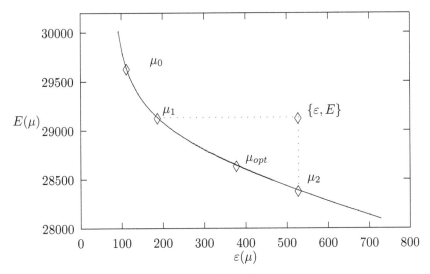

FIGURE 4.14 Plot of the L-curve and of the pair $\{\varepsilon, E\}$ for the example of Figure 4.13. We also indicate the points on the curve corresponding to the different values of the regularization parameter obtained by means of the various methods.

is called the set of *admissible approximate solutions*. This set corresponds to the dashed region represented in Figure 4.6. It is the intersection of the ball of the objects with energy smaller than E^2 and of the ellipsoid of the objects with discrepancy smaller than ε^2. If this intersection is not empty, then the pair $\{\varepsilon, E\}$ is said to be *permissible*.

Then the following question arises: is it possible to establish whether a certain pair $\{\varepsilon, E\}$ is permissible for a given image $g(\mathbf{x})$? The answer to this question is positive and rather simple.

Given $\tilde{f}_1$ or $\tilde{f}_2$, i.e. the solutions introduced in Sections 4.1 and 4.2 respectively, we can check whether these functions belong or not to the set (4.78). If they do, then the set is obviously not empty. If they do not, then the set is empty thanks to the following result: *if the set defined by the conditions (4.78) is not empty, then it contains both $\tilde{f}_1$ and $\tilde{f}_2$.*

The proof is easy. We give it in the case of $\tilde{f}_1$. We observe that $\tilde{f}_1$ is the object which minimizes the discrepancy functional in the set of all objects whose energy does not exceed E^2. Since there exist, by assumption, elements of this set whose discrepancy does not exceed ε^2, it follows that also the discrepancy of $\tilde{f}_1$ must be less than ε^2, i.e. $\tilde{f}_1$ is an admissible approximate solution. A similar argument applies to $\tilde{f}_2$ by exchanging the role of the two functionals.

It is obvious that this condition can be checked numerically. If the computed discrepancy of $\tilde{f}_1$ is greater than ε^2, then the set of the approximate solutions is empty. On the other hand, if the discrepancy of $\tilde{f}_1$ is smaller than ε^2, then also $\tilde{f}_2$ is an admissible approximate solution. Moreover the previous arguments also imply that, if the pair $\{\varepsilon, E\}$ is permissible, then the following inequalities hold true

$$\|\tilde{f}_2\| \leq \|\tilde{f}_1\| , \quad \|A\tilde{f}_1 - g\| \leq \|A\tilde{f}_2 - g\| . \tag{4.79}$$

The first is the inequality of equation (4.43).

Finally it is possible to characterize the set of all permissible pairs. Consider in the plane $\{\varepsilon, E\}$ the curve described by the points $\{\varepsilon(f_\mu; g), E(f_\mu)\}$ when μ varies from 0 to $+\infty$. This curve is a plot of $E(f_\mu)$ versus $\varepsilon(f_\mu; g)$ and it is called the L-curve because, in most cases, its log-log plot has the shape of the letter L. Now, if $\{\varepsilon, E\}$ is an admissible pair, we have shown that $\varepsilon \geq \|A\tilde{f}_1 - g\| = \varepsilon(f_{\mu_1}; g)$ and $E \geq \|\tilde{f}_2\| = E(f_{\mu_2})$. Therefore the point $\{\varepsilon, E\}$ is to the right and above the L-curve. Conversely, any such a point is a permissible pair. In Figure 4.14, we give the L-curve for the example of Figure 4.13 and we indicate the pair $\{\varepsilon, E\}$ corresponding to this example.

Now, if the pair $\{\varepsilon, E\}$ is permissible, it has been shown by Miller [210] that the regularized solution corresponding to the following value of the regularization parameter

$$\mu_0 = \left(\frac{\varepsilon}{E}\right)^2 \tag{4.80}$$

satisfies the conditions (4.78) except for a factor of $\sqrt{2}$ and therefore it is essentially an admissible approximate solution.

In our numerical example we find $\mu_0 = 3.1 \ 10^{-4}$ and this value is much smaller than the values provided by the two other methods. Also the corresponding reconstruction error is much higher, about 21%. The visualization of the reconstructed image shows an important contribution due to the noise propagation. Indeed all numerical experiments seem to indicate that, when the correct values of ε and E are used, the Miller method tends to underestimate the value of the regularization parameter.

(D) *Generalized cross-validation*

The methods considered previously require the knowledge of ε, or of E, or of both. In many cases one does not have a sufficiently accurate estimate of these quantities and therefore it is important to have methods which do not require this kind of information. One such method is the method of *cross-validation* [301] which can only be used in problems with discrete data and is based on the idea of letting the data themselves choose the value of the regularization parameter. In other words it is required that a good value of the regularization parameter should predict missing data values.

In order to simplify the notation, let us assume that a discrete image **g**, represented by an array of numbers $g_{m,n}$, has been rearranged, for instance by means of lexicographic ordering, in order to obtain a vector with N^2 components, denoted by g_m. Then also the imaging matrix **A** is replaced by a matrix with only two entries and its matrix elements are denoted by $A_{m,n}$. We still use the notation **A** for this matrix.

Consider now the problem where the k-th data component is missing. We denote by $\mathbf{f}_{\mu,k}$ the regularized solution for this problem, i.e. the vector which minimizes the functional

$$\Phi_{\mu,k}(\mathbf{f}; \mathbf{g}) = \sum_{m \neq k} |(\mathbf{Af})_m - g_m|^2 + \mu \sum_{n=1}^{N^2} |f_n|^2 . \tag{4.81}$$

By means of $\mathbf{f}_{\mu,k}$ one can compute the missing component, i.e. $(\mathbf{Af}_{\mu,k})_k$. If μ is a good choice, the quantity $(\mathbf{Af}_{\mu,k})_k - g_k$ should, on average, be small. The *cross-validation function* $V_0(\mu)$ is precisely the mean quadratic error we commit when we repeat this procedure for all components of the image

$$V_0(\mu) = \frac{1}{N^2} \sum_{k=1}^{N^2} |(\mathbf{Af}_{\mu,k})_k - g_k|^2 \tag{4.82}$$

and the *cross-validation method* consists in determining the value of μ which minimizes $V_0(\mu)$.

The computation of $V_0(\mu)$ does not require the solution of N^2 minimization problems. Indeed, if we denote by $\mathbf{A}(\mu)$ the following matrix

$$\mathbf{A}(\mu) = \mathbf{AA}^*(\mathbf{AA}^* + \mu\mathbf{I})^{-1}, \tag{4.83}$$

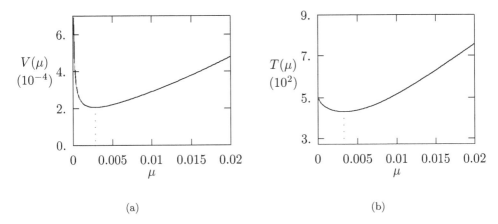

FIGURE 4.15 Plot of the GCV function (a) and of the predictive mean square error (b) in the case of the example of Figure 4.13.

where $\mathbf{A}^*$ is the adjoint matrix and by $\mathbf{f}_\mu$ the regularized solution (4.16), then the following relation holds true [77, 136]

$$V_0(\mu) = \frac{1}{N^2} \sum_{k=1}^{N^2} \frac{|(\mathbf{A}\mathbf{f}_\mu)_k - g_k|^2}{|1 - A_{kk}(\mu)|^2} \qquad (4.84)$$

which can be used for the computation of $V_0(\mu)$.

The main disadvantage of the method is that it is not invariant with respect to a linear transformation of the components of the image and of the object as that provided, for instance, by the DFT. The DFT, indeed, transforms the matrix $\mathbf{A}$ into a diagonal one but, in the case of a diagonal matrix, the cross-validation function is a constant, i.e. $V_0(\mu) = N^{-1}\|g\|^2$.

For this reason, the minimization of the cross-validation function, in spite of its clear meaning, is replaced by the minimization of the *generalized cross-validation function* (GCV function) defined by

$$V(\mu) = \frac{\|\mathbf{A}\mathbf{f}_\mu - \mathbf{g}\|^2}{(Tr[\mathbf{I} - \mathbf{A}(\mu)])^2} \qquad (4.85)$$

where

$$Tr[\mathbf{I} - \mathbf{A}(\mu)] = \sum_{n=1}^{N^2} [1 - A_{n,n}(\mu)] \; , \qquad (4.86)$$

the matrix $\mathbf{A}(\mu)$ being defined in equation (4.83). This function is invariant with respect to rotations of the data vector and is obtained from equation (4.84) by replacing the denominators with their arithmetic mean.

The GCV function has another important property. As it has been shown [77, 136], the minimizer of $V(\mu)$ provides an estimate of the minimizer of the so-called *predictive mean square error* defined by

$$T(\mu) = \|\mathbf{A}\mathbf{f}_\mu - \mathbf{g}^{(0)}\| \; , \qquad (4.87)$$

where $\mathbf{g}^{(0)}$ is the noise-free image, i.e. $\mathbf{g}^{(0)} = \mathbf{A}\mathbf{f}^{(0)}$. Also this function is expected to have a unique minimum as the norm of the reconstruction error investigated in Section 4.3, i.e. $\|\mathbf{f}_\mu - \mathbf{f}^{(0)}\|$. The minimization of $T(\mu)$ provides another criteria of optimality. We denote by μ'_{opt} the minimizer of $T(\mu)$ and by μ_{GCV} the minimizer of $V(\mu)$.

	μ_{opt}	μ_1	μ_2	μ_0	μ'_{opt}	μ_{GCV}
μ	$7.0 \ 10^{-3}$	$1.6 \ 10^{-3}$	$1.2 \ 10^{-2}$	$3.1 \ 10^{-4}$	$3.3 \ 10^{-3}$	$2.8 \ 10^{-3}$
err	13.2%	15.1%	13.5%	21.0%	13.7%	13.9%

TABLE 4.1   The values of the regularization parameter and of the corresponding reconstruction errors obtained by the various methods in the example of Figure 4.13.

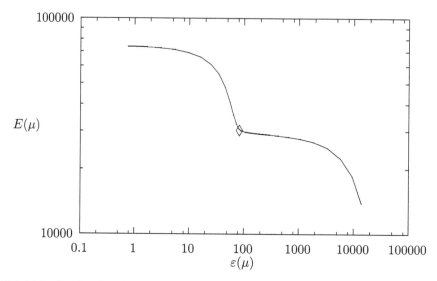

FIGURE 4.16   Log-log plot of the L-curve of Figure 4.14.

In Figure 4.15, we plot the GCV function $V(\mu)$ and the predictive mean square error $T(\mu)$ as functions of μ, in the case of the example of Figure 4.13. The minimum of $T(\mu)$ is $\mu'_{opt} = 3.3 \ 10^{-3}$ and therefore is smaller, by a factor of 2, than the optimal value μ_{opt} but larger than μ_1. The reconstruction error is about 13.7%. On the other hand the minimum of the GCV function is $\mu_{GCV} = 2.8 \ 10^{-3}$ and provides a satisfactory approximation of the minimum of $T(\mu)$. In this case the reconstruction error is 13.9%.

In Table 4.1, we summarize all results obtained in the case of the example of Figure 4.13.

(E) L-curve method

This graphically motivated method, introduced by Hansen [148], is another method which does not require information about the energy of the noise or of the true object. The starting point is that the L-curve, introduced in connection with Miller method, has, in many cases, a rather characteristic L-shaped behavior in a log-log plot. This plot is given in Figure 4.16 for the numerical example of Figure 4.13.

A qualitative explanation of this behavior is the following. We recall that $E(f_\mu)$ is large for small μ and small for large μ while $\varepsilon(f_\mu; g)$ has an opposite behavior. Therefore $E(f_\mu)$ is large when $\varepsilon(f_\mu; g)$ is small and conversely. This is the trade-off between noise-propagation error and approximation error already discussed in Section 4.3. Now the vertical part of the L-curve corresponds to values of the regularization parameter such that f_μ is dominated by the noise propagation error. As a consequence $E(f_\mu)$ is very sensitive to variations of μ while $\varepsilon(f_\mu; g)$ is not. Analogously the horizontal part of the L-curve corresponds to values of the regularization parameter such that f_μ is dominated by the approximation error. As a consequence $\varepsilon(f_\mu; g)$ is very sensitive to variations of μ while $E(f_\mu)$ is not.

Now the L-curve method consists in taking as an estimate of the regularization parameter the value of μ, which we denote by μ_L, corresponding to the corner of the L-curve. In fact this point should correspond to the best compromise between approximation error and noise-propagation error. From the computational point of view a convenient definition of the L-curve corner is the point with maximum curvature. Good performance of the method has been reported in the case of numerical experiments [147]. However in our example we find $\mu_L = 2.3\ 10^{-7}$ and this is an unacceptable value.

Indeed the L-curve method, even if it can be useful in some cases, does not work in all cases and presents some theoretical and practical inconveniences. It has been shown [113, 298] that, in certain cases, it does not provide a regularized solution converging to the exact one when the noise tends to zero. Moreover, examples can be found where the L-curve does not even have an L-shape so that the method cannot be used.

(F) *The interactive method*

The speed and the versatility of the modern digital computers allow to reconstruct images interactively: the user controls the reconstructions obtained by means of several values of the regularization parameter and, by tuning μ, he selects the best reconstruction on the base of his intuition or of the attainment of some specific purpose. Figure 4.7 clearly shows that excellent results can be obtained by means of this method.

5

Iterative regularization methods

Some iterative methods introduced for solving linear algebraic systems can also be used for regularizing the solution of linear ill-posed problems. The basic feature of these methods, when applied to ill-posed problems, is that the number of iterations plays the role of a regularization parameter because semi-convergence holds true in the case of noisy images: when the number of iterations increases, the iterates first approach the unknown object and then go away.

The simplest iterative method having this property is the so-called Landweber method or successive-approximations method. Another one, which is frequently used in practice, is the conjugate gradient method. The main practical difference between the two methods is that the convergence to a sensible approximate solution is faster in the case of the conjugate gradient than in the case of the Landweber method.

However, an interesting feature of the Landweber method is that it can be easily modified to take into account additional *a priori* information about the solution. The resulting projected Landweber method can be used for solving some of the constrained least-squares problems discussed in Section 3.6. The projected Landweber method, steepest descent and conjugate gradient are the first examples we meet of nonlinear regularization methods for the solution of linear inverse problems.

5.1 The van Cittert and Landweber methods

A very simple iterative method for approximating the least-squares solutions of first-kind integral equations has been introduced independently by Landweber [189] and Fridman [123]. For this reason, the method is usually called the *Landweber* or the *Landweber-Fridman method*, at least in the western literature concerning ill-posed problems. In the Russian literature the denomination of *successive-approximations method* is preferred. It is probably more correct to call it the *Jacobi method* because the basic idea is the same as in an iterative method introduced by Jacobi for solving linear algebraic systems.

The convergence of the method was investigated by Bialy [40] in the general case of a linear and ill-posed operator equation. A particular version of the same method was proposed

DOI: 10.1201/9781003032755-5

independently by Gerchberg [130] and Papoulis [227] for the problem of extrapolating a bandlimited signal. This algorithm will be discussed in Chapter 8. It can also be shown that the Cimmino method is a particular case of the Landweber method [178]. Finally the *simultaneous iterative reconstruction technique* (SIRT), which is frequently used in seismic tomography, can also be reduced to the Landweber method if suitable weights are introduced in the data and in the solution space [293].

In the problem of image deconvolution the Landweber method applies to the least-squares equation introduced in Section 3.5, i.e.

$$A^*Af = A^*g \tag{5.1}$$

where A^*A is a convolution operator whose TF is $|\hat{K}(\boldsymbol{\omega})|^2$, if $\hat{K}(\boldsymbol{\omega})$ is the TF of the imaging system. However, in the field of image deconvolution a different method introduced by van Cittert [292] is also considered. The iterative scheme is the same as for the Landweber method but it is applied to the original imaging equation

$$Af = g \tag{5.2}$$

and not to the least-squares equation (5.1). In this context the Landweber method discussed in this Section is also called a *reblurring method* [41] because equation (5.1) can be obtained from equation (5.2) by applying to both members the blurring operator A^*. However, the van Cittert method can be used only in the case of imaging systems having some particular properties, while the Landweber method can be applied to any imaging system. In order to treat simultaneously the two methods we consider a convolution equation of the following form

$$\bar{A}f = \bar{g} \;. \tag{5.3}$$

If $\bar{A} = A^*A$ and $\bar{g} = A^*g$ we get equation (5.1) while if $\bar{A} = A$ and $\bar{g} = g$ we get equation (5.2).

A very simple way for introducing both methods is the following: consider the operator

$$T(f) = f + \tau(\bar{g} - \bar{A}f) \tag{5.4}$$

where τ is the so-called *relaxation parameter* whose choice is discussed in the following. Notice that this operator is nonlinear because of the inhomogeneous term $\tau\bar{g}$. Then any solution of equation (5.3) is also a *fixed point* of the operator T and conversely, i.e. equation (5.3) is equivalent to the following fixed-point equation

$$f = T(f) \;. \tag{5.5}$$

If we use the well-known *method of successive approximations*, which is the most natural way for approximating the fixed points of a given mapping, we obtain the following iterative scheme

$$f_{k+1} = T(f_k) \tag{5.6}$$

or, in a more explicit form

$$f_{k+1} = f_k + \tau(\bar{g} - \bar{A}f_k) \;. \tag{5.7}$$

In order to investigate the convergence of the iterative method, it is convenient to rewrite equation (5.7) as follows

$$f_{k+1} = \tau\bar{g} + (1 - \tau\bar{A})f_k \;. \tag{5.8}$$

If we denote by $\hat{H}(\boldsymbol{\omega})$ the TF of the convolution operator $\bar{A}$ (notice that $H(\boldsymbol{\omega}) = K(\boldsymbol{\omega})$ if $\bar{A} = A$ and $H(\boldsymbol{\omega}) = |K(\boldsymbol{\omega})|^2$ if $\bar{A} = A^*A$), then this equation implies the following equation for the Fourier transforms

$$\hat{f}_{k+1}(\boldsymbol{\omega}) = \tau\hat{\bar{g}}(\boldsymbol{\omega}) + (1 - \tau\hat{H}(\boldsymbol{\omega}))\hat{f}_k(\boldsymbol{\omega}) \;. \tag{5.9}$$

Let us assume, without any loss of generality, that the imaging system has a band $\mathcal{B}$ and let us denote by $\mathcal{B}^C$ is the set of the out-of-band frequencies. Indeed this is not restrictive since $\mathcal{B}^C$ can be empty, if $\mathcal{B}$ coincides with the whole frequency space.
If $\hat{f}_0(\boldsymbol{\omega})$ is the FT of the initial approximation f_0, it is easy to show by induction that

$$\hat{f}_k(\boldsymbol{\omega}) = (1 - \tau\hat{H}(\boldsymbol{\omega}))^k \hat{f}_0(\boldsymbol{\omega}) \tag{5.10}$$
$$+ \quad \tau\left[1 + (1 - \tau\hat{H}(\boldsymbol{\omega})) + (1 - \tau\hat{H}(\boldsymbol{\omega}))^2 + \cdots + (1 - \tau\hat{H}(\boldsymbol{\omega}))^{k-1}\right]\hat{g}(\boldsymbol{\omega}) .$$

When $\boldsymbol{\omega} \in \mathcal{B}^C$ from this equation we obtain

$$\hat{f}_k(\boldsymbol{\omega}) = \hat{f}_0(\boldsymbol{\omega}) + k\tau\hat{g}(\boldsymbol{\omega}) . \tag{5.11}$$

Therefore if $\hat{g}(\boldsymbol{\omega}) = 0$ in $\mathcal{B}^C$, then $\hat{f}_k(\boldsymbol{\omega})$ coincides with the initial approximation $\hat{f}_0(\boldsymbol{\omega})$ for any k. Otherwise the sequence does not converge. We will reconsider this point in the following. On the other hand, for spatial frequencies such that $\hat{H}(\boldsymbol{\omega}) \neq 0$, from the well-known relation

$$1 + \xi + \xi^2 + \cdots + \xi^{k-1} = \frac{1 - \xi^k}{1 - \xi} \quad (\xi \neq 1) , \tag{5.12}$$

with $\xi = 1 - \tau\hat{H}(\boldsymbol{\omega})$, and from equation (5.10) we obtain

$$\hat{f}_k(\boldsymbol{\omega}) = (1 - \tau\hat{H}(\boldsymbol{\omega}))^k \hat{f}_0(\boldsymbol{\omega}) + \left[1 - (1 - \tau\hat{H}(\boldsymbol{\omega}))^k\right]\frac{\hat{g}(\boldsymbol{\omega})}{\hat{H}(\boldsymbol{\omega})} . \tag{5.13}$$

We see from this equation that, for any given $\boldsymbol{\omega}$ such that $\hat{H}(\boldsymbol{\omega}) \neq 0$, the limit of $\hat{f}_k(\boldsymbol{\omega})$, for $k \to \infty$, exists and is precisely $\hat{g}(\boldsymbol{\omega})/\hat{H}(\boldsymbol{\omega})$ if and only if

$$|1 - \tau\hat{H}(\boldsymbol{\omega})| < 1. \tag{5.14}$$

We now discuss the implications of this condition, considering separately the van Cittert and the Landweber method.

(A) *The van Cittert method*

This is the iterative method (5.7) applied to equation (5.2), i.e. with $\bar{A} = A$ and $\bar{g} = g$. If the imaging system is bandlimited and if the image is affected by out-of-band noise, then equation (5.11) implies that the method does not converge. Convergence can only be obtained if the out-of-band noise is suppressed by means of some filtering. On the other hand, when $\boldsymbol{\omega}$ is in $\mathcal{B}$, equation (5.13), with $\hat{H}(\boldsymbol{\omega}) = \hat{K}(\boldsymbol{\omega})$ and $\hat{g}(\boldsymbol{\omega}) = \hat{g}(\boldsymbol{\omega})$, takes the following form

$$\hat{f}_k(\boldsymbol{\omega}) = (1 - \tau\hat{K}(\boldsymbol{\omega}))^k \hat{f}_0(\boldsymbol{\omega}) + \left[1 - (1 - \tau\hat{K}(\boldsymbol{\omega}))^k\right]\frac{\hat{g}(\boldsymbol{\omega})}{\hat{K}(\boldsymbol{\omega})} \tag{5.15}$$

and condition (5.14) becomes

$$|1 - \tau Re\hat{K}(\boldsymbol{\omega})|^2 + \tau^2|Im\hat{K}(\boldsymbol{\omega})|^2 < 1. \tag{5.16}$$

If $Re\hat{K}(\boldsymbol{\omega})$ does not have a definite sign, then it is impossible to find a value of τ such that condition (5.16) is satisfied for any $\boldsymbol{\omega}$. If $Re\hat{K}(\boldsymbol{\omega})$ has a definite sign, then it is not restrictive to assume that

$$Re\hat{K}(\boldsymbol{\omega}) > 0 . \tag{5.17}$$

In such a case τ must be positive and must also satisfy, for any ω, the following condition derived from equation (5.16)

$$\tau|\hat{K}(\omega)|^2 - 2Re\hat{K}(\omega) < 0 \; . \tag{5.18}$$

Such a value of τ exists if and only if

$$\hat{K}_+ = \sup_\omega \frac{|\hat{K}(\omega)|^2}{Re\hat{K}(\omega)} < \infty \; ; \tag{5.19}$$

then the values of τ ensuring the convergence of $\hat{f}_k(\omega)$, equation (5.15), for any ω in $\mathcal{B}$ are given by

$$0 < \tau < \frac{2}{\hat{K}_+} \; . \tag{5.20}$$

Condition (5.17) however is rather restrictive. It is not satisfied, for instance, by the uniform motion blur, out-of-focus blur, etc. In the case of a blur whose TF is real-valued and satisfies condition (5.17) (for instance a Gaussian blur), condition (5.20) becomes

$$0 < \tau < \frac{2}{\hat{K}_{max}} \tag{5.21}$$

where $\hat{K}_{max}$ is the maximum value of $\hat{K}(\omega)$. In such a case the van Cittert method can be used and has a regularization effect analogous but not equivalent to that of the Landweber method which is discussed in the following.

(B) *The Landweber method*

This is the iterative method (5.7) applied to equation (5.1), i.e. with $\bar{A} = A^*A$ and $\bar{g} = A^*g$. In such a case, if the system is bandlimited, then $\hat{\bar{g}}(\omega) = 0$ when ω is in $\mathcal{B}^C$ and therefore from equation (5.11) one obtain $\hat{f}_k(\omega) = \hat{f}_0(\omega)$, for all k's. On the other hand, when ω is in $\mathcal{B}$, equation (5.13), with $\hat{H}(\omega) = |\hat{K}(\omega)|^2$ and $\hat{\bar{g}}(\omega) = \hat{K}^*(\omega)\hat{g}(\omega)$, takes the following form

$$
\begin{aligned}
\hat{f}_k(\omega) \;\; = \;\; & (1 - \tau|\hat{K}(\omega)|^2)^k \hat{f}_0(\omega) \\
+ \;\; & \left[1 - (1 - \tau|\hat{K}(\omega)|^2)^k\right] \frac{\hat{g}(\omega)}{\hat{K}(\omega)}
\end{aligned} \tag{5.22}
$$

and condition (5.14) becomes

$$-1 < 1 - \tau|\hat{K}(\omega)|^2 < 1 \tag{5.23}$$

or also

$$0 < \tau < \frac{2}{|\hat{K}(\omega)|^2} \; . \tag{5.24}$$

This condition is satisfied for any ω in $\mathcal{B}$ if the relaxation parameter is such that

$$0 < \tau < \frac{2}{\hat{K}_{max}^2} \tag{5.25}$$

where now $\hat{K}_{max}$ is the maximum value of $|\hat{K}(\omega)|$. In such a case, equation (5.22) has a limit for any ω and any kind of blur, when $k \to \infty$. For any ω in $\mathcal{B}$ the limit is given by

$$\lim_{k \to \infty} \hat{f}_k(\omega) = \frac{\hat{g}(\omega)}{\hat{K}(\omega)} \tag{5.26}$$

while, when $\boldsymbol{\omega}$ is in $\mathcal{B}^{\mathcal{C}}$, from equation (5.11) we get that $\hat{f}_k(\boldsymbol{\omega})$ is always equal to $\hat{f}_0(\boldsymbol{\omega})$. It follows that, if the generalized solution $f^{\dagger}(\mathbf{x})$ exists, i.e. the image $g(\mathbf{x})$ satisfies condition (3.39), then the limit of the iterates is the following least-squares solution

$$f(\mathbf{x}) = f^{\dagger}(\mathbf{x}) + \frac{1}{(2\pi)^d} \int_{\mathcal{B}^{\mathcal{C}}} \hat{f}_0(\boldsymbol{\omega}) \, e^{i\mathbf{x}\cdot\boldsymbol{\omega}} \, d\boldsymbol{\omega} \ . \tag{5.27}$$

More precisely, from general results on the convergence of the Landweber method [40], one can prove that the convergence is in L^2-norm, i.e. $\|f_k - f\| \to 0$ when $k \to \infty$. If $f_0(\mathbf{x}) = 0$, then the limit is precisely the generalized solution $f^{\dagger}(\mathbf{x})$.

The choice $f_0(\mathbf{x}) = 0$ is the most simple one for obtaining approximations of the generalized solution and, for this reason, it is the most frequently used. We mainly consider this case. As we know from Chapter 3, the generalized solution does not exist in general for noisy data or, if it exists as in the case of discrete images, it is deprived of any physical meaning due to the excessive noise propagation from the data to the solution. In such a case the Landweber method can be used as a regularization algorithm, in the sense defined in Section 4.3. Indeed, from equation (5.22) with $\hat{f}_0(\boldsymbol{\omega}) = 0$ we find that the result of the k-th iteration can be written as follows

$$f_k(\mathbf{x}) = (R^{(k)}g)(\mathbf{x}) = \frac{1}{(2\pi)^d} \int_{\mathcal{B}} \hat{W}^{(k)}(\boldsymbol{\omega}) \, \frac{\hat{g}(\boldsymbol{\omega})}{\hat{K}(\boldsymbol{\omega})} \, e^{i\mathbf{x}\cdot\boldsymbol{\omega}} \, d\boldsymbol{\omega} \tag{5.28}$$

where

$$\hat{W}^{(k)}(\boldsymbol{\omega}) = 1 - (1 - \tau|\hat{K}(\boldsymbol{\omega})|^2)^k \ . \tag{5.29}$$

Notice that, if $\hat{K}(\boldsymbol{\omega}_0) = 0$ for a frequency $\boldsymbol{\omega}_0 \in \mathcal{B}$, then we must set $\hat{W}^{(k)}(\boldsymbol{\omega}_0)/\hat{K}(\boldsymbol{\omega}_0) = 0$, as follows from equation (5.10).

Equation (5.29) has the same structure as equation (4.55) and it is easy to show that the functions $\hat{W}^{(k)}(\boldsymbol{\omega})$ satisfy the conditions, stated in Section 4.4, which characterize the window functions. Therefore equations (5.28) and (5.29) define a regularization algorithm, the parameter being now the number of iterations.

It is interesting to remark that a given number of iterations produces a filtering effect which is similar to that produced by a value of the regularization parameter proportional to the inverse of the number of iterations. Indeed, from equation (4.56) with $\hat{P}(\boldsymbol{\omega}) = 1$, we find that the Tikhonov window function, for small values of $\hat{K}(\boldsymbol{\omega})$, is approximately given by

$$\hat{W}_{\mu}(\boldsymbol{\omega}) \simeq \frac{1}{\mu} \, |\hat{K}(\boldsymbol{\omega})|^2 \ . \tag{5.30}$$

Similarly $\hat{W}^{(k)}(\boldsymbol{\omega})$, for small values of $\hat{K}(\boldsymbol{\omega})$, is given by

$$\hat{W}^{(k)}(\boldsymbol{\omega}) \simeq \tau k \, |\hat{K}(\boldsymbol{\omega})|^2 \tag{5.31}$$

and therefore the two windows take approximately the same values if $\mu = 1/\tau k$. We call $\hat{W}^{(k)}(\boldsymbol{\omega})$ the *Landweber window function*.

In practice, the Tikhonov and Landweber window functions provide rather similar results. The comparison can be done as follows. Without loss of generality we can assume $|\hat{K}_{max}| = 1$ so that we can choose $\tau = 1$. Then the two window functions can be expressed in terms of the following functions

$$\hat{W}_{\mu}(\lambda) = \frac{\lambda}{\lambda + \mu} \ , \quad \hat{W}^{(k)}(\lambda) = 1 - (1 - \lambda)^k \tag{5.32}$$

defined for $0 \leq \lambda \leq 1$. These functions are plotted in Figure 5.1, as functions of $x = -\log_{10} \lambda$, in the case $\mu = 10^{-3}$ and $k = 10^3$, respectively. It is evident that they are similar

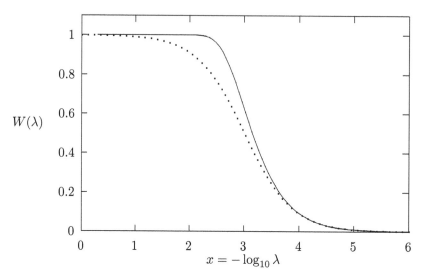

FIGURE 5.1   Comparison of the behavior of the Tikhonov window function with $\mu = 10^{-3}$ (dotted line) and of the Landweber window function with $\tau = 1$ and $k = 10^3$ (full line).

even if $\hat{W}^{(k)}(\lambda)$ is always above $\hat{W}_\mu(\lambda)$ and is practically equal to 1 up to $\lambda \simeq 10^{-2}$. This explains why one obtains rather similar results by means of the Tikhonov and Landweber methods. It is also clear that, if one uses Landweber method as a true iterative method (in deconvolution one can implement the filter (5.29) by means of FT), then one needs in general a very large number of iterations for obtaining reliable results.

This remark applies to the case $\tau = 1$. An acceleration of the method can be obtained by increasing the value of τ. Since the product τk is equivalent to the inverse of the regularization parameter μ, it is obvious that, by increasing τ we can get the same regularizing effect with a smaller number of iterations.

Both the energy of f_k, $E^2(f_k) = \|f_k\|^2$, and the discrepancy of f_k, $\varepsilon(f_k; g) = \|Af_k - g\|^2$ can be easily computed from equation (5.28). We find that $E(f_k)$ is an increasing function of k while $\varepsilon(f_k; g)$ is a decreasing function of k. In particular, $E(f_k)$ increases from 0 to $+\infty$ (or to $E(f^\dagger)$ if $E(f^\dagger) < \infty$) while $\varepsilon(f_k; g)$ decreases from $\|g\|$ to $\|g_{out}\|$ (defined in Section 4.2). The behavior of these functions is plotted in Figure 5.2 in the case of the numerical example of Section 4.6. The value of the relaxation parameter is $\tau = 1$.

We conclude this section by investigating the behavior of the reconstruction error as a function of k when the image g is noisy. As in the case of equation (4.38) the error consists of two terms

$$R^{(k)}g - P^{(\mathcal{B})}f^{(0)} = (R^{(k)}Af^{(0)} - P^{(\mathcal{B})}f^{(0)}) + R^{(k)}w \ . \tag{5.33}$$

The first one is the approximation error while the second one is the error due to noise propagation. The squared norm of the approximation error is given by

$$\|R^{(k)}Af^{(0)} - P^{(\mathcal{B})}f^{(0)}\|^2 = \frac{1}{(2\pi)^d} \int_{\mathcal{B}} |\hat{W}^{(k)}(\boldsymbol{\omega}) - 1|^2 \, |\hat{f}^{(0)}(\boldsymbol{\omega})|^2 \, d\boldsymbol{\omega}$$

$$= \frac{1}{(2\pi)^d} \int_{\mathcal{B}} (1 - \tau|\hat{K}(\boldsymbol{\omega})|^2)^{2k} \, |\hat{f}^{(0)}(\boldsymbol{\omega})|^2 \, d\boldsymbol{\omega} \tag{5.34}$$

and therefore this error is a decreasing function of k when the relaxation parameter satisfies the inequalities 5.25; it takes the value $\|P^{(\mathcal{B})}f^{(0)}\|$ for $k = 0$ and tends to zero for $k \to \infty$. The behavior is plotted in the left panel of Figure 5.3, also in the case of the numerical example discussed is Section 4.6. On the other hand the squared norm of the error due to

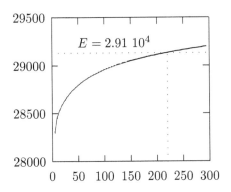
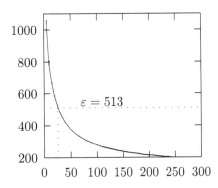

FIGURE 5.2 Behavior of $E(f_k)$ (left panel) and of $\varepsilon(f_k; g)$ (right panel) as functions of the number of iterations k, in the case of the numerical example of Figure 4.13. We also indicate the values of $E = \|f^{(0)}\|$ and $\varepsilon = \|Af^{(0)} - g\|$ corresponding to this numerical example, as well as the numbers of iterations such that $\|f_k\| = E$ and $\|Af_k - g\| = \varepsilon$.

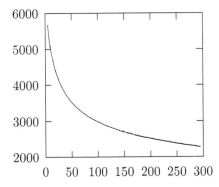
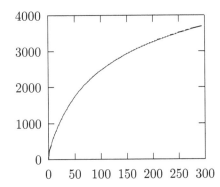

FIGURE 5.3 Left panel: behavior of the norm of the approximation error; right panel: behavior of the norm of the noise-propagation error, both as functions of k. The figures correspond to the numerical example of Figure 4.13.

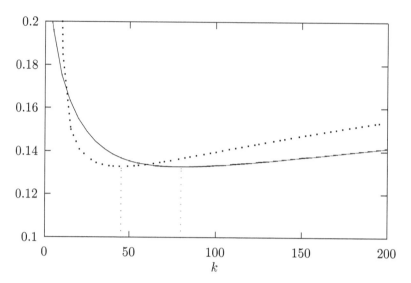

FIGURE 5.4 Behavior of the relative reconstruction error, as a function of the number of iterations k, for two values of the relaxation parameter: $\tau = 1$ (full line) and $\tau = 1.8$ (dotted line). The curves correspond to the numerical example of Figure 4.13. The optimal numbers of iterations are also indicated.

noise propagation is given by

$$\|R^{(k)}w\|^2 = \frac{1}{(2\pi)^d} \int_{\mathcal{B}} \left|1 - (1 - \tau|\hat{K}(\boldsymbol{\omega})|^2)^k\right|^2 \frac{|\hat{w}(\boldsymbol{\omega})|^2}{|\hat{K}(\boldsymbol{\omega})|^2} \, d\boldsymbol{\omega} \qquad (5.35)$$

and this is an increasing function of k, which is zero for $k = 0$ and tends to infinity (or to a very large number) when $k \to \infty$ (see the right panel of Figure 5.3).

By combining the behavior of the approximation error and that of the noise-propagation error, we infer that the norm of the reconstruction error should have a minimum for some suitable value of k. This behavior implies that, when we increase the number of iterations, the iterates should first approach $P^{(\mathcal{B})}f^{(0)}$ and then go away, i.e. that the Landweber method has the *semi-convergence property* which was already observed in the case of the Tikhonov regularization algorithm. Therefore there should exist an optimal value of the number of iterations, k_{opt}, corresponding to an iterate which has minimal distance from $P^{(\mathcal{B})}f^{(0)}$.

In Figure 5.4 we plot, for the numerical example of Section 4.6, the behavior of the relative reconstruction error, already defined in equation (4.42), which is now a function of k, i.e. is defined by

$$\rho_k = \frac{\|f_k - f^{(0)}\|}{\|f^{(0)}\|} \ . \qquad (5.36)$$

Since in this example $\hat{K}_{max} = 1$, the relaxation parameter has to satisfy $0 < \tau < 2$. In Figure 5.4, the two curves correspond to the values $\tau = 1$ and $\tau = 1.8$, respectively. For $\tau = 1.8$ the minimum of the reconstruction error is reached after a number of iterations smaller than for $\tau = 1$. This result is in agreement with the previous remark that the product τk is related to the inverse of the regularization parameter. Indeed for $\tau = 1$ we have $k_{opt} = 80$ and for $\tau = 1.8$ we have $k_{opt} = 45$. The product τk_{opt} is essentially the same in the two cases: 80 for $\tau = 1$ and 81 for $\tau = 1.8$. Since in the case of the Tikhonov method we found $\mu_{opt} = 7. \ 10^{-3}$ (see Section 4.6), it follows $1/\mu_{opt} \simeq 143$ and this value is considerably greater than τk_{opt}. In other words the relationship $\tau k = 1/\mu$ can be used for comparing the high-frequency cut-off of the two methods but not the reconstruction errors.

The minimum reconstruction error is essentially the same for the two values of τ, namely 13.2%. This is also the minimum error provided by the Tikhonov method (see Section 4.6) and therefore the two methods, Tikhonov and Landweber, are essentially equivalent in this particular example.

As concerns the choice of the value of τ, we observe that if we increase τ, then $k_{opt} = k_{opt}(\tau)$ decreases up to a certain value of τ (with a reconstruction error approximately constant) and then suddenly increases, in agreement with the fact that for $\tau = 2$ the method does not converge. This behavior is observed in many examples. It implies that, in general, there exists an optimal value of the relaxation parameter, and that, in practice, τ must be chosen not too close to the upper limit of the interval of allowed values.

In the case of real images it is not possible to compute k_{opt} and therefore one needs methods for estimating the optimal number of iterations. These methods are called *stopping rules*. The simplest stopping rule can be obtained by an extension of the *Morozov discrepancy principle* which is discussed in Section 4.6: let $b \geq 1$ be a given number; then, if there exists a value $\tilde{k}$ of k such that

$$\|Af_{\tilde{k}} - g\| > b\,\varepsilon \;, \quad \|Af_{\tilde{k}+1} - g\| \leq b\,\varepsilon \;, \tag{5.37}$$

then $\tilde{k} = \tilde{k}(\varepsilon)$ can be taken as an estimate of k_{opt} and $\tilde{f} = f_{\tilde{k}}$ can be taken as an estimate of the best approximation to $P^{(\mathcal{B})} f^{(0)}$ provided by the Landweber method. It is possible to prove in a rather simple way [95] that, if $b > 1$ and if the noise on the data tends to zero, so that $\varepsilon \to 0$, then $\|\tilde{f} - P^{(\mathcal{B})} f^{(0)}\|$ also tends to zero.

This stopping rule does not always work well in practice. In the numerical example of Section 4.6 we obtain $\tilde{k} = 25$ and a reconstruction error of 15.8% (a bit too large with respect to the minimum reconstruction error of 13.2%) if we take $b = 1$ and $\tau = 1$, and $\tilde{k} = 20$ and a reconstruction error of 14.1% (not too bad) if we take $b = 1$ and $\tau = 1.8$. If we use larger values of b, then the results are worse as follows from the behavior of $\varepsilon(f_k; g)$ given in the right panel of Figure 5.2. However the condition $b > 1$ is essentially introduced for technical reasons, i.e. in order to prove convergence. In practice, we can take $b = 1$.

If we look at Figure 5.4 we see that the curve of the reconstruction error is rather steep before and rather flat after the minimum. Therefore the stopping rules which tend to underestimate the number of iterations may provide reconstructions which are not sufficiently accurate while methods which tend to overestimate the number of iterations may provide satisfactory reconstructions. For instance, in the numerical example of Section 4.6, the number of iterations such that $\|f_k\| = E$ is rather large ($k = 220$ – see the left panel of Figure 5.2), but the resulting reconstruction error is 14.3%, slightly better than the error corresponding to $\tilde{k}$ (for $\tau = 1$).

5.2 The projected Landweber method

When applied to image deconvolution, the Landweber method discussed in detail in the previous section is not really a genuine iterative method, because it is equivalent to a computable linear filter and therefore, in practice, it is much more convenient and faster to implement the filter rather than the iterative procedure. Obviously this remark does not apply to the general case of linear inverse problems. Anyhow it constitutes a good case example for showing that the number of iterations can play the role of a regularization parameter. Another interesting feature of this method is that it can be used, after some appropriate modifications, for solving constrained least-squares problems such as those introduced in Section 3.6. As shown there, many physical constraints on the unknown object can be expressed by requiring that it belongs to some given closed and convex set $\mathcal{C}$ of a Hilbert space (in particular, of a L^2-space); then the constrained least-squares problem consists in the minimization of the discrepancy functional over this set, i.e.

$$\|Af - g\| = \text{minimum} , \quad f \in \mathcal{C} . \tag{5.38}$$

This problem is still ill posed if the set $\mathcal{C}$ does not satisfy additional conditions. A general discussion of this problem is beyond the scope of this book. We only observe that, from some general mathematical results [169] it follows that if the imaging system is not bandlimited and if the set $\mathcal{C}$ is interior to the set of objects with a prescribed energy E^2, then there exists a unique solution of the problem (5.38) for any image g. Therefore the problem is well posed as concerns uniqueness and existence of the solution.

An example of constraint which is rather natural in many problems of image deconvolution, as well as in some other problems, is the non-negativity of the solution, i.e. the set $\mathcal{C}$ is the set of all nonnegative functions. However, even with this constraint, the problem (5.38) is still ill posed for most applications. Therefore what we need is not a method for obtaining the solution of this problem (which, in general, does not exist for noisy images) but a regularization algorithm, i.e. a family of approximate solutions which converge to the true object in the case of a noise-free image and has the semi-convergence property in the case of noisy images. As we will show, the Landweber method can be modified in order to provide an algorithm which has these properties. This result mainly derives from numerical experiments, hence in a discrete setting. Indeed, with the mathematical results which have been established on the convergence of the modified method in a Hilbert space, neither the convergence nor the semi-convergence property can be proved, as we will briefly discuss. From now on, we will use the following notation, already used in the previous section

$$\bar{A} = A^* A , \quad \bar{g} = A^* g . \tag{5.39}$$

Moreover we will denote by $P_{\mathcal{C}}$ the projection operator onto the set $\mathcal{C}$, as defined in Section 3.6. Then the modified Landweber method, also called *projected Landweber method* goes as follows

$$f_{k+1}^{(\mathcal{C})} = P_{\mathcal{C}} \left[f_k^{(\mathcal{C})} + \tau(\bar{g} - \bar{A}f_k^{(\mathcal{C})}) \right] \tag{5.40}$$

with τ satisfying the conditions (5.25). At each step the method consists in a Landweber iteration followed by a projection onto the convex set $\mathcal{C}$.

In the case of image deconvolution the method can be implemented as follows:

- compute $\hat{f}_k^{(\mathcal{C})}(\boldsymbol{\omega})$ from $f_k^{(\mathcal{C})}(\mathbf{x})$
- compute

$$\hat{h}_{k+1}(\boldsymbol{\omega}) = \tau\hat{K}^*(\boldsymbol{\omega})\hat{g}(\boldsymbol{\omega}) + (1 - \tau|\hat{K}(\boldsymbol{\omega})|^2)\hat{f}_k^{(\mathcal{C})}(\boldsymbol{\omega}) \tag{5.41}$$

- compute $h_{k+1}(\mathbf{x})$ from $\hat{h}_{k+1}(\boldsymbol{\omega})$
- compute $f_{k+1}^{(\mathcal{C})}(\mathbf{x}) = (P_{\mathcal{C}}h_{k+1})(\mathbf{x})$.

We see that, at each iteration, the implementation of the method requires the computation of one direct and one inverse Fourier transform followed by the projection onto the set $\mathcal{C}$. Therefore the iterates $f_k^{(\mathcal{C})}$ are easily computable if this projection is easily computable.

One example is provided by the case where the set $\mathcal{C}$ is the subspace of the functions $f(\mathbf{x})$ whose support is interior to a bounded domain $\mathcal{D}$. Another example is provided by the set of the square-integrable functions $f(\mathbf{x})$ which are nonnegative. In such a case the projection operator is given by equation (3.63). The constraint of non-negativity, however, is a particular example of a more general constraint which consists of upper and lower bounds on $f(\mathbf{x})$. If $\mathcal{C}$ is the set of all square-integrable functions $f(\mathbf{x})$ such that $a \leq f(\mathbf{x}) \leq b$, with *a* and *b* given real numbers, then the projection operator onto this set is given by

$$(P_C f)(\mathbf{x}) = \begin{cases} f(\mathbf{x}) & \text{if } a \le f(\mathbf{x}) \le b \\ a & \text{if } f(\mathbf{x}) < a \\ b & \text{if } f(\mathbf{x}) > b \ . \end{cases} \tag{5.42}$$

The constraint of non-negativity is reobtained when $a = 0$ and $b = +\infty$. Other examples of closed convex sets C are given in Section 3.6.

As concerns the properties of this algorithm, some general results are proved in [111] in the case of Hilbert spaces: if $f^{(0)}$ is an object in the closed and convex set C and $g = g^{(0)} = A f^{(0)}$ is the corresponding noise-free image, then the algorithm (5.40) converges weakly to a solution of the constrained least-squares problem (5.38) (weak convergence is defined in Remark 4.1). If the solution of the problem is unique then the limit is precisely $f^{(0)}$; if the solution is not unique than the limit is a constrained least-squares solution which depends on the choice of the initial approximation f_0. In the same paper it is conjectured that strong convergence, i.e. convergence in norm, should be true. However in [74] this conjecture is disproved by means of a suitable counterexample provided in [165]. In the discrete finite-dimensional case, weak convergence is equivalent to strong convergence and therefore we have convergence of the iterations in the usual quadratic norm.

If g is noisy, as far as we know, no result has been proved ensuring that the iterates $f_k^{(C)}$ first approach $f^{(0)}$ and then go away. However numerical simulations strongly suggest that this result should be true in a discrete setting because the following properties are in general satisfied:

- for any k, $\|f_k^{(C)} - f^{(0)}\| < \|f_k - f^{(0)}\|$ where f_k is the k-th iterate of the unconstrained Landweber method (5.7);

- the reconstruction error $\|f_k^{(C)} - f^{(0)}\|$ has one minimum corresponding to an optimal value k_{opt} of the number of iterations and therefore the algorithm has the semi-convergence property.

These properties (which, we repeat, have not been proved in general) can be verified, for instance, if we apply the projected Landweber method, with the constraint of non-negativity, to the numerical example of Figure 4.13 and if we compare the results with those of the non-projected Landweber method. However, in this case, the error curve for the projected Landweber method (with $f_0 = 0$ and $\tau = 1$) always lies below that of the non-projected method (also with $f_0 = 0$ and $\tau = 1$ – this curve is given by the full line in Figure 5.4) but the improvement in the reconstruction error is not significant. The reason is that, in this example, the non-negativity constraint is not very important, because the optimal solution provided by the Landweber method does not contain large negative values.

An example where the difference between the constrained and unconstrained method is more spectacular is that given in Figure 4.12. In this example the object consists of only two levels: 1/2 for the background and 1 for the letters. As shown in Figure 4.12, the reconstruction provided by the Tikhonov method is affected by ghosts and this is true also for the reconstruction provided by the Landweber method because this is a property of any linear reconstruction method applied to this problem (linear motion blur).

In the application of the projected method we consider two noisy versions of the image of Figure 4.12 both corrupted by additive white Gaussian noise: one with square root of the variance given by $\sigma = 0.008$ (corresponding to a relative RMSE of 1.5%) and the other with $\sigma = 0.055$ (corresponding to a relative RMSE of 10%). In Figure 5.5, we plot the behavior of the relative reconstruction error for the first noisy version in the case $\tau = 1$ and $f_0 = 0$: the full line represents the reconstruction error of the non-projected Landweber method while the dotted line represents the reconstruction error of the projected one, the projection being

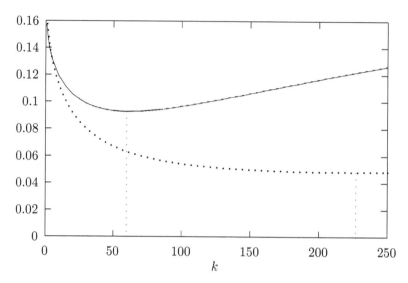

FIGURE 5.5　Comparison of the relative reconstruction errors for the unconstrained Landweber method (full line) and the projected Landweber method (dotted line), with non-negativity constraint. The curves are computed in the case of the image of Figure 4.12, corrupted with white Gaussian noise (square root of the variance $\sigma = 0.008$). Initial approximation: $f_0 = 0$; relaxation parameter: $\tau = 1$.

on the set of the functions with values greater or equal to $1/2$ (see equation (5.42) with $a = 1/2$ and $b = +\infty$). The minimum reconstruction error of the Landweber method is 9.3% corresponding to $k_{opt} = 60$ while the minimum reconstruction error of the projected method is 4.8% corresponding to $k_{opt} = 227$. The improvement is about a factor of 2, even if the minimum is reached after a rather large number of iterations, much larger than that of the usual Landweber method. Nevertheless, the region around the minimum is very flat, so that after 100 iterations one has a reconstruction error which is already acceptable.

The value of k_{opt} and the shape around the minimum, however, depend on the noise level. In the case $\sigma = 0.055$, we have $k_{opt} = 20$ and the reconstruction error is 11.7% (comparable with the RMSE of the image, 10%). The behavior of the relative reconstruction error in the two cases is plotted in Figure 5.6. In the case of larger noise, the Landweber method converges faster.

If we look at the reconstructions provided by the two methods, which are shown in Figure 5.7, we see that the effect of the constraint is the suppression of the ghosts. This result is not in conflict with our analysis of Section 4.5 because the constrained method is nonlinear. Since the presence of ghosts is due to the lack of information around the zeroes of the transfer function (see the discussion of Section 4.5), it follows that the constrained method provides a sort of extrapolation of the Fourier transform of the object in the regions where information is lacking.

The projected Landweber method provides rather good results in many circumstances. The difficulty is that, in general, the number of iterations required for obtaining a satisfactory result is rather large. In the case of the unconstrained algorithm, Section 5.1, a method for increasing the rate of convergence is proposed by Strand [272]. The method consists in modifying the equation for the least-squares solutions as follows

$$D\bar{A}f = D\bar{g} \tag{5.43}$$

where D is a linear and bounded operator with a bounded inverse. An additional requirement is that D commutes with $\bar{A}$. This modification does not change the set of the least-squares

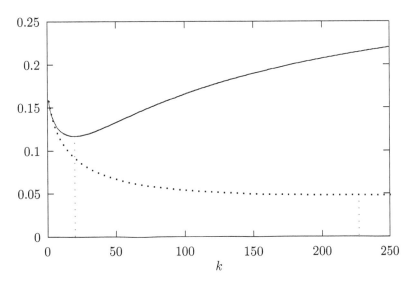

FIGURE 5.6 Behavior of the relative reconstruction error of the projected Landweber method for two noise levels: white noise with $\sigma = 0.008$ (dotted line) and white noise with $\sigma = 0.055$ (full line). The object is the same as that of Figure 5.5.

solutions. It is a form of *preconditioning*, a method introduced in numerical analysis [134] for improving the efficiency of the conjugate gradient method (the application of conjugate gradient to image reconstruction is discussed in Sections 5.4 and 5.5).

In the case of image reconstruction, the assumptions of Strand are satisfied if D is a convolution operator

$$(Df)(\mathbf{x}) = \frac{1}{(2\pi)^d} \int \hat{D}(\boldsymbol{\omega}) \hat{f}(\boldsymbol{\omega}) \, e^{i\mathbf{x}\cdot\boldsymbol{\omega}} \, d\boldsymbol{\omega} \tag{5.44}$$

and the function $\hat{D}(\boldsymbol{\omega})$ is a bounded positive function with a positive lower bound. Moreover $\hat{D}(\boldsymbol{\omega})$ must satisfy the following conditions:

- $\hat{D}(\boldsymbol{\omega})|\hat{K}(\boldsymbol{\omega})|^2 \le 1$ for any $\boldsymbol{\omega}$
- $\hat{D}(\boldsymbol{\omega})|\hat{K}(\boldsymbol{\omega})|^2$ is close to 1 or, at least, not much smaller than 1, for all spatial frequencies $\boldsymbol{\omega}$ where $|\hat{K}(\boldsymbol{\omega})|$ is greater than some suitable threshold value.

An operator satisfying these conditions is called a *preconditioner*. A possible choice for D, which is closely related to the Tikhonov method, is the following one [247]

$$\hat{D}(\boldsymbol{\omega}) = \left(|\hat{K}(\boldsymbol{\omega})|^2 + \gamma \right)^{-1} \tag{5.45}$$

with γ chosen according to the criteria used for estimating the regularization parameter – see Section 4.6.

By applying the Landweber method to equation (5.43) we find

$$f_{k+1} = f_k + \tau D(\bar{g} - \bar{A}f_k) \tag{5.46}$$

with τ such that $0 < \tau < 2$ (this is a consequence of the first condition on $\hat{D}(\boldsymbol{\omega})$). Then, in the case $f_0(\mathbf{x}) = 0$, from equation (5.13) with $\hat{\bar{g}}(\boldsymbol{\omega}) = \hat{D}(\boldsymbol{\omega})\hat{K}^*(\boldsymbol{\omega})\hat{g}(\boldsymbol{\omega})$ and $\hat{H}(\boldsymbol{\omega}) = \hat{D}(\boldsymbol{\omega})|\hat{K}(\boldsymbol{\omega})|^2$ we find

$$\hat{f}_k(\boldsymbol{\omega}) = \left\{ 1 - \left[1 - \tau\hat{D}(\boldsymbol{\omega})|\hat{K}(\boldsymbol{\omega})|^2 \right]^k \right\} \frac{\hat{g}(\boldsymbol{\omega})}{\hat{K}(\boldsymbol{\omega})}. \tag{5.47}$$

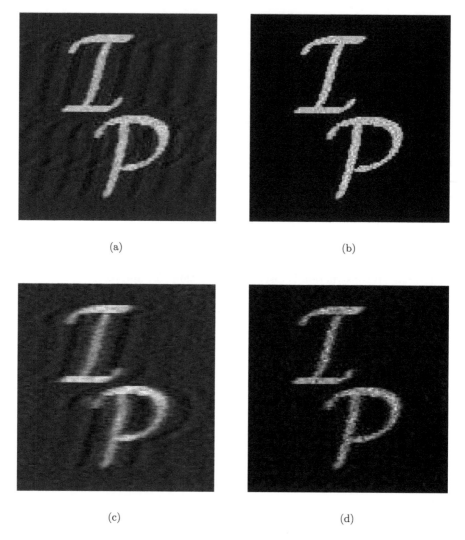

FIGURE 5.7 Optimal reconstructions of noisy versions of the image of Figure 4.12 provided by the Landweber method in (a) and (c) (white noise with $\sigma = 0.008$ in (a) and $\sigma = 0.055$ in (c)) and by the projected Landweber method in (b) and (d) ($\sigma = 0.008$ in (b) and $\sigma = 0.055$ in (d)).

If we compare this equation with equation (5.22) we see that the modification introduced by Strand is equivalent to a modification of the filter. However the choice of D has also effects on the speed of convergence. For instance, if $\hat{D}(\boldsymbol{\omega})$ is given by equation (5.45) and if $\tau = 1$, we have

$$\hat{f}_k(\boldsymbol{\omega}) = \left[1 - \left(\frac{\gamma}{|\hat{K}(\boldsymbol{\omega})|^2 + \gamma} \right)^k \right] \frac{\hat{g}(\boldsymbol{\omega})}{\hat{K}(\boldsymbol{\omega})} \tag{5.48}$$

and therefore the window function becomes approximately 1 after few iterations for those frequencies where $|\hat{K}(\boldsymbol{\omega})|^2$ is greater than or of the order of γ.

The effect on the rate of convergence is especially important in the case of the projected Landweber method. It can be shown [231] that the algorithm obtained by applying the projected Landweber method to equation (5.43), i.e.

$$f_{k+1}^{(\mathcal{C})} = P_{\mathcal{C}} \left[f_k^{(\mathcal{C})} + \tau D(\bar{g} - \bar{A} f_k^{(\mathcal{C})}) \right] \tag{5.49}$$

is the projected Landweber method for the following least-squares problem

$$(D(Af - g), Af - g) = \text{minimum} , \quad f \in \mathcal{C} . \tag{5.50}$$

Since the original least-squares problem, defined in equation (5.38), has been modified, it may be interesting to investigate the effect of this modification by considering, for instance, the preconditioner (5.45). As a rule, one finds that, by decreasing γ, the number of iterations required for reaching the minimum reconstruction error decreases, but the minimum reconstruction error increases.

The statement above can be illustrated by applying the method to the simulated image of a star cluster, produced by AURA/STScI for testing the methods used for the reconstruction of the pre-COSTAR images of the Hubble Space Telescope. The object, a cluster of 470 stars, and the corresponding image are given in Figure 5.8.

The minimum reconstruction error produced by the projected Landweber method, with a non-negativity constraint and $\tau = 1.98/\hat{K}_{max}^2$, is 3.68% and is reached after about 1300 iterations. The good quality of the reconstruction is explained by the fact that the condition number of the problem is rather low ($\alpha = 23.74$). If we use the preconditioner (5.45) with $\gamma = 0.04$, we find the minimum after 56 iterations with a reconstruction error of 5.24%. If we use $\gamma = 0.01$, the minimum is reached after 8 iterations but the reconstruction error is now 7.16%. The reconstructions obtained without preconditioner and with the preconditioner (5.45) corresponding to $\gamma = 0.04$ are given, for comparison, in Figure 5.8. The difference between the two reconstructions is visible. Both reconstructions, however, are much better than the one provided by the Tikhonov regularization method since, in such a case, the minimum reconstruction error is 10.2% (corresponding to $\mu_{opt} = 5 \ 10^{-4}$).

5.3 The projected Landweber method for the computation of constrained regularized solutions

As shown in Section 4.5 the linear deconvolution methods and, in particular, the Tikhonov regularization method, can introduce ringing effects. These artifacts appear, for instance, in the deconvolution of an object consisting of bright spots against a black background or containing sharp intensity variations. Ringing is not desirable because ringing artifacts reduce not only the visual quality of the reconstructed image but also the accuracy in estimating measurable quantities (for instance star intensities in the case of an astronomical image). Since ringing is not compatible with our prior knowledge about the object, it should be

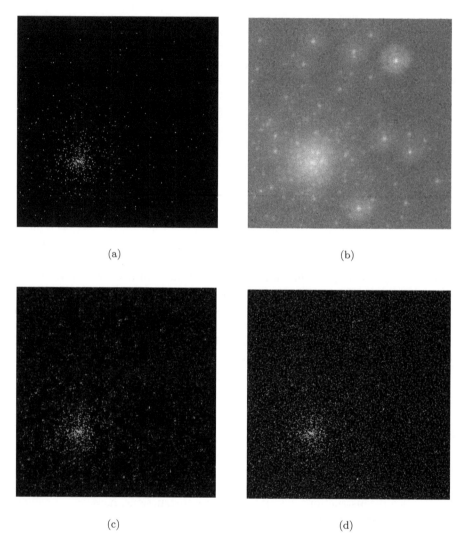

(a)   (b)

(c)   (d)

FIGURE 5.8 Example of a reconstruction obtained by means of the projected Landweber method with preconditioning. In (a) we show the object, a cluster of 470 stars, and in (b) its image (both images are produced by AURA/STScI). In (c) we give the best reconstruction obtained by the projected Landweber method and in (d) the best reconstruction obtained by means of the same method with the preconditioner (5.45) ($\gamma = 0.04$).

suppressed through the use of some suitable *a priori* information. For instance, if the object is positive and the ringing manifests itself through negative values of the reconstructed image, the non-negativity constraint can be used.

The projected Landweber method discussed in the previous section provides a way to improve the situation. Another way, proposed in [187], is to compute constrained regularized solutions. If $\mathcal{C}$ is the closed and convex set expressing the constraints on the unknown object, then one has to solve the following minimization problem

$$\Phi_\mu(f;g) = \|Af - g\|^2 + \mu\|f\|^2 = \text{minimum} , \quad f \in \mathcal{C} . \tag{5.51}$$

It can be proved [221] that the problem (5.51) has a unique solution in $\mathcal{C}$, let us say $f_\mu^{(\mathcal{C})}$, for any $\mu > 0$ and any g. Also in this case there exists an optimal value of μ such that $\|f_\mu^{(\mathcal{C})} - f^{(0)}\|$ is minimum. Therefore, in the case of numerical simulations, one can solve the problem (5.51) for any μ and determine the optimal value of μ by minimizing the reconstruction error. In the case of a real image, one can estimate a value of μ by means of one of the methods of Section 4.6 and then take the solution of the problem (5.51) with this particular value of μ as the optimal solution. The main question is: how can we solve the problem (5.51)?

The projected Landweber method introduced in the previous section provides an answer, even if it is not the most efficient one. Indeed, let us consider the iterative scheme (5.7) for the solution of equation (5.3). If we use the notations (5.39), the Euler-Lagrange equation (4.23) associated with the functional $\Phi_\mu(f;g)$ has the following form

$$(\bar{A} + \mu I)f = \bar{g} \tag{5.52}$$

and therefore it is obtained from equation (5.3) by adding the term μI to $\bar{A}$. Accordingly we can modify the operator T of equation (5.4), and introduce an operator T_μ defined as follows

$$T_\mu(f) = f + \tau\left[\bar{g} - (\bar{A} + \mu I)f\right] . \tag{5.53}$$

so that equation (5.52) is equivalent to the fixed-point equation

$$T_\mu(f) = f . \tag{5.54}$$

The operator T_μ has the nice property to be a *contraction* or *contractive mapping*. Let us recall that an operator T (in general, nonlinear) in a Hilbert space $\mathcal{X}$ is said to be a contraction if there exists a constant $\rho < 1$ such that, for any pair of elements f_1, f_2 of $\mathcal{X}$

$$\|T(f_2) - T(f_1)\| \leq \rho \|f_1 - f_2\| . \tag{5.55}$$

Notice that if the operator T is linear, then condition (5.55) is equivalent to saying that the norm of the operator is strictly smaller than 1. Condition (5.55) implies that the operator T is continuous. Indeed, if f_n is a sequence converging to f, from equation (5.55) we get $\|T(f_n) - T(f)\| \leq \rho \|f_n - f\|$ and therefore the sequence $T(f_n)$ converges to $T(f)$.

If T is a contraction mapping, then there exists at most one fixed point, i.e. an element f of $\mathcal{X}$ such that $T(f) = f$. Indeed, if f_1, f_2 are two fixed points of T, using again condition (5.55), we get $\|f_2 - f_1\| = \|T(f_2) - T(f_1)\| \leq \rho \|f_2 - f_1\|$ and therefore $\|f_2 - f_1\| = 0$ since $\rho < 1$. This implies $f_2 = f_1$. If T is also linear, then the unique fixed point is $f = 0$.

As concerns the existence of the fixed point of the contraction T, this is ensured by the *Banach contraction principle: for any starting element f_0, the sequence of the successive approximations*

$$f_{k+1} = T(f_k) \tag{5.56}$$

is convergent and its limit is the unique fixed point of T.

REMARK 5.1   *The proof is based on the completeness of the Hilbert space $\mathcal{X}$ and essentially consists in proving that the sequence f_k is a Cauchy sequence. First observe that from condition (5.55) we have*

$$\|f_{k+1} - f_k\| \leq \rho \|f_k - f_{k-1}\| \leq \cdots \leq \rho^k \|f_1 - f_0\| \ . \tag{5.57}$$

Then, for any $p > 1$, using the triangular inequality, we obtain

$$
\begin{aligned}
\|f_{k+p} - f_k\| &\leq \|f_{k+p} - f_{k+p-1}\| + \cdots + \|f_{k+1} - f_k\| \\
&\leq (\rho^{k+p-1} + \rho^{k+p-2} + \cdots + \rho^k) \|f_1 - f_0\| < \frac{\rho^k}{1 - \rho} \|f_1 - f_0\|
\end{aligned}
\tag{5.58}
$$

and therefore $\|f_{k+p} - f_k\| \to 0$, $k \to \infty$ for any p. This result implies that the sequence f_k is a Cauchy sequence. Since the space is complete, the sequence has a limit f. If remains only to prove that f is a fixed point of T. Using again the triangular inequality, we have

$$
\begin{aligned}
\|T(f) - f\| &\leq \|T(f) - f_{k+1}\| + \|f_{k+1} - f\| \\
&= \|T(f) - T(f_k)\| + \|f_{k+1} - f\| \leq \rho \|f_k - f\| + \|f_{k+1} - f\| \ .
\end{aligned}
\tag{5.59}
$$

Since the r.h.s. tends to zero when $k \to \infty$, we obtain $T(f) = f$. From the uniqueness of the fixed point we conclude that all sequences of successive approximations converge to this fixed point. Notice that the result still holds true if the contraction is operating on a closed subset of the Hilbert space.

If the relaxation parameter τ satisfies condition (5.14) with $\hat{H}(\boldsymbol{\omega}) = |\hat{K}(\boldsymbol{\omega})|^2 + \mu$, we find

$$0 < \tau < \frac{2}{\hat{K}_{max}^2 + \mu} \tag{5.60}$$

and for these values of τ the following quantity $\rho_\mu(\tau)$ is strictly smaller than one

$$\rho_\mu(\tau) = \min_{\boldsymbol{\omega}} \left| 1 - \tau(|\hat{K}(\boldsymbol{\omega})|^2 + \mu) \right| < 1 \ . \tag{5.61}$$

This implies that the nonlinear operator T_μ defines a contraction mapping. Indeed, from Parseval's equality we get, for any pair of functions $f^{(1)}$, $f^{(2)}$

$$
\begin{aligned}
\|T_\mu(f^{(2)}) - T_\mu(f^{(1)})\|^2 &= \| \left[I - \tau(\bar{A} + \mu I) \right] (f^{(1)} - f^{(2)}) \|^2 \\
&= \frac{1}{(2\pi)^d} \int |1 - \tau(|\hat{K}(\boldsymbol{\omega})|^2 + \mu)|^2 \, |\hat{f}^{(2)}(\boldsymbol{\omega}) - \hat{f}^{(1)}(\boldsymbol{\omega})|^2 d\boldsymbol{\omega} \\
&\leq \frac{\rho_\mu^2(\tau)}{(2\pi)^d} \int |\hat{f}^{(2)}(\boldsymbol{\omega}) - \hat{f}^{(1)}(\boldsymbol{\omega})|^2 d\boldsymbol{\omega}
\end{aligned}
\tag{5.62}
$$

and therefore

$$\|T_\mu(f^{(2)}) - T_\mu(f^{(1)})\| \leq \rho_\mu(\tau) \|f^{(2)} - f^{(1)}\| \ . \tag{5.63}$$

Notice that if $\mu = 0$, i.e. in the case of the operator T given by equation (5.4), since $|\hat{K}(\boldsymbol{\omega})|$ is zero at infinity, we have that $\rho_\mu(\tau) = 1$ and accordingly

$$\|T(f_2) - T(f_1)\| \leq \|f_2 - f_1\| \ . \tag{5.64}$$

An operator T satisfying this condition is said to be a *nonexpansive mapping*. This is a much weaker property than being contractive, which does not allow in general to prove in a simple way the convergence of the iterates to a fixed point of T.

However, for $\mu > 0$, since by inequality (3.64) the projection operator P_C is nonexpansive, the operator $P_C T_\mu$ is the product of a contractive by a nonexpansive mapping and hence it is also a contraction mapping. Indeed

$$\begin{aligned} \|P_C T_\mu(f^{(2)}) - P_C T_\mu(f^{(1)})\| &\leq \|T_\mu(f^{(2)}) - T_\mu(f^{(1)})\| \\ &\leq \rho_\mu(\tau)\, \|f^{(2)} - f^{(1)}\| \end{aligned} \qquad (5.65)$$

and therefore the projected Landweber method

$$f_{k+1}^{(C,\mu)} = P_C T_\mu(f_k^{(C,\mu)}) , \qquad (5.66)$$

which is precisely the successive approximation method for the contraction operator $P_C T_\mu$, converges, for any initial approximation f_0, to the unique fixed point of $P_C T_\mu$. Moreover it can be proved [187] that this fixed point coincides with the unique solution of the problem (5.51).

For image deconvolution, the implementation of the iterative algorithm (5.66) is similar to that of the projected Landweber method. It goes as follows:

- compute $\hat{f}_k^{(C,\mu)}(\boldsymbol{\omega})$ from $f_k^{(C,\mu)}(\mathbf{x})$
- compute $\hat{h}_{k+1}^{(\mu)}(\boldsymbol{\omega}) = \tau \hat{K}^*(\boldsymbol{\omega})\hat{g}(\boldsymbol{\omega}) + [1 - \tau(|\hat{K}(\boldsymbol{\omega})|^2 + \mu)]\, \hat{f}_k^{(C,\mu)}(\boldsymbol{\omega})$
- compute $h_{k+1}^{(\mu)}(\mathbf{x})$ from $\hat{h}_{k+1}^{(\mu)}(\boldsymbol{\omega})$
- compute $f_{k+1}^{(C,\mu)}(\mathbf{x}) = (P_C\, h_{k+1}^{(\mu)})(\mathbf{x})$.

Also in this case, at each iteration we need the computation of one direct and one inverse Fourier transform followed by the projection onto the set C. Since the convergence of the algorithm can be rather slow, one can use the method of Strand discussed in Section 5.2, for increasing the speed of convergence.

5.4 The steepest descent and the conjugate gradient method

The Landweber method is an example of so-called *gradient methods*, i.e. of methods where, at each step, the new approximation is obtained by modifying the old one in the direction of the negative gradient of the discrepancy functional. Another example is provided by the *steepest descent method*.

(A) *The steepest descent method*

Let us first remark that the minimization of the discrepancy functional $\varepsilon^2(f;g)$, as defined in equation (4.2), is equivalent to the minimization of the functional

$$\eta(f;g) = \frac{1}{2}(\bar{A}f, f) - (\bar{g}, f) \qquad (5.67)$$

where the notations (5.39) are used. Indeed, assuming for simplicity here that all functions involved are real-valued, if we expand the square of the norm of $Af - g$, we find that

$$\varepsilon^2(f;g) = 2\eta(f;g) + \|g\|^2 . \qquad (5.68)$$

Now, it is easy to show that the gradient of the function $\eta(f;g)$ is one-half the gradient of the functional $\varepsilon^2(f;g)$ and is given by

$$\nabla_f \eta(f;g) = \bar{A}f - \bar{g} , \qquad (5.69)$$

so that this gradient coincides with minus the residual $\bar{r}$ associated with the least-squares equation

$$\bar{r} = \bar{g} - \bar{A}f \ . \tag{5.70}$$

Given an approximation f_k for the object, in a neighborhood of this point the functional $\eta(f;g)$ decreases most rapidly in the direction of the negative gradient, i.e. in the direction of the residual $\bar{r}$. If we consider the Landweber method and if we put $\bar{r}_k = \bar{g} - \bar{A}f_k$, we see that it can be written as follows

$$f_{k+1} = f_k + \tau \bar{r}_k \tag{5.71}$$

corresponding to modifying the k-th iterate precisely in the direction of the negative gradient. However, if we pick a value of τ satisfying condition (5.25), this value may not be the best one, in the sense that it does not necessarily minimize $\eta(f_{k+1};g)$. By means of simple computations we find that

$$\eta(f_{k+1};g) = \eta(f_k;g) + \frac{1}{2}\tau^2\|A\bar{r}_k\|^2 - \tau\|\bar{r}_k\|^2 \tag{5.72}$$

where the relation $(\bar{A}\bar{r}_k, \bar{r}_k) = (A^*A\bar{r}_k, \bar{r}_k) = (A\bar{r}_k, A\bar{r}_k)$ is used. Hence the value of τ, say τ_k, that minimizes $\eta(f_{k+1};g)$ is given by

$$\tau_k = \frac{\|\bar{r}_k\|^2}{\|A\bar{r}_k\|^2} \ . \tag{5.73}$$

The modification of the Landweber method provided by this choice of the relaxation parameter (which depends on k), i.e.

$$f_{k+1} = f_k + \tau_k \bar{r}_k \tag{5.74}$$

is the so-called *steepest descent method*.
From equations (5.73) and (5.74) we find that the residuals $\bar{r}_k$ have the following properties:

- they can be obtained by means of the iterative scheme

$$\bar{r}_{k+1} = \bar{r}_k - \tau_k \bar{A}\bar{r}_k \tag{5.75}$$

  with τ_k given by equation (5.73)
- they satisfy the orthogonality condition

$$(\bar{r}_{k+1}, \bar{r}_k) = 0 \ . \tag{5.76}$$

In a finite-dimensional case, since $\bar{A}$ is a positive-definite matrix, this method has a very simple interpretation: if f_0 is the initial approximation, then one moves from f_0 in the direction orthogonal to the level set of $\varepsilon^2(f;g)$ passing through f_0 – an ellipsoid, on which $\varepsilon^2(f;g)$ is constant – up to a point f_1 of a level set which is tangent to this direction. Then one moves from f_1 in the direction orthogonal to the new level set up to another level set tangent to this direction and so on. A two-dimensional representation is given in Figure 5.9.

The steepest descent method is a nonlinear method for approximating least-squares solutions. It is also a regularization method. As proved in [132], if $f_0 = 0$, then the iterates f_k converges to $f^\dagger$ in the case of a noise-free image. Moreover they exhibit the semi-convergence behavior in the case of noisy images so that, also for this method, there should exist an optimal number of iterations providing the best approximation of the unknown object.
If we apply the method to the numerical example of Section 4.6, the minimum reconstruction error we obtain is 13.3% with $k_{opt} = 41$. The values of the quantities τ_k, equation (5.73),

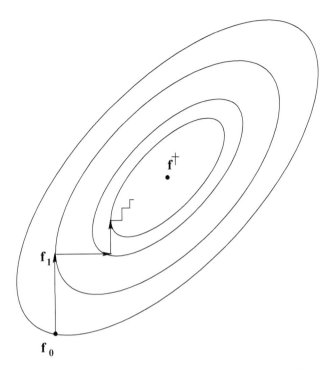

FIGURE 5.9 Two-dimensional representation of the steepest descent method.

oscillate around 2, between 1.8 and 2.1. We recall that in the case of the Landweber method with $\tau = 1.8$ the minimum reconstruction error is 13.2% with $k_{opt} = 45$. We see in this example that the speed of convergence is not significantly improved. The reason can be understood by looking at Figure 5.9. In the case of an ill-conditioned problem, as shown in Section 3.6, the level sets of the discrepancy functional are very elongated ellipsoids and minimization corresponds to finding the lowest point in a very flat and steep-sided valley. The various steps in the steepest-descent method correspond to move back and forth across the valley. Moreover the steps become smaller and smaller. A remedy is provided by the *conjugate gradient* (CG) method, proposed by Hestenes and Stiefel [157], where the steps are made to move down the valley, increasing therefore the speed of convergence.

(B) *The conjugate gradient method*

The origin of the name *conjugate gradient* resides in the following definition: two vectors (or functions or images), f and h, are said to be *conjugate with respect to $\bar{A}$* or also *$\bar{A}$-orthogonal* if

$$(f, \bar{A}h) = 0 . \tag{5.77}$$

Then, given a level set (ellipsoid) of the discrepancy functional and a point of this level set, the direction of the negative conjugate gradient at this point is the direction $\bar{A}$-orthogonal to the tangent plane. In the finite-dimensional case, it is possible to prove that this direction points toward the center of the ellipsoid, as shown in Figure 5.10. In this figure we indicate both the negative gradient $\mathbf{r}$ and the negative conjugate gradient $\mathbf{q}$.

The starting point of the method is the observation that methods such as Landweber or steepest descent provide, at step k, an approximation which is a linear combination of the functions $\bar{g}, \bar{A}\bar{g}, \bar{A}^2\bar{g}, \ldots, \bar{A}^{k-1}\bar{g}$ (we assume here that the initial approximation is $f_0 = 0$). Therefore, whatever the choice of the relaxation parameter, the result of the k-th iteration always lies in the subspace which is the span of these functions. This is the so-called *Krylov*

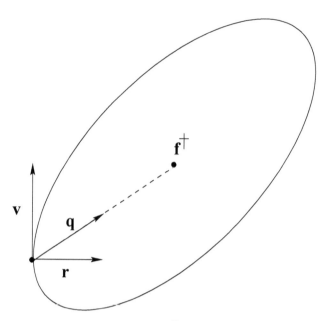

FIGURE 5.10   Two-dimensional representation of $\bar{A}$-orthogonal vectors **v** and **q**. If **v** is tangent to a level curve of the discrepancy functional, then **q** is directed toward the center of the ellipsoid. The direction of the negative gradient **r** is also indicated.

subspace $\mathcal{K}^{(k)}(\bar{A}; \bar{g})$ whose dimension is k if the above-mentioned elements are linearly independent. For the sake of simplicity, we assume that this condition is satisfied for any k.

The Landweber and the steepest descent methods do not provide, at step k, the element of the Krylov subspace which best approximates the generalized solution, i.e. the orthogonal projection of the generalized solution onto $\mathcal{K}^{(k)}(\bar{A}; \bar{g})$. In general, this projection cannot be easily computed and it is not computed by the CG method which is doing something else. It provides the function of $\mathcal{K}^{(k)}(\bar{A}; \bar{g})$ that minimizes the discrepancy functional (or the functional (5.67)). This property can be obtained by showing that the CG method is a *projection method*, in the sense that, at step k, it computes the solution of the least-squares equation projected onto $\mathcal{K}^{(k)}(\bar{A}; \bar{g})$. Denoting by P_k the projection operator onto $\mathcal{K}^{(k)}(\bar{A}; \bar{g})$, the projected equation is

$$P_k \bar{A} P_k f = P_k \bar{g} \,, \tag{5.78}$$

and the solution provided by the CG method satisfies the condition $P_k f = f$. This solution is the minimum of the discrepancy functional restricted to the Krylov subspace $\mathcal{K}^{(k)}(\bar{A}; \bar{g})$. Indeed, if we look for a least-squares solution in $\mathcal{K}^{(k)}(\bar{A}; \bar{g})$, it is a solution of the minimization problem

$$\|A P_k f - g\| = \text{minimum} \,, \quad P_k f = f \,. \tag{5.79}$$

Equation (5.78) is the Euler-Lagrange equation of this problem, as follows from the property that the adjoint of the operator $A P_k$ is the operator $P_k A^*$. We also recall that the solution of the problem (5.79) is the center of the level sets of the discrepancy functional in $\mathcal{K}^{(k)}(\bar{A}; \bar{g})$ and that the conjugate gradient precisely points towards this center.

Projection methods are not iterative methods in a strict sense even if the construction of the projected equation is performed recursively. The *Lanczos method* [293] is another example of a projection method. It provides the construction of a $k \times k$ tridiagonal matrix which is isomorphic to the projected matrix $P_k \bar{A} P_k$. Since the inverse of a tridiagonal matrix is easily computed, the projected equation can be easily solved. The CG method, however, provides the same result in a more stable way, avoiding the explicit construction of the projected system. One can also show that if $\bar{A}$ is a $N \times N$ matrix and $\mathcal{K}^{(N)}(\bar{A}; \bar{g})$ has dimension N, then, in exact arithmetic, the method converges necessarily in N steps. Properties of the CG method are derived in any book of numerical analysis (see, for instance [136] or [293]). Here we only sketch the basic points.

The CG method is based on the following iterative construction of two bases $\bar{r}_k$ and $\bar{p}_k$:

- set

$$\bar{r}_0 = \bar{p}_0 = \bar{g} \tag{5.80}$$

- compute

$$\bar{\alpha}_k = \frac{\|\bar{r}_k\|^2}{(\bar{r}_k, \bar{A}\,\bar{p}_k)}$$

- compute

$$\bar{r}_{k+1} = \bar{r}_k - \bar{\alpha}_k \bar{A}\,\bar{p}_k$$

- compute

$$\bar{\beta}_k = -\frac{(\bar{r}_{k+1}, \bar{A}\,\bar{p}_k)}{(\bar{p}_k, \bar{A}\,\bar{p}_k)}$$

- compute

$$\bar{p}_{k+1} = \bar{r}_{k+1} + \bar{\beta}_k \bar{p}_k \;.$$

Then the iterative scheme for the computation of the approximate solutions is given by

$$\begin{aligned} f_0 &= 0 \\ f_{k+1} &= f_k + \bar{\alpha}_k \bar{p}_k \;. \end{aligned} \tag{5.81}$$

The definition of $\bar{\alpha}_k$ and $\bar{\beta}_k$ as given above is not the usual one but it may be useful for understanding what the CG method is doing. First let us observe that the coefficients $\bar{\alpha}_k$ and $\bar{\beta}_k$ are chosen in such a way that the following orthogonality conditions are satisfied

$$(\bar{r}_{k+1}, \bar{r}_k) = 0 \;, \quad (\bar{p}_{k+1}, \bar{A}\,\bar{p}_k) = 0 \tag{5.82}$$

i.e. $\bar{r}_1$ is orthogonal to $\bar{r}_0$, $\bar{r}_2$ is orthogonal to $\bar{r}_1$, and so on; similarly $\bar{p}_1$ is $\bar{A}$-orthogonal (conjugate) to $\bar{p}_0$, $\bar{p}_2$ is $\bar{A}$-orthogonal to $\bar{p}_1$, and so on. It is not difficult to prove by induction that the following properties hold true:

- the functions $\{\bar{r}_0, \bar{r}_1, \ldots, \bar{r}_{k-1}\}$ form an orthogonal basis of $\mathcal{K}^{(k)}(\bar{A}; \bar{g})$;
- the functions $\{\bar{p}_0, \bar{p}_1, \ldots, \bar{p}_{k-1}\}$ form an $\bar{A}$- orthogonal basis of $\mathcal{K}^{(k)}(\bar{A}; \bar{g})$;
- the $\bar{r}_k$ are precisely the residuals associated with the approximations f_k, i.e.

$$\bar{r}_k = \bar{g} - \bar{A}f_k \tag{5.83}$$

- the following alternative expressions of the coefficients $\bar{\alpha}_k$, $\bar{\beta}_k$ hold true

$$\bar{\alpha}_k = \frac{\|\bar{r}_k\|^2}{(\bar{p}_k, \bar{A}\,\bar{p}_k)} \;, \quad \bar{\beta}_k = \frac{\|\bar{r}_{k+1}\|^2}{\|\bar{r}_k\|^2} \;, \tag{5.84}$$

and therefore the coefficients $\bar{\alpha}_k$, $\bar{\beta}_k$ are positive.

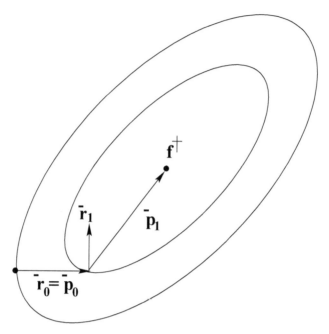

FIGURE 5.11   Two-dimensional representation of the conjugate gradient method.

The first property implies that $\bar{r}_k$ is orthogonal to $\mathcal{K}^{(k)}(\bar{A};\bar{g})$, or also that $P_k\bar{r}_k = 0$. Then, from equation (5.83) we obtain equation (5.78) and we prove in this way that the CG method is a projection method.

We can see now how the CG method works. We start with $f_0 = 0$ and we move in the direction orthogonal to the level set of the discrepancy functional passing through f_0, up to the level set which is tangent to this direction. Next we do not move in the direction of the gradient but we move up to the center of the ellipse which is the intersection of this level set with the plane spanned by the functions $\bar{r}_0, \bar{r}_1$ (the Krylov subspace $\mathcal{K}^{(2)}(\bar{A};\bar{g})$). This is precisely the result we obtain by moving from f_1 in the direction of the conjugate gradient $\bar{p}_1$. Next we consider the level set through f_2 and so on. In Figure 5.11, we represent this process in the two-dimensional case. If we take into account that, as shown in Figure 5.10, the conjugate gradient $\mathbf{q}$ points toward the center of the ellipse, we see that, in this case, we reach the center in two steps.

In the case of the problem of image deconvolution, we can write the iterative process in terms of Fourier transforms. We recognize immediately that if the imaging system is bandlimited, then the FT of $\bar{r}_k, \bar{p}_k$ and f_k is zero outside the band $\mathcal{B}$ of the system and therefore f_k is a bandlimited approximation of the solution of the problem. Indeed we have the following algorithm, where we use the expression (5.84) of the coefficients $\bar{\alpha}_k, \bar{\beta}_k$:

- set

$$\hat{f}_0(\boldsymbol{\omega}) = 0 \quad ; \quad \widehat{\bar{r}}_0(\boldsymbol{\omega}) = \widehat{\bar{p}}_0(\boldsymbol{\omega}) = \hat{K}^*(\boldsymbol{\omega})\hat{g}(\boldsymbol{\omega}) \tag{5.85}$$

- compute

$$\bar{\alpha}_k = \frac{\int_\mathcal{B} |\widehat{\bar{r}}_k(\boldsymbol{\omega})|^2 d\boldsymbol{\omega}}{\int_\mathcal{B} |\hat{K}^*(\boldsymbol{\omega})|^2 |\widehat{\bar{p}}_k(\boldsymbol{\omega})|^2 d\boldsymbol{\omega}} \tag{5.86}$$

- compute

$$\widehat{\bar{r}}_{k+1}(\boldsymbol{\omega}) = \widehat{\bar{r}}_k(\boldsymbol{\omega}) - \bar{\alpha}_k |\hat{K}(\boldsymbol{\omega})|^2 \, \widehat{\bar{p}}_k(\boldsymbol{\omega}) \tag{5.87}$$

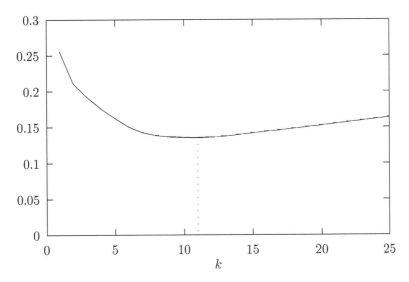

FIGURE 5.12   Behavior of the relative reconstruction error as a function of the number of iterations of the CG method for the reconstruction of the image of Figure 4.13.

- compute

$$\bar{\beta}_k = \frac{\int_{\mathcal{B}} |\widehat{\bar{r}}_{k+1}(\boldsymbol{\omega})|^2 d\boldsymbol{\omega}}{\int_{\mathcal{B}} |\widehat{\bar{r}}_k(\boldsymbol{\omega})|^2 d\boldsymbol{\omega}} \qquad (5.88)$$

- compute

$$\widehat{\bar{p}}_{k+1}(\boldsymbol{\omega}) = \widehat{\bar{r}}_{k+1}(\boldsymbol{\omega}) + \bar{\beta}_k \, \widehat{\bar{p}}_k(\boldsymbol{\omega}) \qquad (5.89)$$

- compute

$$\hat{f}_{k+1}(\boldsymbol{\omega}) = \hat{f}_k(\boldsymbol{\omega}) + \bar{\alpha}_k \, \widehat{\bar{p}}_k(\boldsymbol{\omega}) \ . \qquad (5.90)$$

The convergence properties of the CG method in the case of an ill-posed operator equation are proved in [113]. From these general results one can derive the following convergence properties of the algorithm (5.85)-(5.90):

- in the case of a noise-free image $g^{(0)} = A f^{(0)}$, the iterates f_k converge to the generalized solution $f^\dagger = P^{(\mathcal{B})} f^{(0)}$ when $k \to \infty$, i.e. $\|f_k - f^\dagger\| \to 0$;
- in the case of a noisy image the iterates f_k have the semi-convergence property, i.e. the f_k first approach $f^{(0)}$ and then go away, so that the number of iterations acts as a regularization parameter and the reconstruction error $\|f_k - f^\dagger\|$ typically has a minimum for a certain value k_{opt} of k;
- the optimal number of iterations can be estimated by means of the discrepancy principle (5.37), which provides a finite value $\tilde{k}$ because the norm of the residual of the k-th iterate of the CG method is always smaller than the norm of the corresponding residual of the Landweber method.

The second and the third property are illustrated in the case of the numerical example of Section 4.6. In Figure 5.12, we plot the behavior of the relative reconstruction error $\|f_k - f^{(0)}\|/\|f^{(0)}\|$, as a function of k. It shows a minimum at $k_{opt} = 11$ and the corresponding reconstruction error is 13.6%. By applying the discrepancy principle, as given in equation (5.37) with $b = 1$, we find $\tilde{k} = 7$ and a reconstruction error of 14.2%. Therefore, in this example, the number of iterations required by the CG method is much smaller than that required by the Landweber or the steepest descent method. In some cases, however, it may still be useful to improve the rate of convergence of the CG method and, to this purpose, preconditioning can be used. In other words one can apply the CG method to equation (5.43). In the case of deconvolution problems, one can choose a preconditioner D which is also a convolution operator and an example is given in equation (5.45). In such a case we find that the TF of the operator $D\bar{A}$ is given by $|\hat{K}(\boldsymbol{\omega})|^2(|\hat{K}(\boldsymbol{\omega})|^2 + \gamma)^{-1}$, which is approximately 1 for the space frequencies $\boldsymbol{\omega}$ such that $|\hat{K}(\boldsymbol{\omega})|^2 \gg \gamma$ and approximately zero for spatial frequencies $\boldsymbol{\omega}$ such that $|\hat{K}(\boldsymbol{\omega})|^2 \ll \gamma$. Then the effect on the rate of convergence can be understood by taking into account the filtering properties of the CG method which are discussed in the next section.

The discrete version of the computational scheme (5.85)-(5.90) can be obtained straightforwardly and therefore it is not reported. As concerns the convergence properties, we only observe that, for the reconstruction of a discrete image $N \times N$, in exact arithmetic, the CG method, as any projection method, must terminate in N^2 steps because the dimension of the Krylov subspace cannot exceed N^2. However, the final result is the generalized solution which, as we know, is completely corrupted by noise in the case of an ill-conditioned problem. Therefore, also in the case of discrete images, one must use the semi-convergence property of the algorithm to obtain reliable reconstructions.

5.5   Filtering properties of the conjugate gradient method

The presentation of the CG method given in the previous section does not clarify why this method has regularization properties. In this section we investigate this point by showing that the method has some peculiar filtering properties even if these properties cannot be described in terms of a global PSF because the method is basically nonlinear.

We already know that the approximation f_k, provided by the k-th iteration, is the element of the Krylov subspace $\mathcal{K}^{(k)}(\bar{A}; \bar{g})$ which minimizes the discrepancy functional

$$\|Af_k - g\| = \min_{u \in \mathcal{K}^{(k)}(\bar{A};\bar{g})} \|Au - g\| . \tag{5.91}$$

Moreover, if we observe that any element of $\mathcal{K}^{(k)}(\bar{A}; \bar{g})$ can be written as the result of the application to $\bar{g}$ of a polynomial, of degree $k - 1$, formed with the powers of $\bar{A}$:

$$u = P_{k-1}(\bar{A})\bar{g} = P_{k-1}(\bar{A})A^*g , \tag{5.92}$$

we find that a similar representation holds true also for f_k and we denote by Q_{k-1} the corresponding polynomial

$$f_k = Q_{k-1}(\bar{A})A^*g . \tag{5.93}$$

From equation (5.91) it follows that Q_{k-1} is the polynomial solving the following minimization problem

$$\|AQ_{k-1}(\bar{A})A^*g - g\| = \min_{P_{k-1}} \|AP_{k-1}(\bar{A})A^*g - g\| . \tag{5.94}$$

For the problem of image deconvolution, using Parseval's equality we have

$$\|AP_{k-1}(\bar{A})A^*g - g\|^2 \tag{5.95}$$
$$= \frac{1}{(2\pi)^d} \int |\hat{g}(\boldsymbol{\omega})|^2 \left| 1 - |\hat{K}(\boldsymbol{\omega})|^2 P_{k-1}(|\hat{K}(\boldsymbol{\omega})|^2) \right|^2 d\boldsymbol{\omega}$$

and therefore, if $Q_{k-1}(|\hat{K}(\boldsymbol{\omega})|^2)$ is the polynomial minimizing this functional, we see that $|\hat{K}(\boldsymbol{\omega})|^2 Q_{k-1}(|\hat{K}(\boldsymbol{\omega})|^2)$ must be of the order of 1 for those spatial frequencies such that $|\hat{g}(\boldsymbol{\omega})|$ is large. On the other hand, the function $|\hat{K}(\boldsymbol{\omega})|^2 Q_{k-1}(|\hat{K}(\boldsymbol{\omega})|^2)$ is a filter acting on

the generalized solution since we can write

$$
\begin{aligned}
\hat{f}_k(\boldsymbol{\omega}) &= Q_{k-1}(|\hat{K}(\boldsymbol{\omega})|^2)\hat{K}^*(\boldsymbol{\omega})\hat{g}(\boldsymbol{\omega}) \\
&= |\hat{K}(\boldsymbol{\omega})|^2 Q_{k-1}(|\hat{K}(\boldsymbol{\omega})|^2)\frac{\hat{g}(\boldsymbol{\omega})}{\hat{K}(\boldsymbol{\omega})} \ .
\end{aligned}
\tag{5.96}
$$

This filter is nonlinear because, even if it is not indicated explicitly, the polynomial Q_{k-1} depends on the image g. This point was clarified when we observed that the quantity $|\hat{K}(\boldsymbol{\omega})|^2 Q_{k-1}(|\hat{K}(\boldsymbol{\omega})|^2)$ is of the order of 1 when $|\hat{g}(\boldsymbol{\omega})|$ is large. This also means that the CG method provides a very clever filter. Indeed this filter not only tends to pick up the frequencies corresponding to large values of $|\hat{K}(\boldsymbol{\omega})|$ (as the Tikhonov and Landweber filters do) but also frequencies corresponding to large values of the product $|\hat{K}(\boldsymbol{\omega})\hat{f}^{(0)}(\boldsymbol{\omega})|$ where $f^{(0)}$ is, as usual, the unknown object. The filter (5.96) is often written in the following form

$$
\hat{f}_k(\boldsymbol{\omega}) = \left[1 - R_k(|\hat{K}(\boldsymbol{\omega})|^2)\right]\frac{\hat{g}(\boldsymbol{\omega})}{\hat{K}(\boldsymbol{\omega})}
\tag{5.97}
$$

where R_k, the so-called *Ritz polynomial*, is a polynomial of degree k related to the polynomial Q_{k-1} by

$$
R_k(t) = 1 - t\, Q_{k-1}(t) \ .
\tag{5.98}
$$

The zeroes of the Ritz polynomial are the so-called *Ritz values*, indicated by $\theta_1^{(k)}, \ldots, \theta_k^{(k)}$. Since $R_k(0) = 1$, we have the following representation

$$
R_k(t) = \prod_{j=1}^{k}\left(1 - \frac{t}{\theta_j^{(k)}}\right).
\tag{5.99}
$$

It is possible to prove (see [293] for a discussion and references) that the Ritz values are the eigenvalues of the projected matrix $P_k \bar{A} P_k$. They can be computed by means of the Lanczos method, since this method provides a tridiagonal representation of the matrix, which can then be easily diagonalized.

II

Linear Inverse Problems

6

Examples of linear inverse problems

In the first part of the book the problem of image reconstruction is considered in the case of a space-invariant imaging system. Since the imaging operator is then described by a convolution product, the powerful tool of Fourier transform can be used for obtaining explicit solutions. In this case a simple presentation and discussion of the main ideas and methods used for treating the ill-posedness of the problem is possible.

However, not all imaging systems can be described by a convolution operator. In this chapter we present a few significant examples of imaging systems which can be described, with a sufficient accuracy, by more general linear operators. The corresponding image reconstruction problems are then case examples of more general linear inverse problems.

This chapter introduces the second part of the book where we extend the methods and results presented in the first part to the inversion of a generic linear operator. Moreover, in Chapters 9 and 10, we present new regularization methods which, in a sense, implement different and new kinds of prior information on the solution, and are also able to cure the ill-posedness of the problem.

6.1 Space-variant imaging systems

The first example is strictly related to the problem considered in the first part, i.e. image reconstruction in the case of a space-invariant imaging system, for instance an optical system such as a telescope or a microscope. However, as already observed, this property is seldom satisfied in practice; a moderate space-variance should be taken into account for a more accurate object reconstruction. Therefore, if the system is linear, the general relationship between the object $f^{(0)}(\mathbf{x})$ and the noise-free image $g^{(0)}(\mathbf{x})$ is described by a linear integral operator of the following type

$$g^{(0)}(\mathbf{x}) = \int K(\mathbf{x}, \mathbf{x}') \, f^{(0)}(\mathbf{x}') \, d\mathbf{x}' \ . \qquad (6.1)$$

The function $K(\mathbf{x}, \mathbf{x}')$ is the *space-variant* PSF of the system, already defined in Section 2.1. In the theory of integral equations this function is also called the *kernel* (or integral kernel) of the integral operator. However, we must point out that in other examples considered in

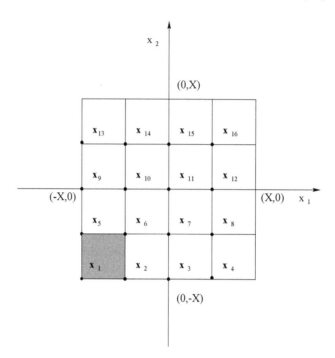

FIGURE 6.1 Partition of a square into 16 pixels. Each pixel is associated with the vertex at bottom left. For instance, the shaded pixel is associated with $\mathbf{x}_1$. These points can be taken as sampling points of the image and of the space-variant PSF. Another possible choice is to take the central points of the pixels as sampling points.

this chapter, such as computed tomography, the kernel is a distribution and not a function. In this section we assume that it is a function.

The problem of image reconstruction consists again in determining an estimate $f(\mathbf{x})$ of $f^{(0)}(\mathbf{x})$ when a noisy version $g(\mathbf{x})$ of $g^{(0)}(\mathbf{x})$ and the space-variant PSF $K(\mathbf{x}, \mathbf{x}')$ are given. The starting point is an equation similar to equation (6.1) with $g(\mathbf{x})$, $f(\mathbf{x})$ in place of $g^{(0)}(\mathbf{x})$, $f^{(0)}(\mathbf{x})$. Such an equation is called a *Fredholm integral equation of the first kind* and the solution of this type of equations is a classical example of ill-posed problem.

Equation (6.1) can be discretized by the method already used for convolution integrals. Here we consider the most frequent case of 2D images and we assume that both the object and the image are defined in the same domain, a square of side $2X$. The domain can be divided into $N_0 = N^2$ square subdomains (pixels) with size $\delta = 2X/N$ and area δ^2 – see Figure 6.1 in the case of $N = 4$. If the area of the pixels is sufficiently small both the object and the image are approximately constant over a single pixel, i.e. they are approximately equal to a value taken in some particular point of the pixel, for instance the vertex at its bottom left (see Figure 6.1) or its central point. If these points are taken in the order indicated in Figure 6.1, and denoted by $\mathbf{x}_1, \mathbf{x}_2, \ldots, \mathbf{x}_{N^2}$, then the integral in equation (6.1) can be approximated by the following sum

$$g^{(0)}(\mathbf{x}_m) = \sum_{n=1}^{N^2} K(\mathbf{x}_m, \mathbf{x}_n) f^{(0)}(\mathbf{x}_n)\, \delta^2 \; ; \quad m = 1, \ldots, N^2 \tag{6.2}$$

so that the effect of the PSF is described by the matrix

$$A_{m,n} = K(\mathbf{x}_m, \mathbf{x}_n)\, \delta^2 \tag{6.3}$$

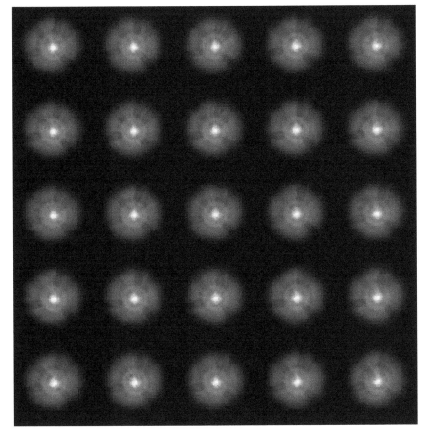

FIGURE 6.2   Grid of 25 PSF's used for the reconstruction of images of the wide-field/planetary camera of the HST (image produced by AURA/STScI).

which can be called the *space-variant discrete* PSF.

In the case of a space-invariant imaging system, i.e. $K(\mathbf{x}_m, \mathbf{x}_n) = K(\mathbf{x}_m - \mathbf{x}_n)$, the matrix (6.3) can be replaced by the block-circulant matrix with circulant blocks associated with the N^2 values $K(\mathbf{x}_n)$ of the space-invariant PSF $(n = 1, 2, \ldots, N^2)$. In the general case of a space-variant system, one needs N^4 values of the space-variant PSF. This number can be exceedingly large. For instance, in the case of an image 512×512, we have $N^4 = (512)^4 \simeq 6.9\,10^{10}$. If the matrix (6.3) does not have some special structure, then too many matrix elements must be stored (even when the matrix is sparse) and the required storage space may be prohibitive. Moreover, the number of operations required for computing the matrix-vector product in equation (6.2) is of the order of N^6 and therefore it is also too large. In other words, the problem is numerically intractable.

For many optical systems, however, the space-variant PSF is slowly varying over the image domain and, in such a case, it can be locally approximated by a space-invariant PSF. For explaining the approach in a simple way, we refer to Figure 6.2. In this picture the image domain is partitioned into 5×5 sub-domains and in each sub-domain the corresponding PSF (more precisely the central part of the PSF) is shown. This representation of a space-variant PSF is that of the wide-field/planetary camera (WF/PC) of the Hubble Space Telescope (HST), before correction of the spherical aberration. The space-variance is due to the vignetting in the internal repeater optics [42]. For the applications to HST the 5×5 grid of the space-variant PSF, shown in the figure, is used for both the computation of synthetic images and the image reconstruction. This example is also frequently used for testing methods designed for image reconstruction in the case of space-variant imaging systems. If we look at the PSF's shown in the figure and corresponding to a given row or column of the grid, we notice small but significant variations, i.e. the PSF is mildly space variant.

More generally, the approach to deal with slowly varying space-variant PSF is the following:

- a computable (exact or approximate) model of the space-variant PSF is available;
- the image domain $\mathcal{D}$ is partitioned into P non-overlapping subdomains $\mathcal{D}_1, ..., \mathcal{D}_P$ with centers $\mathbf{x}_1, ..., \mathbf{x}_P$;
- the subdomains are selected in such a way that the difference between the PSF's computed in adjacent subdomains is not too large (let us say, about 15 % or less);
- the size d_i of the subdomain $\mathcal{D}_i$ is also selected in such a way that it is greater than the width of the PSF computed in the center of the domain, i.e. $K(\mathbf{x}, \mathbf{x}_i)$;
- finally, d_i is also selected in such a way that, for any point $\mathbf{x}' \in \mathcal{D}_i$, the shape of the PSF computed in $\mathbf{x}'$, i.e $K(\mathbf{x}, \mathbf{x}')$, is similar to that of $K(\mathbf{x}, \mathbf{x}_i)$.

The last condition implies that $K(\mathbf{x}, \mathbf{x}')$ can be approximately obtained by translating $K(\mathbf{x}, \mathbf{x}_i)$ from $\mathbf{x}_i$ to $\mathbf{x}'$, i.e.

$$K(\mathbf{x}, \mathbf{x}') \simeq K(\mathbf{x} - (\mathbf{x}' - \mathbf{x}_i), \mathbf{x}_i), \ \mathbf{x}, \mathbf{x}_i \in \mathcal{D}_i \ . \tag{6.4}$$

Therefore, it is possible to introduce in $\mathcal{D}_i$ a space-invariant PSF $K_i(\mathbf{x}) = K(\mathbf{x} + \mathbf{x}_i, \mathbf{x}_i)$ such that

$$\int K(\mathbf{x}, \mathbf{x}') f(\mathbf{x}') \, d\mathbf{x}' \simeq \int K_i(\mathbf{x} - \mathbf{x}') f(\mathbf{x}') \, d\mathbf{x}', \ \mathbf{x} \in \mathcal{D}_i \ . \tag{6.5}$$

The first approach based on this approximation of a space-variant PSF is the *sectioning approach* [287, 286, 12]. The idea is to enlarge the non-overlapping subdomains considered above to render them partially overlapping and to apply to each of these enlarged subdomains a deconvolution method based on the corresponding PSF. The amount of overlapping is dictated by the width of the PSF. Finally, the reconstructions obtained on the non-overlapping subdomains are extracted and reassembled so that the final reconstructed object is the mosaic of these local reconstructions. The drawback of this approach is mainly due to possible boundary artifacts generated by the use of FFT and by discontinuities due to the differences between adjacent PSF's. A refinement of this approach, where deconvolution methods with boundary effect correction are used, is proposed in [185].

A second approach, also based on the representation of the space-variant PSF in terms of locally space-invariant PSF's, as exemplified in Figure 6.2, can be called an *interpolation* approach. More precisely, a first interpolation approach is proposed in [217] while a different interpolation approach is proposed in [131, 158]. The first approach consists in interpolating the images provided by the different space-invariant PSF while the second one consists in interpolating the PSF's. To describe both cases, let us introduce a function $\phi(\mathbf{x})$ which is approximately 1 over a domain with the size of the subdomains $\mathcal{D}_i$ and which tends to 0 outside. Then, in the approach proposed in [217] the model of the image is given by

$$g(\mathbf{x}) = \sum_{i=1}^{P} \phi(\mathbf{x} - \mathbf{x}_i)(K_i * f)(\mathbf{x}) \tag{6.6}$$

corresponding to the PSF

$$K(\mathbf{x}, \mathbf{x}') = \sum_{i=1}^{P} \phi(\mathbf{x} - \mathbf{x}_i) \, K_i(\mathbf{x} - \mathbf{x}') \ , \tag{6.7}$$

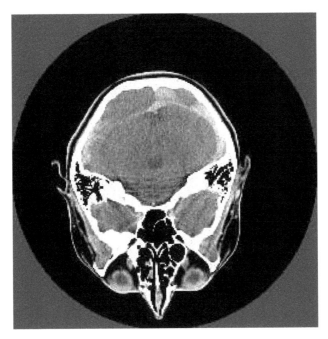

FIGURE 6.3   CT image of a slice of the head of one of the authors of the first edition of this book.

where K_i is the space-invariant PSF associated with the domain $\mathcal{D}_i$. In the second interpolation approach, one interpolates directly the space-invariant PSF's by introducing the model

$$K(\mathbf{x}, \mathbf{x}') = \sum_{i=1}^{P} K_i(\mathbf{x} - \mathbf{x}')\, \phi(\mathbf{x} - \mathbf{x}') . \qquad (6.8)$$

Both models provide an approximation of the space-variant PSF such that the reconstruction problem is numerically tractable since they can be used in conjunction with any deconvolution method. A comparison of the two methods with application to astronomy is discussed in [100] and the conclusion of the authors is that PSF interpolation provides more accurate results than image interpolation.

A similar conclusion is reached in [27] where the problem is considered in the case of confocal microscopy. More precisely, in this problem, the PSF is depth-variant when optical sectioning is used for 3D imaging of biological samples.

6.2   X-ray tomography

The word *tomography* comes from the Greek and means cross-sectional representation of a 3D object, i.e. representation of a 3D object by means of 2D cross-sections or slices. In this sense the word only indicates a particular way of representing 3D objects and therefore can also be applied, for instance, to the case of 3D confocal microscopy, discussed in Section 2.6. However, with reference to the first case where this name was introduced, it is now generally used for denominating imaging methods which are based on the knowledge of integrals of the object over lines or surfaces or other manifolds.

The first important example is the X-ray tomography, introduced by Hounsfield in 1971, which is usually called *computed tomography* (CT). A machine producing CT images is called a CT *scanner*. Several generations of CT scanners have been already designed, corresponding essentially to different ways of collecting data. A short description of the collecting schemes

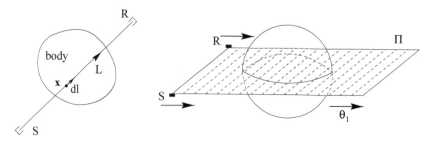

FIGURE 6.4 Left panel: schematic representation of the elementary process of X-ray tomography; right panel: the planar slice defined by the movement of the source-receiver pair.

of the various machines can be found in [304].

A CT image can be described as follows: it looks as though a planar slice of the body had been physically removed and then radiographed by passing X-rays through it in a direction perpendicular to its plane. Such an image shows the human anatomy with a spatial resolution of about 1 mm and a density discrimination of about 1%. In Figure 6.3, an example of CT image is shown.

The elementary process at the basis of CT is the X-ray absorption by the tissues of the human body. This absorption is described by a function whose value at point $\mathbf{x}$ is the *linear attenuation coefficient* of the tissue at that point. This function, which is here denoted by $f^{(0)}(\mathbf{x})$, is roughly proportional to the density of the body and is the function to be imaged, i.e. it is the object of the CT imaging system.

Let us consider a finely collimated source S emitting a pencil beam of X-rays which propagates through the body along a straight line L up to a well-collimated detector R (see the left panel of Figure 6.4). If $I(\mathbf{x})$ is the intensity of the beam at the point $\mathbf{x}$ of the straight line L, then the loss of intensity in the corresponding element dl of L is given by

$$dI(\mathbf{x}) = -f^{(0)}(\mathbf{x})\, I(\mathbf{x})\, dl \ . \tag{6.9}$$

By integrating this equation we obtain

$$I = I_0 \exp\left(-\int_L f^{(0)}(\mathbf{x})\, dl\right) \tag{6.10}$$

where I_0 is the intensity of the beam at the exit of the source S and I is the intensity of the beam detected by the receiver R. Therefore, if we know both I_0 and I, by taking the logarithm of the ratio I_0/I we obtain the integral of the linear attenuation coefficient $f^{(0)}(\mathbf{x})$ along L.

By repeating the measurement outlined above for several different positions of the source-receiver pair, the basic information we obtain is a set of line integrals of $f^{(0)}(\mathbf{x})$ and therefore the problem is the recovery of $f^{(0)}(\mathbf{x})$ from these data. To this purpose the strategy followed in the collection of the data is important. We describe the one chosen in the first CT scanner, which is also basic for understanding other methods of data collection.

As shown in the right panel of Figure 6.4, the source and the receiver are moved simultaneously along two parallel straight lines which define a plane Π, in practice a planar slice of the body under consideration. The common direction of the two straight lines is denoted by $\boldsymbol{\theta}_1$. By measuring the ratio I_0/I for all positions of the source-receiver pair, we obtain the integrals of $f^{(0)}(\mathbf{x})$ along all parallel lines defined by these positions. These integrals provide the so-called projection of $f^{(0)}(\mathbf{x})$ in the direction $\boldsymbol{\theta}_1$.

Once these data are collected, the source-receiver system is rotated by an angle $\Delta\phi$ around an axis orthogonal to the plane Π and then moved along two parallel straight lines with

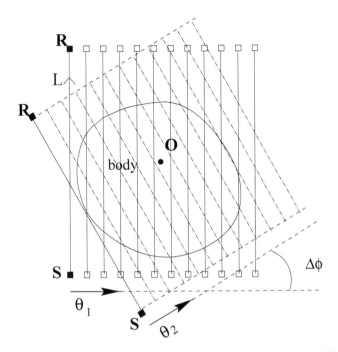

FIGURE 6.5 Scheme of the scanning used for measuring various projections of $f^{(0)}(\mathbf{x})$.

direction $\boldsymbol{\theta}_2$. As shown in Figure 6.5, $\Delta\phi$ is the angle between the directions $\boldsymbol{\theta}_1$ and $\boldsymbol{\theta}_2$. The intersection O between the rotation axis and the plane Π is also indicated in the same figure. By measuring the ratio I_0/I for all positions of the source-receiver pair along the direction $\boldsymbol{\theta}_2$, we obtain the projection of $f^{(0)}(\mathbf{x})$ in this direction. By repeating this measurement for several directions $\boldsymbol{\theta}_1, \boldsymbol{\theta}_2, \ldots, \boldsymbol{\theta}_p$, we obtain the corresponding projections of $f^{(0)}(\mathbf{x})$.

The procedure outlined above is a double scanning consisting of an angular scanning and of a linear scanning for each angle. This is the origin of the name CT scanner used for a machine producing CT images. Moreover the geometry of the scanning procedure outlined above is the so-called *parallel beam* geometry because one collects data corresponding to bunches of parallel lines. This was used in the first-generation scanners. In more recent scanners, the scanning procedure is based on the so-called *fan beam* geometry. The name derives from the fact that, in this case, a single source illuminates a ring of detectors with a broad fan beam of X-rays, so that one collects data corresponding to bunches of lines emanating from a point source. The difference between the parallel beam and the fan beam geometry is shown in Figure 6.6. The important point is that in both cases one collects data which provide the integrals of $f^{(0)}(\mathbf{x})$ along lines lying in a well-defined plane Π.

The inverse problem consists now in reconstructing the function $f^{(0)}(\mathbf{x})$ in the plane Π from the knowledge of its line integrals. Since, as shown above, the data provided by a CT scanner can be grouped into projections of the function $f^{(0)}(\mathbf{x})$, this problem is frequently called *image reconstruction from projections* [155] while in the mathematical literature, the name *Radon transform inversion* is preferred. Even if, in practice, parallel beam geometry is no longer used in commercial scanners, in this book we mainly consider this case because it provides the best framework for introducing the basic ideas underlying CT imaging. The modifications required in the case of fan beam geometry can be found in [176].

Consider the plane Π corresponding to the section of the body to be imaged. Then, it is quite natural to take the intersection of this plane with the rotation axis of the machine as the origin O of a coordinate system. The orientation of the Cartesian axes is arbitrary

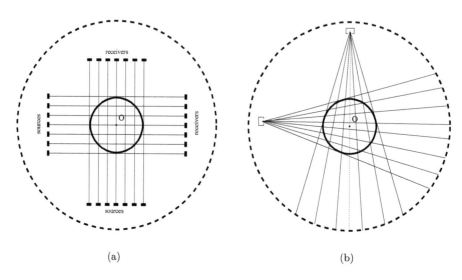

FIGURE 6.6   Schematic representation of parallel beam geometry (a) and of fan beam geometry (b).

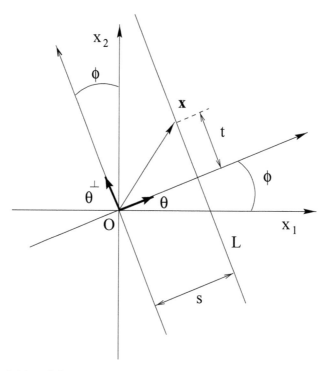

FIGURE 6.7   Definition of the geometrical variables for the Radon transform.

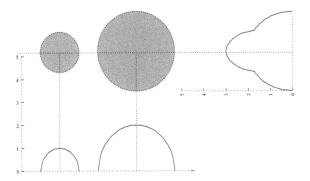

FIGURE 6.8   Two projections of an object which consists of a pair of cylinders.

but one can take, for instance, the x_1-axis along the initial direction of movement of the source-receiver pair.

Given a direction $\boldsymbol{\theta}$ in the plane, the straight lines orthogonal to $\boldsymbol{\theta}$ are characterized by their signed distance s from the origin (see Figure 6.7). If ϕ is the angle between $\boldsymbol{\theta}$ and the x_1-axis, then $\boldsymbol{\theta} = \{\cos\phi, \sin\phi\}$ while $\boldsymbol{\theta}^\perp = \{-\sin\phi, \cos\phi\}$ is the unit vector orthogonal to $\boldsymbol{\theta}$. In terms of these unit vectors the equation of the straight line L, orthogonal to $\boldsymbol{\theta}$ with signed distance s from the origin, is given by $\mathbf{x} = s\boldsymbol{\theta} + t\boldsymbol{\theta}^\perp$, where t is the variable defined in Figure 6.7. Then the *projection of $f^{(0)}$ in the direction $\boldsymbol{\theta}$* is the function of s defined by

$$\left(R_{\boldsymbol{\theta}} f^{(0)}\right)(s) = \int_{-\infty}^{+\infty} f^{(0)}(s\boldsymbol{\theta} + t\boldsymbol{\theta}^\perp)\, dt \ . \tag{6.11}$$

For a given $\boldsymbol{\theta}$, the parallel beam method described above provides sampled and noisy values of the projection $\left(R_{\boldsymbol{\theta}} f^{(0)}\right)(s)$.

As an example, in Figure 6.8, we show two projections of a function $f^{(0)}$ which is the sum of the characteristic functions of two disjoint cylinders, one with radius $1/2$ and the other with radius 1. It is obvious that when the direction $\boldsymbol{\theta}$ is parallel to the straight line joining the centers of the two cylinders, their projections are separated. When the direction $\boldsymbol{\theta}$ is orthogonal to this line, the projections of the two cylinders are superimposed. For other directions one can have intermediate situations.

If we consider all projections of $f^{(0)}$, corresponding to all possible directions $\boldsymbol{\theta}$, then a function of two variables, the angle ϕ characterizing the direction $\boldsymbol{\theta}$ and the signed distance s, is defined. This function is denoted by

$$\left(R f^{(0)}\right)(s, \boldsymbol{\theta}) = \left(R_{\boldsymbol{\theta}} f^{(0)}\right)(s) \tag{6.12}$$

and is called the *Radon transform* of $f^{(0)}$, in honor of the mathematician Johann Radon who first investigated the problem of recovering a function of two variables from its line integrals. The result of Radon was even more general, since he proved formulas for the reconstruction of a function of q variables from knowledge of its integrals over all hyperplanes with dimension $q - 1$ [238].

The Radon transform defines a mapping R which transforms a function of two space variables into a function of one space and one angular variable. The domain of definition of this function in the $\{s, \phi\}$ plane is the strip defined by: $-\infty < s < \infty$, $-\pi < \phi \leq \pi$. In this strip the Radon transform has the symmetry property

$$\left(R f^{(0)}\right)(-s, -\boldsymbol{\theta}) = \left(R f^{(0)}\right)(s, \boldsymbol{\theta}) \tag{6.13}$$

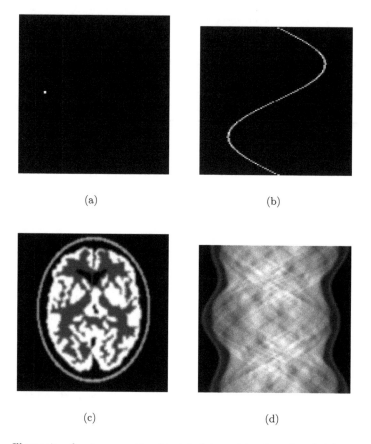

(a)   (b)

(c)   (d)

FIGURE 6.9   Illustrating the sinogram: the object in (a) consists of a unique bright spot and the corresponding sinogram is the white curve in (b). Each point of this curve corresponds to a straight line passing through the bright spot in (a). In (c) a more structured object is given and its sinogram is shown in (d).

which easily follows from the explicit representation of $Rf^{(0)}$ in terms of the variables s, ϕ

$$\left(Rf^{(0)}\right)(s, \boldsymbol{\theta}) = \int_{-\infty}^{\infty} f^{(0)}(s\cos\phi - t\sin\phi, s\sin\phi + t\cos\phi)\, dt \ . \tag{6.14}$$

The $\{s, \phi\}$-plot of the Radon transform of $f^{(0)}$, obtained by representing its values as gray levels, is called the *sinogram* of $f^{(0)}$ [156]. This picture is, in a sense, the image of $f^{(0)}$ provided by a CT scanner, so that the sinogram obtained from the measured values of the Radon transform is a discrete and noisy image of $f^{(0)}$.

If the support of the function $f^{(0)}$ is interior to the disc of radius a, then the support of the sinogram is the rectangle: $|s| \le a$, $|\phi| \le \pi$. Each point in the rectangle corresponds to a straight line crossing the disc of radius a in the $\{x_1, x_2\}$-plane. The points corresponding to straight lines through a fixed point in the $\{x_1, x_2\}$-plane describe a sinusoidal curve in the $\{s, \phi\}$-plane, given by: $s = x_1\cos\phi + x_2\sin\phi$. As a consequence, if the object consists of a bright spot over a black background, then its sinogram is precisely a bright sinusoidal curve over a black background (see Figure 6.9). This is the origin of the name. In Figure 6.9, we also give an example of sinogram of a complex object.

We conclude this section by remarking that, in the last generation scanners, the acquisition process is based on a concept able to reduce the acquisition time, hence also the dose of radiation absorbed by the patient. The source of X-rays emits a cone beam and rotates around the patient while the detectors form a fixed ring consisting of several layers. The rotation of the source is combined with a slow translation of the patient along the axis of the scanner so that the trajectory of the source, from the point of view of the patient, is a spiral. For this reason the mode is called *spiral tomography*, or also helical tomography. The treatment of this mode is more complex than that of parallel beam tomography and it is beyond the scope of this book.

6.3 Emission tomography

X-ray tomography is also called *transmission computed tomography* (TCT) because the image is obtained by detecting the X-rays transmitted by the body. This technique provides information about anatomical details of the body organs: the map of the linear attenuation function is essentially the map of the density of the tissues.

A quite different type of information is obtained by the so-called *emission computed tomography* (ECT). This technique is based on the administration, either by injection or by inhalation, of radionuclide-labeled agents known as radiopharmaceuticals. Their distribution in the body of the patient depends on factors such as blood flow and metabolic processes. Then a map of this distribution is obtained by detecting the γ-rays produced by the decay of the radionuclides. Therefore ECT yields functional information, in the sense that the images produced by ECT show the function of the biological tissues of the organs.

Two different modalities of ECT have been developed and are used in the practice of medical imaging:

- *single-photon emission tomography* (SPECT), which makes use of radioisotopes such as ^{99m}Tc, where a single γ-ray is emitted per nuclear disintegration;
- *positron emission tomography* (PET), which makes use of β^+-emitters such as $^{11}C, ^{15}O, ^{13}N$ and ^{18}F, where the final result of a nuclear disintegration is a pair of γ-rays, propagating in opposite directions, produced by the annihilation of the emitted positron with an electron of the tissue.

In both cases it is necessary to detect the γ-rays coming from well-defined regions of the body and, to this purpose, different methods are used in the two cases.

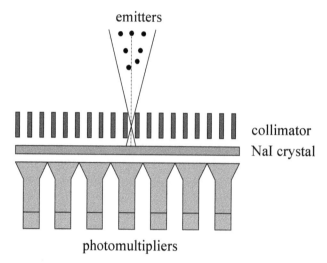

FIGURE 6.10   Scheme of the detection system in SPECT . The γ-rays, transmitted by the collimator, cause scintillations in the NaI crystal. The energy and position of the interaction of the γ-quantum in the crystal are determined by an array of photomultipliers. The acceptance cone of a hole of the collimator is also indicated.

In SPECT the discrimination of the γ-rays is obtained by a *collimator* which is a large slab covering the crystal detector face, consisting of holes separated by lead septa (see Figure 6.10). The axes of the holes are perpendicular to the detector face. Therefore each hole mainly collects the γ-quanta which are emitted in the direction of its axis by the radionuclides located in voxels also crossed by this axis. These γ-quanta are counted by the detection system and their number (in a given time τ) is approximately proportional to the number of radionuclides contained in the voxels discriminated by the hole. In this way we obtain a number which is roughly proportional to the integral of the radionuclides concentration along the axis of the hole. We can denote again the unknown concentration by $f^{(0)}(\mathbf{x})$ since this is the object to be imaged by the SPECT machine.

If we consider now a line of holes of the collimator and the plane Π, defined by this line and the axis of the holes, then each hole provides a sample of the projection (in the direction of the line) of the concentration function $f^{(0)}(\mathbf{x})$ in the plane Π. By rotating the collimator – detector system around the body in a way similar to that of the parallel-beam scanning in X-ray tomography, we obtain information which roughly consists in a set of projections of $f^{(0)}(\mathbf{x})$. These data can also be represented in the form of a sinogram. The basic reconstruction technique of TCT, i.e. the filtered backprojection which is discussed in Chapter 8, can be used in the reconstruction of SPECT images, even if, in modern scanners, the iterative methods discussed in Chapter 13 are frequently used. Indeed, these methods can be applied to models of image formation which can take into account several corrections. The principal ones are due to the following effects.

- *Collimator blur* – The holes of the collimator are not infinitely narrow so that photons moving in directions other than that perpendicular to the detector plane can also be detected (see Figure 6.10). In other words, each hole of the collimator collects photons coming from the radionuclides interior to a small cone around its axis. This effect causes a substantial loss of resolution as shown in Figure 6.11.

- *Attenuation and scatter* –The photons emitted by the radionuclides may be absorbed or scattered by the body tissues before reaching the detectors. Because of the attenuation produced by these effects, the radionuclides concentration is

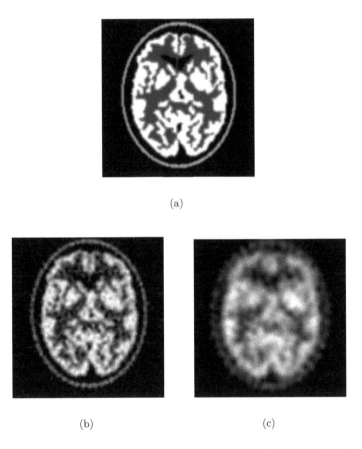

(a)

(b)   (c)

FIGURE 6.11 The reconstruction of a digital phantom (shown in (a)) as provided by the filtered back-projection in the case of TCT data (shown in (b)) and in the case of SPECT data (shown in (c)). In this example, the SPECT projections are computed by taking into account only the collimator blur and the Poisson noise.

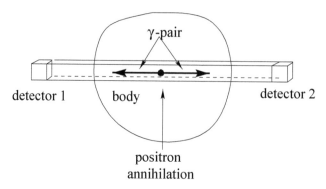

FIGURE 6.12   Scheme of the electronic collimation in PET.

underestimated, the effect being not uniform over the cross-section of the body (it is greater at the center than at the boundary). Moreover, due to scattering effects, a photon detected in a certain direction may correspond to a radionuclide which in not located in the acceptance cone of the hole.

- *Noise* – The amount of radiopharmaceutical administered to the patient as well as the acquisition time must be subjected to severe limitations. As a consequence, at each position in the detector plane the number of counts is relatively low, i.e. the data are affected by a rather large amount of noise which is accurately described by Poisson statistics. As a consequence a filtering of the projections before reconstruction or the use of statistical methods may be required.

The last point is very important. Indeed, nowadays, the image reconstruction methods in SPECT and PET scanners are iterative methods derived in the framework of models which take into account the Poisson statistics of the photon counting process. It is interesting to point out that these methods can also be applied to fluorescence microscopy and to astronomy; they will be discussed in Chapters 11 and 13.

In PET the collimation is obtained by the use of pairs of detectors in coincidence. This technique, which is also called *electronic collimation* is more accurate than the physical collimation (described above) and can be used for designing a detection system with a great efficiency. The procedure is illustrated in Figure 6.12: annihilation photons, emitted in a voxel intersected by a line joining the centers of the two detectors, are detected if their direction of emission is parallel, or close, to the previous line. Since the number of positrons (and therefore of γ-pairs) emitted in a given time and in a given volume is proportional to the concentration of the radionuclides, the number of detected photons is proportional to the total number of radionuclides interior to the line defined by the two detectors and shown in Figure 6.12, and therefore to the integral of their concentration over this line. Therefore the fundamental information we obtain is of the same nature as for X-ray tomography.

The description of the various geometries of the PET scanners is outside the scope of this book. We only observe that data are, in general, organized into a sinogram so that, in principle, the basic algorithm of tomography can be used also in this case. However, nowadays more accurate reconstruction methods are preferred [96].

Several corrections are applied also to PET data, some of them being due to effects which are also present in SPECT. However, the effects of collimator blur are not as important as in SPECT, thanks to the great precision and efficiency of the electronic collimation. Moreover the effect of attenuation can be corrected fairly easily.

The brief discussion of the principles of SPECT and PET imaging should make clear that the basic approximation, which consists in considering SPECT and PET data as line integrals of the radionuclides concentration, is rather rough. In order to improve the quality of the reconstructed images, an improved physical model of the acquisition process should be used. Such a model must take into account the features of the specific scanner which is considered. Therefore we can only give here a rather general description of these models.

In SPECT scanners but not in PET scanners, the geometry of the detection system is designed in order to define planar cross-sections of the body to be imaged; however, in all cases the reconstruction problem is always 3D both as a consequence of the scatter effect and also as a consequence of the collimation blur (especially in SPECT). Therefore, from the beginning, the region of interest is a certain domain $\mathcal{D}$ in 3D space. This domain is replaced by a fine grid consisting of N voxels, characterized by the index n. Moreover the detection system defines M detectors characterized by the index m.

The basic physical quantity describing the acquisition process is the probability that a decay occurring at voxel n is detected by the detector m. This probability is denoted by $P_{m,n}$. If we denote by $f_n^{(0)}$ the average number of decays occurring in the voxel n, and by $g_m^{(0)}$ the average number of γ-rays detected by the detector m, then the relationship between the quantities $f_n^{(0)}$ and $g_m^{(0)}$ is given by

$$g_m^{(0)} = \sum_{n=1}^{N} P_{m,n} f_n^{(0)} . \tag{6.15}$$

As we see the probabilities $P_{m,n}$ are the elements of a matrix which is called the *transition matrix*. Since not all decays occurring at voxel n are detected, these matrix elements must satisfy the conditions

$$\sum_{m=1}^{M} P_{m,n} < 1 \tag{6.16}$$

for $n = 1, 2, \ldots, N$.

The quantities $g_m^{(0)}$ are the noise-free data. Since $g_m^{(0)}$ is the average number (with respect to a large number of realizations) of the detector counts m, the measured number of counts g_m is the value of a Poisson random variable whose expectation value is $g_m^{(0)}$. The difference $g_m - g_m^{(0)}$ represents the so called Poisson noise. We come back to this point in Chapter 11. From this scheme it comes out that the key problem, in the modeling of a SPECT or PET scanner, is the determination or approximation of the matrix elements $P_{m,n}$. This computation must take into account the geometry of the detection system as well as the physical processes affecting the propagation of the photons through the tissues, namely attenuation and scattering. Since these effects depend on the density of the tissues, their estimation requires the knowledge of the linear attenuation function, i.e. the function imaged by the CT scanners. As already observed, one of the advantages of the PET scanners with respect to the SPECT scanners, is that they allow to estimate the corrections of the transition matrix due to attenuation by means of two separate measurements which precede the imaging scan. Anyway, modern PET scanners are usually coupled with a CT scanner, the so-called PET-CT machines. In such a way a co-registration of PET and CT images allows a great improvement of the anatomical localization of the main features (in particular tumors) of PET images.

Nowadays, one of the most important clinical use of SPECT and PET is oncology since, according to the tracer used, they allow to estimate and quantify, in a non-invasive way, numerous patho-physiological parameters. Other clinical applications are: cerebral or myocardial perfusion studies, epilepsy as well as others brain pathologies and trauma.

6.4   Diffusion MRI and spherical deconvolution

Magnetic Resonance Imaging (MRI) is a widely used modality able to image anatomy or physiological processes of the human body. It is not based on the use of X-rays or γ-rays, as in the previously discussed imaging modalities, but on the use of a strong magnetic field combined with magnetic field gradients and radio-waves.

The physical phenomenon used in MRI is *nuclear magnetic resonance*: when immersed in a strong magnetic field the magnetic nuclear spins of the atoms, which constitute the body, align with the field; this alignment can be perturbed significantly by an oscillating electromagnetic signal if its frequency is close to a specific frequency of the nuclei, the so-called Larmor frequency, which depends on the field and on specific properties of the nuclei; when the probing signal is removed the nuclei emit a radio-frequency signal which can be detected by a coil around the patient [57]. The idea of using this effect for the imaging of the human body was proposed by Paul Lauterbur [195] and further developed by Peter Mansfield [207]. Both were awarded the Nobel Prize for Physiology and Medicine in 2003.

MRI is able to image both the anatomy and the physiological processes of the human body. In general, the strong magnetic field and the radio-frequency are tailored to match the intrinsic frequency of the magnetic moment of the hydrogen nucleus, the proton, because of the abundance of water in the human body; therefore, MRI images provide information about the water content of the organs, tissues, etc. MRI is producing 3D images, in general displayed as planar sections; they look similar to CT images but their information content is different.

The understanding of image formation in MRI is much more complex than that of transmission or emission tomography and for this reason, is beyond the scope of this book. We only remark that MRI is a very flexible tool and can be used for obtaining different images of the same part of the body. This flexibility is due to the great variety of the so-called MRI *sequences* which can be designed. Roughly speaking, an MRI sequence is a particular setting of pulsed radio-frequency signals and pulsed field gradients, resulting in a particular image appearance. In any case they always contain, together with the Larmor frequency pulse, a pulse of field gradients to select the slice (plane) of the body to be imaged and two encoding pulses to differentiate signals with respect to their phase and frequency. Finally, a Fourier transform of the measured data is performed and its modulus is displayed as the MR image.

Thanks to its flexibility MRI can be used for very different tasks, among them, functional imaging. In particular, it can be used for mapping the diffusion of molecules, in particular water, in biological tissues. This MRI methodology is called *Diffusion Magnetic Resonance Imaging* (dMRI) and is relevant in many applications. For instance it is known that in brain ischemia water diffusion drops very early during an ischemic event [303]. For instance, in the first version of dMRI, a quantitative evaluation of the diffusion coefficient in the various voxels of the region of interest was achieved. In this way early diagnosis of ischemia is possible, allowing for a reduction of the damage produced by the stroke. This application also explains why dMRI is mainly used to investigate regions of the *White Matter* (WM) in the brain.

However, diffusion of water is anisotropic because it interacts with the structure of the surrounding tissues. In neuronal tissues, as WM, the tissue is organized in bundles – called *fibers* in the following – formed by axonal fibers running parallel to each other. Then water moves preferentially along the direction of a fiber and less in the orthogonal direction.

The first technique introduced for detecting this effect is *Diffusion Tensor Imaging* (DTI), proposed in the mid 80's [196]. It is able to detect the main direction of the fibers in each voxel of the observed region and in this way a so-called *fiber tractography* is possible, i.e. a 3D reconstruction technique to assess neural tracts. However, the important limit of DTI is that it is based on the assumption that only one fiber is crossing each voxel while it is known that about 90 % of the white-matter voxels contains fiber crossings. To this purpose the technique known as *High Angular Resolution Diffusion Imaging* (HARDI) was developed; it consists in sampling the diffusivity signal along a large number of directions uniformly distributed on the sphere. This makes possible the discrimination of multiple fiber populations within a single voxel [288] and, consequently, the reconstruction of fiber crossings in WM . This information is of major importance for improving tractographic techniques which must establish the 3D connections between the fibers of the different voxels.

A relationship between the detected signals and the fiber orientations within each voxel is required to reconstruct the number and the orientation of the fibers. Here we consider the one based on the *Gaussian mixture model* proposed in [288]. Let us consider, for instance, a certain volume of WM and a generic voxel in that volume and let us denote by $g(\mathbf{s})$ the signal arriving from that voxel in the direction $\mathbf{s}$ of the diffusion gradient. It can be written in the following form

$$g(\mathbf{s}) = g_0 \sum_{i=1}^{M} f_i \exp\left[-b\,\left(\beta_i + \alpha_i(\mathbf{s}\cdot\mathbf{s}_i)^2\right)\right] \tag{6.17}$$

where g_0 denotes the non-diffusion weighted MRI signal; $\mathbf{s}$ and $\mathbf{s}_i$ are unit vectors in the directions of the diffusion gradient and of the i-th fiber, respectively, and $\mathbf{s}\cdot\mathbf{s}_i$ denotes the usual scalar product; f_i is the *partial volume* of the i-th fiber; b is the diffusion weighting factor depending on the strength and duration of the diffusion gradient; finally, α_i and β_i are parameters related to the diffusion tensor associated with the i-th fiber. A typical value of b used in HARDI acquisition is $b = 3 \times 10^3$ s/mm^2 while typical values of the other parameters are $\alpha = 1.5 \times 10^{-3}$ mm^2/s and $\beta = 0.2 \times 10^{-3}$ mm^2/s, consistent with physiological values [232].

The vector $\mathbf{f}$ with components f_i is usually called *fiber orientation function* (FOF). Moreover, the partial volumes satisfy the constraints

$$f_i \geq 0\ , \quad \sum_{i=1}^{M} f_i = 1\ . \tag{6.18}$$

The basic problem is the estimation of the FOF from given (noisy) signal values. We introduce the FOF density which, in the notation used for the exact solution, can be written as follows

$$f^{(0)}(\mathbf{s}') = \sum_{i=1}^{M} f_i\,\delta(\mathbf{s}' - \mathbf{s}_i)\ , \tag{6.19}$$

where δ denotes the Dirac delta measure on the sphere. Moreover, we assume that all the fibers have the same diffusivity characteristics α, β; then equation (6.17) takes the following form (we use the notation for noise-free data)

$$g^{(0)}(\mathbf{s}) = \int_{S^2} H\,(\mathbf{s}\cdot\mathbf{s}')\,f^{(0)}(\mathbf{s}')\,\mu(d\mathbf{s}')\ , \tag{6.20}$$

with $H(t) = g_0 \exp[-b\,(\beta + \alpha\,t^2)]$ and $\mu(d\mathbf{s}')$ the standard measure on the unit sphere S^2, i.e. $\mu(d\mathbf{s}) = \sin\theta\,d\theta\,d\phi$. The r.h.s. of this equation is a particular case of a *spherical convolution* with a specific kernel H. It defines a linear operator in $L^2(S^2)$ commuting with the group of rotations; if the kernel is bounded, the operator is an integral operator of the Hilbert-Schmidt class, defined in Section 7.4; it is also self-adjoint if the kernel is real-valued. Both conditions are satisfied in our case. The interesting point is that its eigenfunctions are the well-known *spherical harmonics*. They are defined by

$$Y_{l,m}(\mathbf{s}) = \sqrt{\frac{2l+1}{4\pi}\frac{(l-m)!}{(l+m)!}}\ P_l^m(\cos\theta)\,e^{im\phi}\ ; \quad l = 0,1,2,\dots\ ; \quad m = -l,..,0,..,l\ ; \tag{6.21}$$

where the P_l^m are the *associated Legendre polynomials*, and they form an orthonormal basis in $L^2(S^2)$

$$\int_{S^2} Y_{l,m}(\mathbf{s})\, Y_{l',m'}^*(\mathbf{s})\, \mu(d\mathbf{s}) = \delta_{l,l'}\, \delta_{m,m'} \, . \tag{6.22}$$

Therefore any function $f \in L^2(S^2)$ can be represented as follows

$$f^{(0)}(\mathbf{s}) = \sum_{l=0}^{\infty} \sum_{m=-l}^{l} f_{l,m}^{(0)}\, Y_{l,m}(\mathbf{s}) \, , \quad f_{l,m}^{(0)} = \int_{S^2} f^{(0)}(\mathbf{s})\, Y_{l,m}^*(\mathbf{s})\, \mu(d\mathbf{s}) \, . \tag{6.23}$$

Now, let us represent the function H as a series of Legendre polynomials ($\cos \xi = \mathbf{s} \cdot \mathbf{s}'$)

$$H(\cos \xi) = \sum_{l=0}^{\infty} H_l\, P_l(\cos \xi) \, , \quad H_l = \frac{2l+1}{2} \int_{-\pi}^{\pi} H(\cos \xi) P_l(\cos \xi)\, \sin \xi\, d\xi \, . \tag{6.24}$$

Then, if A is the linear operator defined in equation (6.20), from the *addition theorem for spherical harmonics*

$$P_l(\mathbf{s} \cdot \mathbf{s}') = \frac{4\pi}{2l+1} \sum_{m=-l}^{l} Y_{l,m}(\mathbf{s})\, Y_{l,m}^*(\mathbf{s}') \, , \tag{6.25}$$

and from the expansion of H in Legendre polynomials, we obtain

$$(AY_{l,m})(\mathbf{s}) = \frac{4\pi}{2l+1} H_l\, Y_{l,m}(\mathbf{s}) \, , \tag{6.26}$$

so that the eigenvalues of the operator A are given by

$$\lambda_l = \frac{4\pi}{2l+1} H_l \, , \tag{6.27}$$

with multiplicity $2l+1$ since the eigenfunctions associated with λ_l are the spherical harmonics $Y_{l,m}$, $m = -l, ..., 0, ...l$.

In the application to Diffusion-MRI the function H is given by

$$H(\cos \xi) = C\, e^{-ba\cos^2 \xi} \, , \quad C = e^{-b\beta} \, ; \tag{6.28}$$

since H is an even function of $\cos \xi$, its coefficients of odd order are zero as well as the eigenvalues of the corresponding convolution operator. It follows that the null space of the operator is the subspace of the odd functions.

The coefficients $C^{-1}H_l$ of the kernel $C^{-1}H$ uniquely depend on the parameter $a = b\alpha$ and are given by Anderson [7] up to $l = 8$. If we take into account their multiplicity, we can compute 45 eigenvalues of the corresponding operator as given in equation (6.27); they are alternately positive and negative, the first one being positive.

In order to give a flavor of their behavior, we compute these eigenvalues for the values of the parameters $\{b, \alpha, \beta\}$ given after equation (6.17); they are obtained from the coefficients given by Anderson multiplied by $4\pi \exp(-b\beta)/(2l+1)$. We plot their absolute values in Figure 6.13, each one being represented as many times as its multiplicity, generating the flat regions with increasing length appearing in the figure. As it can be expected, the eigenvalues are decreasing for increasing values of l but their decrease is not dramatic. If we use the previous 45 eigenvalues for an approximate solution of the inversion problem, the condition number, i.e. the ratio between the first and last eigenvalue (see Section 8.1, equation (8.22)), is about 31, hence not very large. We expect that, in general, the problem is not extremely ill-posed, even if, as far as we know, the asymptotic behavior for large values of l is not available.

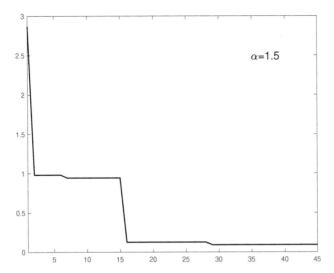

FIGURE 6.13 Plot of the absolute values of the first five non-zero eigenvalues of the kernel H in the case of the following values of the parameters: $b = 3 \times 10^3$ s/mm^2, $\alpha = 1.5 \times 10^{-3}$ mm^2/s and $\beta = 0.2 \times 10^{-3}$ mm^2/s. They are plotted as many times as their multiplicity $2l + 1$ for $l = 0, 2, 4, 6, 8$, generating the flat regions of increasing length which appear in the plot.

6.5 Inverse diffraction and inverse source problems

Inverse diffraction and *inverse source problems* are sometimes considered as intermediate steps in the solution of inverse scattering problems. However, they may also have some direct application. For instance, as we mentioned in Section 2.7, inverse diffraction is the basic problem of acoustic holography. In that section only the case of planar surfaces was considered. Here we shortly discuss the more general case of arbitrary surfaces. As concerns inverse source problem, an example is the problem of determining the charge-current distribution of a radiating antenna from the knowledge of its radiation pattern. These problems have been investigated both in the scalar case (acoustic waves) and in the vector case (electromagnetic waves). Here we only consider here the simple case of scalar waves.

As a general comment we observe that, while the solution of an inverse diffraction problem is unique, inverse source problems are always affected by a rather serious non-uniqueness. This property is related to the existence of the so-called *non-radiating sources*, i.e. to the existence of sources (for instance an oscillating body or an oscillating charge-current distribution) which do not produce any radiation field. A survey of this subject can be found in [18].

Before stating the two problems, we describe the physical situation we are considering. We assume that a scalar field is generated by oscillating sources contained in a bounded domain $\mathcal{D}$, with a regular boundary Σ. If the sources oscillate with a frequency ν, then the field is monochromatic and can be represented by

$$U(\mathbf{x}, t) = u(\mathbf{x}) \, e^{-i2\pi\nu t} \tag{6.29}$$

where $\mathbf{x} = \{x_1, x_2, x_3\}$. The free propagation of the field outside Σ is described by the Helmholtz equation

$$\Delta u + k^2 u = 0 \tag{6.30}$$

where $k = 2\pi\nu/c$ is the wavenumber of the monochromatic field, c being the velocity of the propagating waves. To the wavenumber k one can also associate the wavelength λ, which is given by

$$\lambda = \frac{2\pi}{k} = \frac{c}{\nu} \tag{6.31}$$

and is a characteristic length of the problem under consideration.

Moreover, the field amplitude $u(\mathbf{x})$ satisfies the *Sommerfeld radiation condition* at infinity

$$\lim_{r \to \infty} \left[r \left(\frac{\partial u}{\partial r} - ik \right) \right] = 0 \ , \quad r = |\mathbf{x}| \tag{6.32}$$

which expresses the fact that we are considering an outgoing wave, i.e. a wave propagating from the sources, and not an ingoing wave, i.e. a wave propagating toward the sources.

(A) Inverse diffraction

Consider two closed surfaces, Σ_1 and Σ_2, both surrounding the domain $\mathcal{D}$ of the sources, such that Σ_1 is interior to Σ_2 (see Figure 6.14). Then the problem of *inverse diffraction* associated with the surfaces Σ_1, Σ_2 is the following: given the values of the propagating field $u(\mathbf{x})$ on the surface Σ_2, estimate its values on the surface Σ_1. The solution of this problem provides a map of the field over Σ_1; if Σ_1 is close to the domain $\mathcal{D}$ of the sources, this map provides information on the internal sources.

Before giving the mathematical formulation of the inverse problem, we first solve the direct one: determine the field amplitude $u(\mathbf{x})$ which satisfies Helmholtz equation in the region exterior to Σ_1, takes prescribed values, let us say $f^{(0)}(\mathbf{x})$, on Σ_1

$$u(\mathbf{x})|_{\Sigma_1} = f^{(0)}(\mathbf{x}) \tag{6.33}$$

and satisfies the Sommerfeld radiation condition at infinity. If $G(k; \mathbf{x}, \mathbf{x}')$ is the Green function of this problem, then $u(\mathbf{x})$ can be expressed as follows

$$u(\mathbf{x}) = \int_{\Sigma_1} G(k; \mathbf{x}, \mathbf{x}') \, f^{(0)}(\mathbf{x}') \, \mu(d\mathbf{x}') \tag{6.34}$$

where $\mu(d\mathbf{x}')$ is the measure on the surface Σ_1. In simple cases, the Green function $G(k; \mathbf{x}, \mathbf{x}')$ can be determined analytically; otherwise it must be computed numerically.

Coming back now to the inverse problem, let $g^{(0)}(\mathbf{x})$ be the value of $u(\mathbf{x})$ on Σ_2

$$u(\mathbf{x})|_{\Sigma_2} = g^{(0)}(\mathbf{x}) \ . \tag{6.35}$$

Then the problem consists in looking for boundary values $f^{(0)}(\mathbf{x})$ such that $u(\mathbf{x})$, as given by equation (6.34), satisfies condition (6.35). We obtain

$$g^{(0)}(\mathbf{x}) = \int_{\Sigma_1} G(k; \mathbf{x}, \mathbf{x}') \, f^{(0)}(\mathbf{x}') \, \mu(d\mathbf{x}') \ , \quad \mathbf{x} \in \Sigma_2 \ . \tag{6.36}$$

Therefore, the problem is reduced to the solution of a first-kind Fredholm integral equation. In a practical situation, we do not have the exact values $g^{(0)}(\mathbf{x})$ of the field on Σ_2 but only measured values $g(\mathbf{x})$, affected by noise or experimental errors, i.e. $g(\mathbf{x}) = g^{(0)}(\mathbf{x}) + w(\mathbf{x})$, where $w(\mathbf{x})$ is the term describing the noise.

The operator (6.36) belongs to the class of integral operators discussed in Section 7.4 and therefore, if the kernel is square integrable, it has a singular value decomposition which can be used in the investigation of the problem. The computation of this singular system, however, may be difficult if the surfaces Σ_1, Σ_2 are irregular.

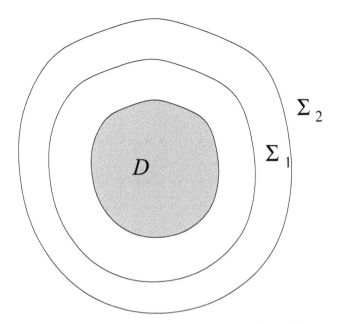

FIGURE 6.14 Geometry of the problem of inverse diffraction. The dashed area corresponds to the domain $\mathcal{D}$ containing the sources of the field. Σ_2 is the surface where data are recorded.

The case of spherical surfaces is simple because spherical harmonic expansions can be used for solving the problem. Let Σ_1, Σ_2 be spheres with center at the origin and let a_1, a_2 be their radii, $a_1 < a_2$. Moreover, let $r, \mathbf{s}$ be the variables

$$r = |\mathbf{x}| \ , \quad \mathbf{s} = \frac{\mathbf{x}}{r} \ , \tag{6.37}$$

$\mathbf{s}$ defining a point on the sphere S^2 in the 3D space. The boundary values of $u(\mathbf{x})$ on Σ_1 define a function on S^2, denoted by $f^{(0)}(\mathbf{s}) = u(a_1, \mathbf{s})$, which can be represented by means of its expansion in terms of spherical harmonics $Y_{l,m}$, equation (6.23). Then, in terms of the spherical Hankel functions of the first kind,

$$h_l^{(1)}(r) = \sqrt{\frac{\pi}{2r}} \, H_{l+1/2}^{(1)}(r) \ , \tag{6.38}$$

the solution of the direct problem is given by

$$u(r, \mathbf{s}) = \sum_{l=0}^{\infty} \sum_{m=-l}^{l} f_{l,m}^{(0)} \frac{h_l^{(1)}(kr)}{h_l^{(1)}(ka_1)} \, Y_{l,m}(\mathbf{s}) \ . \tag{6.39}$$

Consider now the inverse problem. The exact data define a function on S^2, denoted by $g^{(0)}(\mathbf{s}) = u(a_2, \mathbf{s})$. Then, by taking into account equation (6.39), the relationship between $g^{(0)}(\boldsymbol{\theta})$ and $f^{(0)}(\boldsymbol{\theta})$ can be written as follows

$$g^{(0)}(\mathbf{s}) = \int_{S^2} G(\mathbf{s}, \mathbf{s}') \, f^{(0)}(\mathbf{s}') \, \mu(d\mathbf{s}') \tag{6.40}$$

where

$$G(\mathbf{s}, \mathbf{s}') = \sum_{l=0}^{\infty} \sum_{m=-l}^{l} \lambda_l \, Y_{l,m}(\mathbf{s}) \, Y_{l,m}^*(\mathbf{s}') \tag{6.41}$$

and

$$\lambda_l = \frac{h_l^{(1)}(ka_2)}{h_l^{(1)}(ka_1)} \; . \tag{6.42}$$

Equation (6.40) is precisely equation (6.36) in the particular case of spherical surfaces. The integral operator defined by equations (6.40-6.42) has some simple and interesting properties which can be summarized as follows.

- The numbers λ_l are the eigenvalues of the integral operator with the kernel (6.41).
- Each λ_l has multiplicity $2l + 1$.
- All eigenvalues are different from zero so that, thanks to the completeness of the spherical harmonics, no function $f^{(0)}(\mathbf{s}) \neq 0$ exists such that the corresponding $g^{(0)}(\mathbf{s})$ is exactly zero.

The last property implies that the null space of the integral operator (6.40) is trivial and therefore the solution of the inverse diffraction problem is unique. However, from the asymptotic behavior of the Hankel functions for large l we get

$$\lambda_l \simeq \exp\left[-(l + 1) \ln\left(\frac{a_2}{a_1}\right)\right] \tag{6.43}$$

so that the eigenvalues tend to zero, when $l \to \infty$, with an exponential decay. This behavior implies that the problem is ill posed and this point is discussed in Chapter 8. We also remark that the decay of the eigenvalues increases when the distance between the two spheres increases; therefore also the ill-posedness of the problem increases.

The previous analysis may be appropriate in the case of *near-field data*, i.e. when the difference $a_2 - a_1$ is not much larger than the wavelength λ. When this condition is not satisfied, then we are in the case of *far-field data* and one can consider the asymptotic limit $a_2 \to \infty$.

From the asymptotic behavior of the spherical Hankel functions for $r \to \infty$

$$h_l^{(1)}(r) \simeq (-i)^{l+1} \frac{e^{ikr}}{r} \tag{6.44}$$

we obtain the following asymptotic behavior of the field amplitude (6.39)

$$u(r, \mathbf{s}) \simeq \frac{e^{ikr}}{r} u_\infty(\mathbf{s}) \tag{6.45}$$

where

$$u_\infty(\mathbf{s}) = \sum_{l=0}^{\infty} \sum_{m=-l}^{l} (-i)^{l+1} \frac{f_{l,m}^{(0)}}{h_l^{(1)}(ka_1)} Y_{l,m}(\mathbf{s}) \; . \tag{6.46}$$

The function $u_\infty(\mathbf{s})$ is usually called the *radiation pattern*. In the inverse-diffraction problem with far-field data, the exact data are precisely the values of the radiation pattern, i.e.

$$g^{(0)}(\mathbf{s}) = u_\infty(\mathbf{s}) \; . \tag{6.47}$$

Then, from equation (6.46) we can derive an equation analogous to (6.40)

$$g^{(0)}(\mathbf{s}) = \int_{S^2} G_\infty(\mathbf{s}, \mathbf{s}') \, f^{(0)}(\mathbf{s}') \, \mu(ds) \tag{6.48}$$

where

$$G_\infty(\mathbf{s}, \mathbf{s}') = \sum_{l=0}^{\infty} \sum_{m=-l}^{l} \lambda_l^{(\infty)} Y_{l,m}(\mathbf{s}) \, Y_{l,m}^*(\mathbf{s}') \tag{6.49}$$

and

$$\lambda_l^{(\infty)} = \frac{(-i)^{l+1}}{h_l^{(1)}(ka_1)} \, . \tag{6.50}$$

Also in the case of the integral operator (6.48) the null space does not contain functions $f^{(0)} \neq 0$, so that the solution of the problem is unique. The eigenvalues $\lambda_l^{(\infty)}$, however, tend to zero more rapidly than the eigenvalues λ_l of the problem with near-field data. Their asymptotic behavior, indeed, is given by $|\lambda_l^{(\infty)}| \simeq \exp\left[-l \ln\left(2l/ka_1\right)\right]$. Since the components of $g^{(0)}(\mathbf{s})$ with respect to the spherical harmonics are given by $g_{m,l}^{(0)} = \lambda_l \, f_{m,l}^{(0)}$ in the near-field case and by $g_{m,l}^{(0)} = \lambda_l^{(\infty)} f_{m,l}^{(0)}$ in the far-field case, it follows that, for a given $f^{(0)}(\mathbf{s})$, the number of components significantly different from zero for far-field data is smaller than that number for near-field data. In other words, and in agreement with physical intuition, far-field data contain less information than near-field data because, in the presence of noise, they allow the recovery of a smaller number of components of $f^{(0)}(\mathbf{s})$.

(B) *Inverse source problem*

Assume that the sources of the field, located in the domain $\mathcal{D}$, are described by a function which is again denoted by $f^{(0)}(\mathbf{x})$. The precise meaning of this function is that it represents the inhomogeneous term in the Helmholtz equation: in the region $\mathcal{D}$ the field amplitude $u(\mathbf{x})$ is a solution of the following inhomogeneous equation

$$\Delta u + k^2 u = -4\pi f^{(0)} \tag{6.51}$$

while, outside $\mathcal{D}$, it is a solution of the homogeneous equation (6.30) and satisfies the Sommerfeld radiation condition, equation (6.32).

With reference to Figure 6.14, the *inverse source problem* can be formulated as the problem of estimating $f^{(0)}(\mathbf{x})$ in $\mathcal{D}$ from given values of the radiation field $u(\mathbf{x})$ on a closed surface Σ' (i.e. Σ_1 or Σ_2) such that $\mathcal{D}$ is interior to Σ'.

It is clear that this problem is highly underdetermined because we wish to recover a function of three variables from the knowledge of a function of two variables, i.e. the values of the field on a 2D surface. This underdetermination, however, cannot be removed by taking the values of the field in a volume because, as shown before in the case of a spherical surface (but the result is also true for more general surfaces), the knowledge of the field on a closed surface determine uniquely the field everywhere outside $\mathcal{D}$. This intrinsic underdetermination of the problem is due to the existence of the so-called non-radiating sources which are discussed in the following.

As usual, in order to derive the mathematical formulation of the inverse problem, we first state the direct problem, i.e. the problem of determining for a given source function $f^{(0)}(\mathbf{x})$ a function u satisfying equation (6.51) inside $\mathcal{D}$, equation (6.30) outside $\mathcal{D}$ as well as condition (6.32). This solution can be obtained by introducing the Green function

$$G_0(r) = \frac{e^{ikr}}{r} \tag{6.52}$$

and is given by

$$u(\mathbf{x}) = \int_{\mathcal{D}} G_0(|\mathbf{x} - \mathbf{x}'|) \, f^{(0)}(\mathbf{x}') \, d\mathbf{x}' \, . \tag{6.53}$$

If we assume that the values of $u(\mathbf{x})$ are given on the surface Σ', i.e.

$$u(\mathbf{x})|_{\Sigma'} = g^{(0)}(\mathbf{x}) \tag{6.54}$$

then the inverse source problem consists in looking for a function $f^{(0)}(\mathbf{x})$ such that $u(\mathbf{x})$, as given by equation (6.53), satisfies condition (6.54). In this way we obtain the following integral equation for $f^{(0)}(\mathbf{x})$

$$g^{(0)}(\mathbf{x}) = \int_{\mathcal{D}} G_0(|\mathbf{x} - \mathbf{x}'|)\, f^{(0)}(\mathbf{x}')\, d\mathbf{x}' \ . \tag{6.55}$$

The null space of this integral operator, i.e. the subspace of the non-radiating sources, consists of the solutions $f^{(0)}(\mathbf{x})$ of the following homogeneous equation for any $\mathbf{x}$ outside $\mathcal{D}$

$$\int_{\mathcal{D}} G_0(|\mathbf{x} - \mathbf{x}'|) f^{(0)}(\mathbf{x}')\, d\mathbf{x}' = 0 \ . \tag{6.56}$$

From equations (6.52) and (6.53) it follows that, for large values of $r = |\mathbf{x}|$, $u(\mathbf{x})$ has the following asymptotic behavior

$$u(\mathbf{x}) \simeq \frac{e^{ikr}}{r}\, u_\infty(\mathbf{s}) \tag{6.57}$$

with $\mathbf{s} = \mathbf{x}/r$ and

$$u_\infty(\mathbf{s}) = \int_{\mathcal{D}} e^{-ik(\mathbf{s}\cdot\mathbf{x}')} f^{(0)}(\mathbf{x}')\, d\mathbf{x}' \ . \tag{6.58}$$

This function, which is the radiation pattern of any solution of equation (6.53), is also the radiation pattern of the solutions of equation (6.56) and therefore is zero in the case of a non-radiating source. Conversely, if the radiation pattern is zero, then, thanks to the uniqueness of the inverse diffraction problem, it is the radiation pattern of a field which is zero everywhere outside $\mathcal{D}$. Hence a source is non-radiating if and only if its radiation pattern $u_\infty(\mathbf{s})$ is zero, i.e. if and only if its Fourier transform $\hat{f}^{(0)}(\boldsymbol{\omega})$ vanishes on the sphere with center the origin and radius k: $\hat{f}^{(0)}(k\mathbf{s}) = 0$. This sphere is the so-called *Ewald sphere*. In other words, *the set of the non-radiating sources $f^{(0)}(\mathbf{x})$ with support in $\mathcal{D}$ is the set of all functions whose Fourier transform vanishes on the Ewald sphere.*

We conclude by formulating the inverse source problem from far-field data in terms of a first-kind Fredholm integral equation. If $g^{(0)}(\mathbf{s})$ are the given values of the radiation pattern, then from equation (6.58) we see that $f^{(0)}(\mathbf{x})$ can be obtained by solving the integral equation

$$g^{(0)}(\mathbf{s}) = \int_{\mathcal{D}} K(\mathbf{s}, \mathbf{x}')\, f^{(0)}(\mathbf{x}')\, d\mathbf{x}' \tag{6.59}$$

where

$$K(\mathbf{s}, \mathbf{x}') = e^{-ik(\mathbf{s}\cdot\mathbf{x}')} \tag{6.60}$$

Let us assume that $\mathcal{D}$ is the sphere of radius a. If we use the well-known expansion of a plane wave into spherical harmonics

$$e^{-ikr'(\mathbf{s}\cdot\mathbf{s}')} = 4\pi \sum_{l=0}^{\infty} \sum_{m=-l}^{l} (-i)^l j_l(kr')\, Y_{l,m}(\mathbf{s})\, Y_{l,m}^*(\mathbf{s}') \ , \tag{6.61}$$

where the $j_l(x)$ are the spherical Bessel functions, and if we introduce the functions

$$v_{l,m}(\mathbf{x}') = \frac{4\pi}{\sigma_l}\, i^l\, j_l(kr')\, Y_{l,m}(\mathbf{s}') \ , \quad u_{l,m}(\mathbf{s}) = Y_{l,m}(\mathbf{s}) \ , \tag{6.62}$$

where

$$\sigma_l = 4\pi \left(\int_0^a r^2\, j_l^2(kr)\, dr \right)^{1/2} , \tag{6.63}$$

we obtain the following representation of the integral kernel (6.60)

$$K(\mathbf{s}, \mathbf{x}') = \sum_{l=0}^{\infty} \sum_{m=-l}^{l} \sigma_l \, u_{l,m}(\mathbf{s}) \, v_{l,m}^*(\mathbf{x}') \, . \tag{6.64}$$

This representation is the singular value decomposition of the integral operator (6.59), as follows from the results of Chapter 7. We observe that, from the asymptotic behavior of the spherical Bessel functions for large l, it follows that $\sigma_l \to 0$ when $l \to \infty$, a result which implies the ill-posedness of the inverse source problem.

Moreover, we remark that the solution obtainable from the previous representation of the kernel, for a given diffraction pattern, is orthogonal to the subspace of the non-radiating sources and therefore it is not unique because other solutions of the same problem can be obtained by the addition of an arbitrary non-radiating source.

6.6 Linearized inverse scattering problems

Inverse scattering problems are very important in many different domains, such as quantum mechanics [61], acoustics and electromagnetism [73]. These problems arise from the attempts at obtaining information on a body, the scatterer, by illuminating this body with waves of various wavelengths and directions and by recording the waves scattered by the body.

We assume that the scatterer is contained in a bounded domain $\mathcal{D}$ of the 3D space. Then in general two different kinds of inverse scattering problems are considered: the inverse medium problem and the inverse obstacle problem.

In the inverse medium problem the scatterer is an inhomogeneous medium characterized by one or more physical quantities, varying in a continuous manner, and the inverse problem consists in estimating these parameters from scattering data.

In the inverse obstacle problem, the scatterer is a homogeneous body and the problem is to estimate the shape of the body from scattering data and given boundary conditions on the surface of the body.

For the sake of simplicity we consider again the scalar case. Then, outside the bounded domain $\mathcal{D}$ containing the scatterer, the field amplitude $u(\mathbf{x})$ satisfies the Helmholtz equation (6.30). As concerns the boundary condition at infinity, we consider the most frequent case where the body is illuminated by a plane wave propagating in the direction $\mathbf{s}_0$. Then $u(\mathbf{x})$ can be written in the following form

$$u(\mathbf{x}) = e^{ik\mathbf{s}_0 \cdot \mathbf{x}} + u_{sc}(\mathbf{x}) \, . \tag{6.65}$$

The scattered wave $u_{sc}(\mathbf{x})$ is also a solution of the Helmholtz equation outside $\mathcal{D}$, and satisfies Sommerfeld's radiation condition (6.32) at infinity. As follows from the analysis of Section 6.5, $u_{sc}(\mathbf{x})$ behaves as an outgoing spherical wave at infinity

$$u_{sc}(\mathbf{x}) \simeq u_\infty(k; \mathbf{s}_0, \mathbf{s}) \, \frac{e^{ikr}}{r} \tag{6.66}$$

where $\mathbf{s} = \mathbf{x}/r$. The function $u_\infty(k; \mathbf{s}_0, \mathbf{s})$ is called the *scattering amplitude* and is the quantity which is measured in the case of far-field scattering data. It is a function of five variables and therefore a complete determination of this function provides, in general, redundant data. Thanks to this redundancy many different experimental situations can be considered. We only mention a few of them: *backward scattering*, which corresponds to measuring $u_\infty(k; \mathbf{s}_0, \mathbf{s})$ in the case $\mathbf{s} = -\mathbf{s}_0$, for various values of k and $\mathbf{s}$; *forward scattering*, which corresponds to measuring $u_\infty(k; \mathbf{s}_0, \mathbf{s})$ in the case $\mathbf{s} = \mathbf{s}_0$, for various values of k and $\mathbf{s}$; *fixed-frequency scattering*, which corresponds to measuring $u_\infty(k; \mathbf{s}_0, \mathbf{s})$ for a fixed value of k and various values of $\mathbf{s}_0$ and $\mathbf{s}$, etc. In some cases, problems with near-field data are also considered.

Inverse scattering problems are nonlinear and ill-posed and therefore are difficult problems both from the mathematical and from the computational point of view. A wide literature exists on these problems which are outside the scope of this book. However, under some circumstances, it is possible to introduce physical approximations which allow for a linearization of the nonlinear problem. A well-known case is that of a weak scatterer since, in this case, the Born approximation can be used. Another kind of approximation also leading to a linear problem, is the *Rytov approximation*, which is valid in the case of a slowly varying scatterer, i.e. in the case where the fluctuation length of the properties of the scatterer is large compared to the wavelength $\lambda = 2\pi/k$ of the incident radiation. These approximations apply to the inverse medium problem. An approximation leading to the linearization of the inverse obstacle problem is the so-called *physical-optics approximation*.

(A) *Born approximation*

We consider the inverse medium problem in the case of a semitransparent body characterized by a refraction index $n(\mathbf{x})$. If we introduce the function $f^{(0)}(\mathbf{x}) = 1 - n^2(\mathbf{x})$, then the total field $u(\mathbf{x})$ (incident plus scattered) is a solution, in $\mathcal{D}$, of the equation

$$\Delta u + k^2 u = k^2 f^{(0)} u . \tag{6.67}$$

Since the plane wave is a solution of the Helmholtz equation, from equation (6.65) we see that the scattered wave $u_{sc}(\mathbf{x})$ is a solution in $\mathcal{D}$ of the equation

$$\Delta u_{sc} + k^2 u_{sc} = k^2 f^{(0)} u \tag{6.68}$$

while outside $\mathcal{D}$ it satisfies the Helmholtz equation and the Sommerfeld radiation condition at infinity. Then, from the solution (6.53) of the inhomogeneous equation (6.51), we obtain the following representation of the scattered wave

$$u_{sc}(\mathbf{x}) = -\frac{1}{4\pi} \int_{\mathcal{D}} G_0(|\mathbf{x} - \mathbf{x}'|) \, f^{(0)}(\mathbf{x}') \, u(\mathbf{x}') \, d\mathbf{x}' , \tag{6.69}$$

the Green function $G_0(r)$ being defined in equation (6.52). Finally, from the asymptotic behavior (6.57), by identifying the scattering amplitude (6.66) with the radiation pattern (6.58), we obtain the following representation of the scattering amplitude

$$u_\infty(k; \mathbf{s}_0, \mathbf{s}) = -\frac{k^2}{4\pi} \int_{\mathcal{D}} e^{-ik\mathbf{s}\cdot\mathbf{x}'} f^{(0)}(\mathbf{x}') \, u(\mathbf{x}') \, d\mathbf{x}' \tag{6.70}$$

where $u(\mathbf{x})$ is the total field.

The Born approximation, which applies to so-called weak scatterers, consists in replacing the total field $u(\mathbf{x})$ in equation (6.70) by the incident field (the plane wave). The result is

$$u_\infty(k; \mathbf{s}_0, \mathbf{s}) = -\frac{k^2}{4\pi} \int_{\mathcal{D}} e^{-ik(\mathbf{s}-\mathbf{s}_0)\cdot\mathbf{x}'} f^{(0)}(\mathbf{x}') \, d\mathbf{x}' . \tag{6.71}$$

This equation shows that, if we measure the scattering amplitude for all values of $\mathbf{s}$, with k and $\mathbf{s}_0$ fixed, then we obtain the Fourier transform of $f^{(0)}(\mathbf{x})$ on the surface of the Ewald sphere with center $\mathbf{s}_0$ and radius k. It follows that, from the mathematical point of view, this problem is analogous to the inverse source problem discussed in the previous section. However, by varying the direction of incidence $\mathbf{s}_0$, a (theoretically infinite) number of experiments would allow to determine $\hat{f}^{(0)}(\boldsymbol{\omega})$ within the sphere with center at the origin and radius $2k$ (the limiting Ewald sphere). In this case, since $\hat{f}^{(0)}(\boldsymbol{\omega})$ is an analytic function thanks to the boundedness of the support of $f^{(0)}(\mathbf{x})$, the uniqueness of the solution of the inverse problem is ensured. This inverse problem is an example of the general problem of out-of-band extrapolation, which is investigated in Section 8.5. The scattering data, indeed, provide a band-limited approximation of the unknown object $f^{(0)}(\mathbf{x})$.

The case of data collected over planes not intersecting $\mathcal{D}$ is also considered in the framework of Born approximation [309], and the same conclusions as outlined above are reached. Moreover, the Rytov approximation leads to the same mathematical problem as the Born approximation [101].

(B) *Physical-optics approximation*

We consider the inverse obstacle problem in the case of a perfectly reflecting body (a sound-soft obstacle in acoustic or a perfectly conducting obstacle in electromagnetism). In such a case the total field u, as given by equation (6.65), satisfies the boundary condition

$$u(\mathbf{x})|_{\Sigma} = 0 \tag{6.72}$$

where Σ is the unknown surface of the scatterer, which is assumed to be a convex body. From equation (6.65) and condition (6.72) we obtain that the scattered wave satisfies the following boundary condition on Σ

$$u_{sc}(\mathbf{x})|_{\Sigma} = -e^{ik\mathbf{s}_0 \cdot \mathbf{x}}|_{\Sigma} . \tag{6.73}$$

Moreover, $u_{sc}(\mathbf{x})$ is a solution of the Helmholtz equation outside the domain $\mathcal{D}$ occupied by the body, and satisfies Sommerfeld's radiation condition at infinity. Then, by means of the second Green theorem one can easily derive the following representation of the scattered wave

$$u_{sc}(\mathbf{x}) = -\frac{1}{4\pi} \int_{\Sigma} G_0(|\mathbf{x} - \mathbf{x}'|) \frac{\partial}{\partial \nu(\mathbf{x}')} u(\mathbf{x}') \, \mu(d\mathbf{x}') \tag{6.74}$$

where $\nu(\mathbf{x}')$ is the unit vector orthogonal to Σ at the point $\mathbf{x}'$ and directed towards the exterior of $\mathcal{D}$. The Green function $G_0(r)$ is defined in equation (6.52). Equation (6.74) provides the foundation of the *Huygens principle*, because it represents the scattered wave as a superposition of spherical waves emitted by a double layer located at the surface of the body.

From equation (6.74) it follows that the behavior at infinity of $u_{sc}(\mathbf{x})$ is given by equation (6.66) with a scattering amplitude now given by

$$u_{\infty}(k; \mathbf{s}_0, \mathbf{s}) = -\frac{1}{4\pi} \int_{\Sigma} e^{-ik\mathbf{s} \cdot \mathbf{x}'} \frac{\partial}{\partial \nu(\mathbf{x}')} u(\mathbf{x}') \, \mu(d\mathbf{x}') . \tag{6.75}$$

The *physical-optics approximation* provides a simple expression of this scattering amplitude if the wavelength $\lambda = 2\pi/k$ is much smaller than the diameter of the convex body $\mathcal{D}$.

To this purpose, for a given $\mathbf{s}_0$, we introduce the illumination region $\Sigma_+(\mathbf{s}_0)$ and the shadow region $\Sigma_-(\mathbf{s}_0)$ of the surface Σ. The meaning of these regions is clearly illustrated in Figure 6.15. The physical-optics approximation consists first in neglecting the contribution of the shadow region. Then, as concerns the illumination region, the surface is locally approximated by a plane and the scattered wave in a point $\mathbf{x}'$ of the surface is computed as the

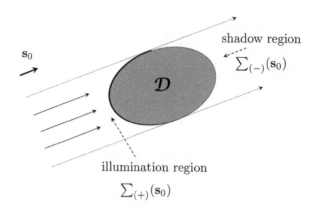

FIGURE 6.15   Illustrating the illumination and the shadow region of the surface Σ, for a given incidence direction $\mathbf{s}_0$.

wave reflected by the tangent plane. In conclusion the following approximations are used

$$\left.\frac{\partial}{\partial\nu(\mathbf{x}')}u(\mathbf{x}')\right|_{\Sigma_-(\mathbf{s}_0)} = 0 \tag{6.76}$$

$$\left.\frac{\partial}{\partial\nu(\mathbf{x}')}u(\mathbf{x}')\right|_{\Sigma_+(\mathbf{s}_0)} = 2\left.\frac{\partial}{\partial\nu(\mathbf{x}')}e^{ik\mathbf{s}_0\cdot\mathbf{x}'}\right|_{\Sigma_+(\mathbf{s}_0)} \tag{6.77}$$

so that from equation (6.75) we obtain

$$u_\infty(k;\mathbf{s}_0,\mathbf{s}) = -\frac{ik}{2\pi}\int_{\Sigma_+(\mathbf{s}_0)}(\mathbf{s}_0\cdot\nu(\mathbf{x}'))\,e^{ik\mathbf{s}_0-\mathbf{s}\cdot\mathbf{x}'}\,\mu(d\mathbf{x}')\;. \tag{6.78}$$

This formula can be very useful in the case of backscattering (for instance, its application to radar imaging has been suggested), i.e. in the case $\mathbf{s} = -\mathbf{s}_0$. From equation (6.78) we get

$$u_\infty(k;\mathbf{s},-\mathbf{s}) = -\frac{1}{4\pi}\int_{\Sigma_+(\mathbf{s})}\frac{\partial}{\partial\nu(\mathbf{x}')}e^{2ik\mathbf{s}\cdot\mathbf{x}'}\mu(d\mathbf{x}')\;. \tag{6.79}$$

If we exchange $\mathbf{s}$ and $-\mathbf{s}$, by observing that the illumination region is now $\Sigma_-(\mathbf{s})$, we also obtain

$$u_\infty(k;-\mathbf{s},\mathbf{s}) = -\frac{1}{4\pi}\int_{\Sigma_-(\mathbf{s})}\frac{\partial}{\partial\nu(\mathbf{x}')}e^{-2ik\mathbf{s}\cdot\mathbf{x}'}\mu(d\mathbf{x}')\;. \tag{6.80}$$

By adding the complex conjugate of equation (6.80) to equation (6.79) and by using the first Green theorem, we finally obtain the *Bojarski identity* [45]

$$u_\infty(k;\mathbf{s},-\mathbf{s}) + u_\infty^*(k;-\mathbf{s},\mathbf{s}) = \frac{k^2}{\pi}\int\chi_{\mathcal{D}}(\mathbf{x})\,e^{2ik\mathbf{s}\cdot\mathbf{x}}\,d\mathbf{x} \tag{6.81}$$

where $\chi_{\mathcal{D}}(\mathbf{x})$ is the characteristic function of the domain $\mathcal{D}$.

This formula implies that, in the physical-optics approximation, the Fourier transform of the characteristic function of the scatterer (and therefore the shape of the scatterer) can be obtained from the knowledge of the backscattering amplitude for all incident directions $\mathbf{s}$ and all wavenumbers k. In practice, the backscattering amplitudes can be measured only for a restricted region of values of k, i.e. for $k_{min} \le k \le k_{max}$. In other words one obtains the image of $\chi_{\mathcal{D}}(\mathbf{x})$ provided by a band-pass filter, so that a special feature of this problem is the lack of information both at low and at high frequencies. We see here another instance of the general problem of out-of-band extrapolation, which will be discussed in Chapter 8.

7

Singular value decomposition (SVD)

One of the most fruitful tools in the theory of linear inverse problems is the singular value decomposition (SVD) of a matrix and its extension to certain classes of linear operators. Indeed, SVD is basic both for understanding the ill-posedness of linear inverse problems and for describing the effect of the regularization methods.

The SVD of a matrix was essentially introduced in 1873 by the Italian mathematician Eugenio Beltrami for solving the problem of the diagonalization of a bilinear form [26]. A more complete treatment, published independently one year later, is due to Camille Jordan [173]. For a short and interesting history of SVD we refer to [269].

Nowadays efficient and stable algorithms to compute the SVD of a matrix are available and in many problems, when the number of data is not too large, they can be used for obtaining stable and approximate solutions. However, in imaging problems, the size of the matrices is so large that it may be difficult to use these algorithms in practice.

In this chapter, after a description of linear imaging systems in terms of linear operators, we derive the SVD of an arbitrary matrix by assuming that the reader is familiar with the diagonalization of a symmetric (self-adjoint) matrix. Then we extend the SVD to a class of linear operators related to problems where the image is discrete but the object is assumed to be a function. Finally we give the SVD of certain integral operators in Hilbert spaces and that of the Radon transform in two dimensions.

7.1 Mathematical description of linear imaging systems

Based on the examples of the previous chapter, we provide now a general mathematical description of a linear imaging system. The object function $f^{(0)}$ is an element of a Hilbert space $\mathcal{X}$, very frequently a space of square-integrable functions, which will be called the *object space*. Analogously the image function (or data function) $g^{(0)}$ is an element of another Hilbert space $\mathcal{Y}$, which will be called the *image space*.

The two spaces $\mathcal{X}$ and $\mathcal{Y}$ may coincide. This is the case, for instance, of the imaging systems considered in the first part of the book. However $\mathcal{X}$ and $\mathcal{Y}$ may be different as in the case of scattering problems or in X-ray tomography. In the latter case the object, i.e. the linear attenuation coefficient, is a function of two space variables while the image function, i.e. the

DOI: 10.1201/9781003032755-7

Radon transform of $f^{(0)}$, is a function of the space variable s and of the angular variable ϕ, so that the object and image spaces are different.

Once the object and image spaces are defined, the imaging process is described by a *linear operator*, i.e. a linear mapping which associates functions $f^{(0)}$ of $\mathcal{X}$ to functions $g^{(0)}$ of $\mathcal{Y}$

$$g^{(0)} = Af^{(0)} . \tag{7.1}$$

In general, it is also assumed that A is a bounded, hence a continuous operator, from $\mathcal{X}$ into $\mathcal{Y}$; we write $A: \mathcal{X} \to \mathcal{Y}$. As in Chapter 2, $g^{(0)}$ is called the *noise-free image* of the object $f^{(0)}$. In the case of a space-variant imaging system, the linear operator A is the integral operator defined in equation (6.1), while in the case of X-ray tomography the operator A is the Radon transform R defined in equations (6.11)-(6.12).

As discussed in Chapter 2, the recorded image g is affected by the noise introduced by the recording process. As in equation (2.3) and the related discussion, this effect can be described by an additional term w, which is also assumed to be an element of $\mathcal{Y}$, so that

$$g = Af^{(0)} + w . \tag{7.2}$$

The function g is called the *noisy image*.

In Section 3.1 we consider not only the integral equation (3.1) but also the discretized equation (3.10), where both the object and the image are discrete. For more general inverse problems, it is sometimes convenient to consider an intermediate model where the image is discrete but the object is assumed to be a function (*semi-discrete problems*). It is obvious that an experimental image is always described by a finite set of numbers: one or several detectors measure variations in space (and, possibly, also in time) of the emitted or scattered radiation; the output of the detectors is digitized and the final result is precisely a set of numbers stored in the computer. We denote the elements of this set by $g_1, g_2, \ldots, g_M$. As already mentioned many times, in the case of a 2D image, they can be obtained by a lexicographic ordering of the numbers corresponding to the gray levels associated with the pixels of the image. Therefore an image can also be viewed as a vector which will be called the *image vector*, or the *discrete image*, and denoted by $\mathbf{g}$.

The components g_m of $\mathbf{g}$ are proportional (but, for simplicity, we omit the proportionality constant) to sampled values of the image g:

$$g_m = g(\mathbf{x}_m) ; \quad m = 1, 2, \ldots, M . \tag{7.3}$$

More realistically, however, since any detector integrates over some region around a sampling point $\mathbf{x}_m$, the g_m are proportional to weighted averages of g:

$$g_m = \int p_m(\mathbf{x}) \, g(\mathbf{x}) \, d\mathbf{x} ; \quad m = 1, 2, \ldots, M . \tag{7.4}$$

The (real) averaging function $p_m(\mathbf{x})$, which represents the response of the detector, has typically a peak centered at a point $\mathbf{x} = \mathbf{x}_m$ of the image plane. Then equation (7.3) is obtained from equation (7.4) in the special case where $p_m(\mathbf{x}) = \delta(\mathbf{x} - \mathbf{x}_m)$.

The image vectors form a M-dimensional vector space $\mathcal{Y}_M$. In this vector space we can introduce the canonical Euclidean scalar product or, more frequently, a weighted scalar product characterized by a positive definite weighting matrix. This weighting matrix can be related to statistical properties of the noise (see Part III) or to a quadrature formula used for the discretization of the scalar product of $\mathcal{Y}$ [36].

If we assume, for simplicity, that $\mathcal{Y}$ is a space of square-integrable functions, we can view the integral in equation (7.4) as the scalar product of the functions p_m and g, i.e. $g_m = (g, p_m)_\mathcal{Y}$.

From equation (7.2) and from the definition of adjoint operator it follows that

$$
\begin{aligned}
(g, p_m)_{\mathcal{Y}} &= (A f^{(0)}, p_m)_{\mathcal{Y}} + (w, p_m)_{\mathcal{Y}} \\
&= (f^{(0)}, A^* p_m)_{\mathcal{X}} + (w, p_m)_{\mathcal{Y}} .
\end{aligned}
\tag{7.5}
$$

Therefore, if we introduce the functions of $\mathcal{X}$ defined by

$$
\varphi_m(\mathbf{x}') = (A^* p_m)(\mathbf{x}') ; \quad m = 1, 2, \ldots, M
\tag{7.6}
$$

we obtain

$$
g_m = g_m^{(0)} + w_m ; \quad m = 1, 2, \ldots, M
\tag{7.7}
$$

where $w_m = (w, p_m)_{\mathcal{Y}}$ and

$$
g_m^{(0)} = \left(f^{(0)}, \varphi_m \right)_{\mathcal{X}} ; \quad m = 1, 2, \ldots, M .
\tag{7.8}
$$

When $\mathcal{X}$ is also a space of square-integrable functions equipped with the canonical scalar product of L^2, equation (7.8) takes the following form

$$
g_m^{(0)} = \int_{\mathcal{D}} \varphi_m^*(\mathbf{x}') \, f^{(0)}(\mathbf{x}') \, d\mathbf{x}' ; \quad m = 1, 2, \ldots, M
\tag{7.9}
$$

where $\mathcal{D}$ is the object domain. Therefore the components of the noise-free discrete image are suitable weighted averages of the object $f^{(0)}$, also referred to as generalized moments of $f^{(0)}$.

The weighting functions φ_m must be determined for the specific problem at hand. For instance, if we consider the space-variant imaging system described by equation (6.1) and if $p_m(\mathbf{x}) = \delta(\mathbf{x} - \mathbf{x}_m)$, then

$$
\varphi_m^*(\mathbf{x}') = K(\mathbf{x}_m, \mathbf{x}') \quad m = 1, 2, \ldots, M .
\tag{7.10}
$$

In such a case the functions φ_m are given directly by the integral kernel of the imaging operator.

Equation (7.9) describes a linear mapping which transforms a function of the Hilbert space $\mathcal{X}$ into a vector, i.e. an element of the M-dimensional vector space $\mathcal{Y}_M$. For this reason it is referred to as a *semi-discrete mapping*, which can be denoted by A_M and is defined by the rule

$$
(A_M f)_m = (f, \varphi_m)_{\mathcal{X}} ; \quad m = 1, 2, \ldots, M .
\tag{7.11}
$$

In this approach the image is discrete while the object is still considered as an element of a space of functions, typically a Hilbert space. Therefore only when an estimate of the object has been obtained, one can perform a fine discretization of this estimate in order to produce a numerical or graphical result. In other words, this approach points out that in a practical inverse problem one has two distinct types of discretization. The first one is the discretization of the image, which is related to the design of the experiment or of the imaging system and therefore can be affected by rather strong instrumental restrictions. The second one is the discretization of the object which depends essentially on the approximation method used by the mathematician or by the practitioner. In the semi-discrete approach outlined above this second kind of discretization is not performed.

In practical problems, however, the usual approach consists in a complete discretization of the problem. If we start from equation (7.8), the discretization of this equation is obtained by assuming that the object $f^{(0)}$ can be reliably approximated by a linear combination of suitable and linearly independent basis functions $\psi_1, \psi_2, \ldots, \psi_N$, so that we can write

$$
f^{(0)}(\mathbf{x}') = \sum_{n=1}^{N} f_n^{(0)} \, \psi_n(\mathbf{x}') .
\tag{7.12}
$$

In the case of a 2D object, for instance, the functions ψ_n can be the characteristic functions of the pixels; in that case equation (7.12) is equivalent to assuming that $f^{(0)}$ can be approximated by a piecewise constant function. On the other hand, if $f^{(0)}(\mathbf{x})$ is band-limited, then the ψ_n can be suitable sampling functions. By substituting equation (7.12) into equation (7.8), we get

$$g_m^{(0)} = \sum_{n=1}^{N} A_{m,n} \, f_n^{(0)} \quad ; \quad m = 1, 2, \ldots, M \tag{7.13}$$

where

$$A_{m,n} = (\psi_n, \varphi_m)_{\mathcal{X}} \; . \tag{7.14}$$

An expression of $A_{m,n}$ in terms of the linear operator A can be obtained by taking into account equation (7.6) and using again the relationship between A and A^*

$$A_{m,n} = (\psi_n, A^* p_m)_{\mathcal{X}} = (A\psi_n, p_m)_{\mathcal{Y}} \tag{7.15}$$

so that $A_{m,n}$ is the m-th component of the discrete image of ψ_n.

If we denote by $\mathbf{f}^{(0)}$ the vector with components $f_1^{(0)}, \ldots, f_N^{(0)}$, and by $\mathbf{A}$ the matrix (in general non-square) whose elements $(\mathbf{A})_{m,n} = A_{m,n}$ are given by equation (7.14) or equation (7.15), then from equations (7.7) and (7.13) we obtain the discrete imaging equation

$$\mathbf{g} = \mathbf{A}\mathbf{f}^{(0)} + \mathbf{w} \tag{7.16}$$

which is the discrete version of equation (7.2).

In conclusion, let us say a few words about the scalar product in the space $\mathcal{X}_N$ of the vectors $\mathbf{f}$. Equation (7.12) defines a linear subspace of $\mathcal{X}$ which is finite-dimensional. If f and h are two functions in this subspace, their scalar product is given by

$$(f, h)_{\mathcal{X}} = \sum_{n, n'=1}^{N} (\psi_n, \psi_{n'})_{\mathcal{X}} \, f_n \, h_{n'}^* \; . \tag{7.17}$$

Therefore, if we require that the scalar product of two vectors of $\mathcal{X}_N$, $\mathbf{f}$ and $\mathbf{h}$, coincides with the scalar product of the functions f and h they represent, i.e. $(\mathbf{f}, \mathbf{h})_N = (f, h)_{\mathcal{X}}$, we must introduce in $\mathcal{X}_N$ a weighted scalar product, with a weighting matrix given by

$$C_{n,n'} = (\psi_n, \psi_{n'})_{\mathcal{X}} \; . \tag{7.18}$$

The matrix $C_{n,n'}$ is the *Gram matrix* of the basis functions ψ_n. If they form an orthonormal set, then $C_{n,n'} = \delta_{n,n'}$ and $\mathcal{X}_N$ can be equipped with the canonical scalar product. Otherwise we have to deal with a weighted scalar product. Indeed, the matrix (7.18) is positive definite because if $\mathbf{f}$ is an element of $\mathcal{X}_N$ and $f(\mathbf{x})$ the corresponding function of $\mathcal{X}$ given by (7.12), then equation (7.17) implies that

$$\sum_{n, n'=1}^{N} C_{n,n'} \, f_n \, f_{n'}^* = \|f\|_{\mathcal{X}}^2 > 0 \; . \tag{7.19}$$

In conclusion the matrix $\mathbf{A}$ defines a linear mapping from a vector space $\mathcal{X}_N$ into a vector space $\mathcal{Y}_M$, both vector spaces being equipped, in general, with weighted scalar products. Notice that a change of variables, based on the so-called Choleski factorization of the weighting matrices, allows to transform a problem formulated in weighted spaces into a problem formulated in canonical vector spaces.

7.2   SVD of a matrix

We consider first the case of a completely discretized imaging problem. This is characterized by an image space $\mathcal{Y}_M$ of dimension M, an object space $\mathcal{X}_N$ of dimension N and a matrix $\mathbf{A}$, $M \times N$, transforming a vector of $\mathcal{X}_N$ into a vector of $\mathcal{Y}_M$. Thanks to the remark at the end of the previous section, we assume, without loss of generality, that both $\mathcal{X}_N$ and $\mathcal{Y}_M$ are equipped with the canonical scalar product and in the following we denote by $\mathcal{E}_N$, and $\mathcal{E}_M$ the corresponding Euclidean vector spaces and indicate with a subscript N or M the respective canonical scalar product. This point is important because the theory and the algorithms for the singular value decomposition of a matrix are based on this assumption, even if the usual way of formulating SVD in numerical analysis does not mention at all the structure of the vector spaces.

The standard formulation is as follows: *let* $\mathbf{A}$ *be a matrix* $M \times N$, *with rank* p; *then there exists a* $p \times p$ *diagonal matrix* $\mathbf{\Sigma}$, *with positive diagonal elements, and two isometric matrices* $\mathbf{U}$ *and* $\mathbf{V}$, *respectively* $M \times p$ *and* $N \times p$, *such that*

$$\mathbf{A} = \mathbf{U}\mathbf{\Sigma}\mathbf{V}^* . \tag{7.20}$$

Here $\mathbf{V}^*$ denotes the adjoint of the matrix $\mathbf{V}$. We also recall that a non-square matrix $\mathbf{V}$ is isometric if it satisfies the condition $\mathbf{V}^*\mathbf{V} = \mathbf{I}$ ($\mathbf{I}$ is the $p \times p$ identity matrix). A square isometric matrix is unitary (orthogonal). We must also mention that the standard algorithms for SVD apply to the case of a real matrix and, in such a case, the matrix $\mathbf{V}$ is also real, so that the adjoint matrix $\mathbf{V}^*$ coincides with the transposed matrix $\mathbf{V}^T$.

For our applications to inverse problems, however, it is more convenient to write the decomposition (7.20) in a different way. Our starting point is the diagonalization of a self-adjoint matrix: *let* $\mathbf{A}$ *be a self-adjoint matrix* $N \times N$, *i.e. a matrix such that* $\mathbf{A}^* = \mathbf{A}$; *then there exists an* $N \times N$ *diagonal matrix* $\mathbf{\Lambda}$ *and a unitary (orthogonal) matrix* $\mathbf{V}$ *such that*

$$\mathbf{A} = \mathbf{V}\mathbf{\Lambda}\mathbf{V}^* . \tag{7.21}$$

As it is well known, this representation is a synthetic formulation of the basic results on the eigenvalue problem for the matrix $\mathbf{A}$. A self-adjoint $N \times N$ matrix has always N real eigenvalues, $\lambda_1, \lambda_2, \ldots, \lambda_N$, if each eigenvalue is counted as many times as its multiplicity. They can be ordered in such a way that: $|\lambda_1| \geq |\lambda_2| \geq \cdots \geq |\lambda_N|$. Moreover eigenvectors associated with distinct eigenvalues are automatically orthogonal while m orthogonal eigenvectors can always be associated to each eigenvalue with multiplicity m. If all eigenvectors are normalized, then one can conclude that the solution of the eigenvalue problem for a self-adjoint matrix $\mathbf{A}$ provides a set of eigenvectors $\mathbf{v}_1, \mathbf{v}_2, \ldots, \mathbf{v}_N$ which constitute an orthonormal basis in $\mathcal{E}_N$. Here the eigenvector $\mathbf{v}_k$ is associated with the eigenvalue $\lambda_k : \mathbf{A}\mathbf{v}_k = \lambda_k\mathbf{v}_k$. Finally the representation (7.21) is related to the solution of the eigenvalue problem as follows: the eigenvalues λ_k are the diagonal elements of the diagonal matrix $\mathbf{\Lambda}$ while the eigenvectors $\mathbf{v}_k$ are the columns of the unitary matrix $\mathbf{V}$.

If $\mathbf{f}$ is an arbitrary vector of $\mathcal{E}_N$, then $\mathbf{V}^*\mathbf{f}$ is the vector whose components are the scalar products of $\mathbf{f}$ with the eigenvectors $\mathbf{v}_k$, i.e. $(\mathbf{f} \cdot \mathbf{v}_k)_N$. Hence the representation (7.21) implies the following equation

$$\mathbf{A}\mathbf{f} = \sum_{k=1}^{N} \lambda_k \, (\mathbf{f} \cdot \mathbf{v}_k)_N \, \mathbf{v}_k \tag{7.22}$$

(remember that the suffix of the scalar product indicates the dimension of the vector space). If the matrix $\mathbf{A}$ has rank $p < N$, then $\mathbf{A}$ has the eigenvalue $\lambda = 0$ with multiplicity $N - p$ so that, taking into account the ordering of the eigenvalues, we conclude that $\lambda_{p+1} = \lambda_{p+2} =$

$\cdots = \lambda_N = 0$. It follows that in equation (7.22) the summation extends only up to p

$$\mathbf{A f} = \sum_{k=1}^{p} \lambda_k \, (\mathbf{f} \cdot \mathbf{v}_k)_N \, \mathbf{v}_k \ . \tag{7.23}$$

This expansion is called the *spectral representation* of the self-adjoint matrix $\mathbf{A}$. Notice that a similar representation holds true for any circulant matrix, even when the matrix is not self-adjoint.

The representation (7.23) is equivalent to the representation (7.21). Indeed, in (7.21) we can assume that $\boldsymbol{\Lambda}$ is a $p \times p$ diagonal matrix formed with the non-zero eigenvalues of $\mathbf{A}$ and that $\mathbf{V}$ is an $N \times p$ isometric matrix whose columns are the p eigenvectors associated with these non-zero eigenvalues.

We consider now the case of an arbitrary matrix $\mathbf{A}$ with M rows and N columns. If the matrix is non-square, i.e. $M \neq N$, then the eigenvalue problem is meaningless. If the matrix is square, i.e. $M = N$, but not self-adjoint, it may have eigenvalues and eigenvectors but these do not have, in general, the nice properties which hold true for self-adjoint matrices. The circulant matrices considered for the discretization of the convolution product represent a very particular case because they always have N orthonormal eigenvectors. This is not true in general and therefore, for an arbitrary matrix the eigenvalue problem may not be interesting for the solution of inverse problems because it does not provide a representation of the matrix similar to the spectral representation (7.23). The singular value representation of the matrix provides the appropriate generalization of equation (7.23).

In the case of an arbitrary $M \times N$ matrix $\mathbf{A}$, with rank $p \leq \min\{M, N\}$, it is always possible to form two self-adjoint matrices, namely

$$\bar{\mathbf{A}} = \mathbf{A}^* \mathbf{A} \ , \quad \tilde{\mathbf{A}} = \mathbf{A} \mathbf{A}^* \ . \tag{7.24}$$

The first is $N \times N$ while the second is $M \times M$. They have the following properties which can be easily proved:

- both matrices are self-adjoint (more precisely symmetric if real-valued) and positive semi-definite;
- both matrices have rank p.

Since both matrices have exactly p positive eigenvalues, the zero eigenvalue has multiplicity $N - p$ for the matrix $\bar{\mathbf{A}}$ and $M - p$ for the matrix $\tilde{\mathbf{A}}$. If $M \neq N$, at least one of the two matrices has the zero eigenvalue and precisely the matrix which has the largest dimension. A first basic result is the following: *the matrices $\bar{\mathbf{A}}$ and $\tilde{\mathbf{A}}$ have exactly the same positive eigenvalues with the same multiplicity*. The proof of this result is easy. Let us consider the matrix $\bar{\mathbf{A}}$. Since it is symmetric and positive semi-definite, from the properties mentioned above it follows that it has p positive eigenvalues $\sigma_1^2 \geq \sigma_2^2 \geq \cdots \geq \sigma_p^2$; we denote by $\mathbf{v}_1, \mathbf{v}_2, \ldots, \mathbf{v}_p$ the eigenvectors associated with these eigenvalues. These eigenvectors form an orthonormal basis in the orthogonal complement of the null space of the matrix $\mathbf{A}$, $\mathcal{N}(\mathbf{A})$, which coincides with $\mathcal{N}(\bar{\mathbf{A}})$. To each eigenvector $\mathbf{v}_k$, which is a vector of $\mathcal{E}_N$, we can associate a vector of $\mathcal{E}_M$ as follows

$$\mathbf{u}_k = \frac{1}{\sigma_k} \mathbf{A} \mathbf{v}_k \ . \tag{7.25}$$

All vectors $\mathbf{u}_k$ are different from zero and they are eigenvectors of $\tilde{\mathbf{A}}$ associated with the eigenvalue σ_k^2. Indeed

$$\begin{aligned} \tilde{\mathbf{A}} \mathbf{u}_k &= \mathbf{A} \mathbf{A}^* \mathbf{u}_k = \frac{1}{\sigma_k} (\mathbf{A} \mathbf{A}^*) \mathbf{A} \mathbf{v}_k = \\ &= \frac{1}{\sigma_k} \mathbf{A} (\mathbf{A}^* \mathbf{A}) \mathbf{v}_k = \frac{1}{\sigma_k} \mathbf{A} (\sigma_k^2 \, \mathbf{v}_k) = \sigma_k^2 \, \mathbf{u}_k \ . \end{aligned} \tag{7.26}$$

This simple computation implies that all positive eigenvalues of $\bar{\mathbf{A}}$ are also positive eigenvalues of $\tilde{\mathbf{A}}$. If all σ_k^2's have multiplicity one, the two matrices have precisely the same eigenvalues because also $\tilde{\mathbf{A}}$ can have only p positive eigenvalues. The proof is not complete in the case of eigenvalues with multiplicity > 1. However, if we compute the scalar products of the vectors $\mathbf{u}_k$, we obtain

$$
\begin{aligned}
(\mathbf{u}_k \cdot \mathbf{u}_j)_M &= \frac{1}{\sigma_k \sigma_j} (\mathbf{A}\mathbf{v}_k \cdot \mathbf{A}\mathbf{v}_j)_M \qquad &(7.27)\\
&= \frac{1}{\sigma_k \sigma_j} (\mathbf{A}^*\mathbf{A}\mathbf{v}_k \cdot \mathbf{v}_j)_N = \frac{\sigma_k}{\sigma_j} (\mathbf{v}_k \cdot \mathbf{v}_j)_N = \delta_{k,j} \ .
\end{aligned}
$$

This result implies that the eigenvectors $\mathbf{u}_k$ are orthonormal, so that they are linearly independent even when they are associated with the same positive eigenvalue. Since we have obtained p linearly independent eigenvectors and since $\tilde{\mathbf{A}}$ has precisely p linearly independent eigenvectors associated with positive eigenvalues, we have proved that equation (7.25) provides all the eigenvectors of $\tilde{\mathbf{A}}$.

If we multiply by $\mathbf{A}^*$ both sides of equation (7.25) we obtain

$$
\frac{1}{\sigma_k} \mathbf{A}^* \mathbf{u}_k = \frac{1}{\sigma_k^2} \mathbf{A}^* \mathbf{A} \mathbf{v}_k = \mathbf{v}_k \ . \qquad (7.28)
$$

Therefore equations (7.25) and (7.28) show that the eigenvectors $\mathbf{u}_k, \mathbf{v}_k$ are solutions of the following *shifted eigenvalue problem*, a term introduced by Lanczos [188],

$$
\mathbf{A}\mathbf{v}_k = \sigma_k \mathbf{u}_k \ , \quad \mathbf{A}^* \mathbf{u}_k = \sigma_k \mathbf{v}_k \ . \qquad (7.29)
$$

These equations can be written in the form of a standard eigenvalue problem if we define the following vectors with dimension $M + N$

$$
\begin{pmatrix} \mathbf{u}_k \\ \mathbf{v}_k \end{pmatrix} \qquad (7.30)
$$

and the following symmetric matrix with dimension $(M + N) \times (M + N)$

$$
\begin{pmatrix} 0 & \mathbf{A} \\ \mathbf{A}^* & 0 \end{pmatrix}. \qquad (7.31)
$$

Indeed, it is easy to verify that the equations (7.29) imply that

$$
\begin{pmatrix} 0 & \mathbf{A} \\ \mathbf{A}^* & 0 \end{pmatrix} \begin{pmatrix} \mathbf{u}_k \\ \mathbf{v}_k \end{pmatrix} = \sigma_k \begin{pmatrix} \mathbf{u}_k \\ \mathbf{v}_k \end{pmatrix} \qquad (7.32)
$$

i.e. the vectors (7.30) are eigenvectors of the matrix (7.31) associated with the eigenvalues σ_k. Since the matrix (7.31) has rank $2p$, it has $2p$ non-zero eigenvalues. If we consider the following vectors

$$
\begin{pmatrix} \mathbf{u}_k \\ -\mathbf{v}_k \end{pmatrix} \qquad (7.33)
$$

it is easy to verify that they are also eigenvectors of the matrix (7.31) associated with the eigenvalues $-\sigma_k$.

The positive numbers σ_k are called the *singular values* of the matrix $\mathbf{A}$. This term comes from the theory of integral equations and presumably it was used for the first time by Smithies [269]. The vectors $\mathbf{u}_k, \mathbf{v}_k$ are called the *singular vectors* of the matrix $\mathbf{A}$ and the set of the triples $\{\sigma_k; \mathbf{u}_k, \mathbf{v}_k\}$ is called the *singular system* of the matrix $\mathbf{A}$. As in the case

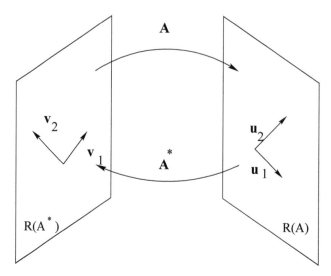

FIGURE 7.1   Schematic representation of the relationship between the two bases of singular vectors.

of the diagonalization of a self-adjoint matrix, each singular value is counted as many times as its multiplicity (which coincides with the multiplicity of σ_k^2 as an eigenvalue of $\bar{\mathbf{A}}$ and $\tilde{\mathbf{A}}$) and the following ordering is used: $\sigma_1 \geq \sigma_2 \geq \cdots \geq \sigma_p$.

The singular vectors $\mathbf{v}_k$ are orthogonal to the null space of $\mathbf{A}$, as one can verify directly in the following way. Let $\mathbf{v}$ be an element of $\mathcal{N}(\mathbf{A})$; then from the second relation of equation (7.29) we get

$$(\mathbf{v}_k \cdot \mathbf{v})_N = \frac{1}{\sigma_k} (\mathbf{A}^* \mathbf{u}_k \cdot \mathbf{v})_N = \frac{1}{\sigma_k} (\mathbf{u}_k \cdot \mathbf{A}\mathbf{v})_M = 0 \ . \tag{7.34}$$

In a similar way, we can prove that the $\mathbf{u}_k$ are orthogonal to the null space of $\mathbf{A}^*$. Therefore the $\mathbf{v}_k$ form an orthonormal basis in $\mathcal{N}(\mathbf{A})^\perp = \mathcal{R}(\mathbf{A}^*)$ while the $\mathbf{u}_k$ form an orthonormal basis in $\mathcal{N}(\mathbf{A}^*)^\perp = \mathcal{R}(\mathbf{A})$. The matrix $\mathbf{A}$ transforms the basis $\mathbf{v}_k$ into the basis $\mathbf{u}_k$, except for the scaling factors σ_k, and analogously the matrix $\mathbf{A}^*$ transforms the basis $\mathbf{u}_k$ into the basis $\mathbf{v}_k$. We have represented this situation in Figure 7.1.

We can now obtain the singular value decomposition of the matrix $\mathbf{A}$ in a form similar to the spectral representation (7.23) of a self-adjoint matrix. Since the singular vectors $\mathbf{u}_k$ form a basis in $\mathcal{R}(\mathbf{A})$, for any $\mathbf{f}$ in $\mathcal{E}_N$ we can write $\mathbf{A}\mathbf{f}$ as a linear combination of the $\mathbf{u}_k$

$$\mathbf{A}\mathbf{f} = \sum_{k=1}^{p} (\mathbf{A}\mathbf{f} \cdot \mathbf{u}_k)_M \, \mathbf{u}_k \ . \tag{7.35}$$

Then, by the use of the second relation of equation (7.29) we have

$$(\mathbf{A}\mathbf{f} \cdot \mathbf{u}_k)_M = (\mathbf{f} \cdot \mathbf{A}^* \mathbf{u}_k)_N = \sigma_k (\mathbf{f} \cdot \mathbf{v}_k)_N \tag{7.36}$$

which put into equation (7.35) yields

$$\mathbf{A}\mathbf{f} = \sum_{k=1}^{p} \sigma_k (\mathbf{f} \cdot \mathbf{v}_k)_N \, \mathbf{u}_k \ . \tag{7.37}$$

This is precisely the *singular value decomposition* of the matrix $\mathbf{A}$; in a similar way we obtain the SVD of $\mathbf{A}^*$

$$\mathbf{A}^* \mathbf{g} = \sum_{k=1}^{p} \sigma_k (\mathbf{g} \cdot \mathbf{u}_k)_M \, \mathbf{v}_k \ . \tag{7.38}$$

In order to show that the representation (7.37) is equivalent to the representation (7.20) it is sufficient to remark that the isometric matrix $\mathbf{U}$ is the matrix whose columns are the p orthonormal vectors $\mathbf{u}_k$, while the isometric matrix $\mathbf{V}$ is the matrix whose columns are the p orthonormal vectors $\mathbf{v}_k$. Then equation (7.37) follows from equation (7.20) precisely in the same way as equation (7.23) follows from equation (7.21).

By means of the representation (7.37) it is easy to show that the maximum singular value σ_1 is the norm of the matrix $\mathbf{A}$. Indeed, from the orthogonality and normalization of the vectors $\mathbf{u}_k$ – equation (7.27) – we obtain

$$\|\mathbf{A}\mathbf{f}\|_M^2 = \sum_{k=1}^{p} \sigma_k^2 \left|(\mathbf{f} \cdot \mathbf{v}_k)_N\right|^2 . \tag{7.39}$$

Since $\sigma_k \leq \sigma_1$ it follows that for any $\mathbf{f}$

$$\|\mathbf{A}\mathbf{f}\|_M^2 \leq \sigma_1^2 \sum_{k=1}^{p} \left|(\mathbf{f} \cdot \mathbf{v}_k)_N\right|^2 \leq \sigma_1^2 \|\mathbf{f}\|_N^2 \tag{7.40}$$

and therefore $\|\mathbf{A}\| \leq \sigma_1$. If we observe that equality holds true in equation (7.40) when $\mathbf{f} = \mathbf{v}_1$, we conclude that

$$\|\mathbf{A}\| = \sigma_1 . \tag{7.41}$$

In a similar way, we can prove that $\|\mathbf{A}^*\| = \sigma_1$.

In conclusion, let us comment on the analogies and differences between the spectral representation – equation (7.23) – of a self-adjoint matrix and the SVD – equation (7.37) – of an arbitrary matrix. In both cases there exists an orthonormal basis in the orthogonal complement of $\mathcal{N}(\mathbf{A})$, i.e. that formed by the vectors $\mathbf{v}_k$, such that the action of the matrix on a vector $\mathbf{f}$ consists in multiplying the components of $\mathbf{f}$ (with respect to this basis) by scaling factors (the eigenvalues in the case of a self-adjoint matrix, the singular values in the case of an arbitrary matrix). In this way one obtains the components of $\mathbf{A}\mathbf{f}$. The difference is that in the self-adjoint case these are the components of $\mathbf{A}\mathbf{f}$ with respect to the same basis, i.e. that formed by the vectors $\mathbf{v}_k$, while in the general case they are the components of $\mathbf{A}\mathbf{f}$ with respect to a different basis, i.e. that formed by the vectors $\mathbf{u}_k$. In other words only one basis is needed in the self-adjoint case while two bases are needed in the general case.

We also observe that the spectral representation (7.23) can be written in the form (7.37) if we put $\sigma_k = |\lambda_k|$ and $\mathbf{u}_k = \text{sgn}(\lambda_k)\mathbf{v}_k$, without modifying the definition of the $\mathbf{v}_k$'s. In particular, we find that $\mathbf{u}_k = \mathbf{v}_k$, for any k, only in the case of a symmetric and positive semi-definite matrix.

Efficient algorithms for the computation of the SVD of a matrix are available. They are based on a method developed by Golub and Reinsch [135] (see also [236]). The input is a matrix $M \times N$ with $M \geq N$ (if this condition is not satisfied one takes the transposed matrix). The output is a vector of length N (the singular values) and two matrices $\mathbf{U}$ and $\mathbf{V}$, respectively $M \times N$ and $N \times N$, i.e. the matrices of equation (7.20). If $p < N$, then $N - p$ singular values are zero or of the order of the machine precision. It is also important to point out that the usual routines apply to the case of a real-valued matrix and cannot be trivially extended to the case of a complex-valued one. Routines specially designed for complex-valued matrices are also available (for instance in the LAPACK package).

7.3   SVD of a semi-discrete mapping

In Section 7.1 we introduced the case where the image is discrete while the object is a function of continuous variables, for instance an element of L^2. This imaging problem can be formulated in terms of a semi-discrete mapping, transforming functions into vectors. In this section we show that such an operator has a singular value decomposition and we provide a method which can be used, in principle, for computing its singular system. To this purpose we need results given in Section 7.2 because, as we will show, it is always possible to reduce the problem to the diagonalization of a self-adjoint matrix.

We assume, for simplicity, that $\mathcal{Y}_M$ is the usual vector space $\mathcal{E}_M$, while $\mathcal{X}$ is the Hilbert space of the square-integrable functions defined on a domain $\mathcal{D}$. It is not difficult to extend the analysis if $\mathcal{Y}_M$ is a vector space equipped with a weighted scalar product [36].

The semi-discrete mappings introduced in Section 7.1 have the following general structure

$$(A_M f)_m = (f, \varphi_m)_{\mathcal{X}} \; ; \quad m = 1, \ldots, M \tag{7.42}$$

where $\varphi_1, \varphi_2, \ldots, \varphi_M$ are given functions of $\mathcal{X}$ which describe the imaging process. Then, for investigating the singular system of A_M we need the adjoint operator A_M^* which is defined by the usual rule (we denote again with a suffix M the scalar product of $\mathcal{E}_M$)

$$(A_M f \cdot \mathbf{g})_M = (f, A_M^* \mathbf{g})_{\mathcal{X}} \; . \tag{7.43}$$

From this definition it is evident that A_M^* is a mapping which transforms a vector into a function. By taking into account the definition (7.42) of A_M we get

$$
\begin{aligned}
(A_M f \cdot \mathbf{g})_M &= \sum_{m=1}^{M} (A_M f)_m \, g_m^* = \sum_{m=1}^{M} (f, \varphi_m)_{\mathcal{X}} \, g_m^* = \\
&= \left(f, \sum_{m=1}^{M} g_m \, \varphi_m \right)_{\mathcal{X}} = (f, A_M^* \mathbf{g})_{\mathcal{X}}
\end{aligned}
\tag{7.44}
$$

and therefore

$$(A_M^* \mathbf{g})(\mathbf{x}) = \sum_{m=1}^{M} g_m \, \varphi_m(\mathbf{x}) \; . \tag{7.45}$$

We conclude that the range of A_M^* is the finite-dimensional subspace spanned by the functions $\varphi_1, \varphi_2, \ldots, \varphi_M$. If these functions are linearly independent then this subspace has dimension M. In the following, we assume that this condition is satisfied.

From the general rule $\mathcal{R}(A_M^*)^\perp = \mathcal{N}(A_M)$, we get that the null space of A_M, i.e. the subspace of the invisible objects, is the infinite-dimensional subspace of all functions which are orthogonal to the subspace spanned by the functions $\varphi_1, \varphi_2, \ldots, \varphi_M$. This result can also be obtained directly from the definition (7.42). On the other hand, if the φ_m are linearly independent, the null space of A_M^* contains only the zero vector and therefore the relationship $\mathcal{N}(A_M^*) = \mathcal{R}(A_M)^\perp$ implies that the range of A_M coincides with $\mathcal{E}_M$.

As in the case of a matrix, we investigate the operators $\bar{A}_M = A_M^* A_M$ and $\tilde{A}_M = A_M A_M^*$. Since A_M maps functions into vectors while A_M^* maps vectors into functions, the operator $\bar{A}_M$ maps functions into functions, i.e. it is an operator in $\mathcal{X}$. In the particular case where $\mathcal{X}$ is a space of square-integrable functions, we have

$$(f, \varphi_m)_{\mathcal{X}} = \int_{\mathcal{D}} f(\mathbf{x}) \, \varphi_m^*(\mathbf{x}) \, d\mathbf{x} \tag{7.46}$$

and from equation (7.45) with $g_m = (A_M f)_m = (f, \varphi_m)_{\mathcal{X}}$ we get that $\bar{A}_M$ is the integral operator

$$(\bar{A}_M f)(\mathbf{x}) = \int \bar{K}_M(\mathbf{x}, \mathbf{x}') f(\mathbf{x}') d\mathbf{x}' \tag{7.47}$$

with

$$\bar{K}_M(\mathbf{x}, \mathbf{x}') = \sum_{m=1}^{M} \varphi_m(\mathbf{x}) \, \varphi_m^*(\mathbf{x}') \, . \tag{7.48}$$

An integral operator with this structure is called a *finite-rank integral operator* because its range is finite-dimensional (and has dimension M). It is a particular case of the integral operators considered in the next section.

As concerns the operator $\tilde{A}_M$, it maps vectors into vectors and therefore is an operator in $\mathcal{E}_M$, which can be represented by a matrix. From equation (7.45) we get

$$\left(\tilde{A}_M \mathbf{g}\right)_m = (A_M^* \mathbf{g}, \varphi_m)_{\mathcal{X}} = \sum_{m'=1}^{M} (\varphi_{m'}, \varphi_m)_{\mathcal{X}} \, g_{m'} \tag{7.49}$$

so that the matrix associated with $\tilde{A}_M$ is given by

$$\left(\tilde{\mathbf{A}}_M\right)_{m,m'} = (\varphi_{m'}, \varphi_m)_{\mathcal{X}} \, . \tag{7.50}$$

This is the *Gram matrix* of the functions φ_m. Its rank is M since the φ_m are linearly independent. Moreover it is a positive-definite matrix, as follows from the previous property of the φ_m and the relationship

$$\begin{aligned}
\left(\tilde{\mathbf{A}}_M \mathbf{g} \cdot \mathbf{g}\right)_M &= \sum_{m,m'}^{M} (\varphi_m, \varphi_{m'})_{\mathcal{X}} \, g_m \, g_{m'}^* \tag{7.51} \\
&= \left\| \sum_{m=1}^{M} g_m \varphi_m \right\|_{\mathcal{X}}^2 > 0
\end{aligned}$$

if $\mathbf{g}$ is not the null vector. Since $\tilde{\mathbf{A}}_M$ is a self-adjoint and positive-definite matrix, it has exactly M positive eigenvalues.

By repeating the arguments used in Section 7.2 in the case of a matrix we conclude that the operators $\bar{A}_M$ and $\tilde{A}_M$ have the same non-zero eigenvalues which are, precisely, the M non-zero eigenvalues of the Gram matrix of the functions φ_m. Let us denote by $\sigma_1^2 \geq \sigma_2^2 \geq \cdots \geq \sigma_M^2$ its eigenvalues and by $\mathbf{u}_1, \mathbf{u}_2, \ldots, \mathbf{u}_M$ the corresponding eigenvectors. Eigenfunctions $v_1(\mathbf{x}), v_2(\mathbf{x}) \ldots, v_M(\mathbf{x})$ of the operator $\bar{A}_M$ (i.e. of the integral operator (7.47) if $\mathcal{X}$ is a space of square-integrable functions) can be associated to the eigenvectors $\mathbf{u}_k$ by means of the same procedure as in the case of matrices: $v_k = \sigma_k^{-1} A_M^* \mathbf{u}_k$. In conclusion we find that the singular system of the operator A_M consists of the solutions of the shifted eigenvalue problem

$$A_M v_k = \sigma_k \mathbf{u}_k \, , \quad A_M^* \mathbf{u}_k = \sigma_k v_k \, . \tag{7.52}$$

A possible procedure for the computation of this singular system is the following:

- compute the Gram matrix of the functions φ_m;
- compute the eigenvalues σ_k^2 and the eigenvectors $\mathbf{u}_k$ of the Gram matrix; the (positive) square roots of the eigenvalues are the singular values while the $\mathbf{u}_k$ are the singular vectors of A_M in the image space;
- compute the singular functions $v_k(\mathbf{x})$ by means of equation (7.45) and the second relation of equation (7.52); the result is

$$v_k(\mathbf{x}) = \frac{1}{\sigma_k} \sum_{m=1}^{M} (\mathbf{u}_k)_m \, \varphi_m(\mathbf{x}) \, . \tag{7.53}$$

In the case of a large image, the Gram matrix is also large and the computation of the singular system may be difficult in practice. However, the method can be very useful for the solution of inverse problems with few data, as of those where functions of only one variable are involved. Examples of applications to instrumental physics can be found in [37].

We conclude by deriving the singular value decomposition of the operators A_M and A_M^* in a form analogous to the one obtained for the matrices $\mathbf{A}$ and $\mathbf{A}^*$, equations (7.37) and (7.38) respectively. Indeed, since $A_M f$ is a vector of $\mathcal{E}_M = \mathcal{Y}_M$ and since the singular vectors $\mathbf{u}_k$ form a basis in $\mathcal{E}_M$, we can write

$$A_M f = \sum_{k=1}^{M} (A_M f \cdot \mathbf{u}_k)_M \, \mathbf{u}_k \ . \tag{7.54}$$

From the second relation of equation (7.52) we get

$$(A_M f \cdot \mathbf{u}_k)_M = (f, A_M^* \mathbf{u}_k)_\mathcal{X} = \sigma_k \, (f, v_k)_\mathcal{X} \tag{7.55}$$

and the substitution of this relation in equation (7.54) provides the SVD of A_M:

$$A_M f = \sum_{k=1}^{M} \sigma_k \, (f, v_k)_\mathcal{X} \, \mathbf{u}_k \ . \tag{7.56}$$

Moreover, equation (7.45) implies that the range of A_M^* is the subspace spanned by the (linearly independent) functions $\varphi_m(\mathbf{x})$. The singular functions $v_k(\mathbf{x})$, which are linear combinations of the $\varphi_m(\mathbf{x})$, form an orthonormal basis in this subspace so that

$$A_M^* \mathbf{g} = \sum_{k=1}^{M} (A_M^* \mathbf{g}, v_k)_\mathcal{X} \, v_k \ . \tag{7.57}$$

From the first relation of equation (7.52), we obtain

$$(A_M^* \mathbf{g}, v_k)_\mathcal{X} = (\mathbf{g} \cdot A_M v_k)_M = \sigma_k \, (\mathbf{g} \cdot \mathbf{u}_k)_M \tag{7.58}$$

so that

$$A_M^* \mathbf{g} = \sum_{k=1}^{M} \sigma_k \, (\mathbf{g} \cdot \mathbf{u}_k)_M \, v_k \ , \tag{7.59}$$

which is the SVD of the adjoint operator A_M^*.

7.4 SVD of an integral operator with a square-integrable kernel

In Section 6.1 a space-variant imaging system is described in terms of an integral operator of the following form

$$(Af)(\mathbf{x}) = \int_{\mathcal{D}} K(\mathbf{x}, \mathbf{x}') \, f(\mathbf{x}') \, d\mathbf{x}' \ , \quad \mathbf{x} \in \mathcal{D}' \ , \tag{7.60}$$

where $\mathcal{D}$ and $\mathcal{D}'$ are the object and the image domain, respectively. Other examples are described in Sections 6.5 and 6.6 (see, for instance, equations (6.40) and (6.55)). In the case of functions defined on the sphere, another example is the spherical convolution introduced in Section 6.4, equation (6.20). Finally, an example of a finite-rank integral operator is given in Section 7.3 (see equations (7.47) and (7.48)). This operator is self-adjoint. A more general type of finite-rank integral operators corresponds to integral kernels with the following structure

$$K(\mathbf{x}, \mathbf{x}') = \sum_{m=1}^{M} \varphi_m(\mathbf{x})\, \psi_m^*(\mathbf{x}') \tag{7.61}$$

the functions φ_m and ψ_m being, in general, different.

The analysis of this section applies also to the case of space-invariant imaging systems, i.e. systems described by convolution operators, when the symmetry of the system with respect to translations is destroyed by the fact that the objects are localized in a bounded domain $\mathcal{D}$. In such a case the integral operator takes the following form

$$(Af)(\mathbf{x}) = \int_{\mathcal{D}} K(\mathbf{x} - \mathbf{x}')\, f(\mathbf{x}')\, d\mathbf{x}' , \quad \mathbf{x} \in \mathcal{D}' , \tag{7.62}$$

where $\mathcal{D}'$ can possibly be the whole image plane.

We assume, for simplicity, that both the object and the image are square-integrable functions of the space variables, i.e. we assume that $\mathcal{X}$ and $\mathcal{Y}$ are the Hilbert spaces $\mathcal{X} = L^2(\mathcal{D})$ and $\mathcal{Y} = L^2(\mathcal{D}')$. Then equation (7.60) defines an operator from $L^2(\mathcal{D})$ into $L^2(\mathcal{D}')$. This operator is continuous if the integral kernel is square-integrable, i.e.

$$\|K\|^2 = \int_{\mathcal{D}'} d\mathbf{x} \int_{\mathcal{D}} d\mathbf{x}'\, |K(\mathbf{x}, \mathbf{x}')|^2 < \infty . \tag{7.63}$$

An integral operator satisfying this condition is called an operator of the *Hilbert-Schmidt class*. In order to show that this operator is continuous, we apply the Schwarz inequality to the r.h.s. of equation (7.60) which, for $\mathbf{x}$ fixed, is the scalar product of two square-integrable functions. It follows that

$$|(Af)(\mathbf{x})|^2 \le \left(\int_{\mathcal{D}} |K(\mathbf{x}, \mathbf{x}')|^2 d\mathbf{x}' \right) \left(\int_{\mathcal{D}} |f(\mathbf{x}')|^2 d\mathbf{x}' \right) . \tag{7.64}$$

If we integrate both sides of this inequality with respect to $\mathbf{x}$, and we take the square-root of the result, we get

$$\|Af\|_{\mathcal{Y}} \le \|K\|\, \|f\|_{\mathcal{X}} , \tag{7.65}$$

i.e. the operator is bounded. As it is well known, this property implies the continuity of the operator.

REMARK 7.1 *Condition (7.63) is satisfied in the case of the integral operators of Sections 6.4, 6.5 and 6.6 as well as in the case of the finite rank operator (7.61) (if the functions $\varphi_m(\mathbf{x})$ and $\psi_m(\mathbf{x})$ are square integrable). Moreover, it is also satisfied in the case of the operator (7.62) if the domain $\mathcal{D}$ or the domain $\mathcal{D}'$ is bounded and measurable, with a finite non-zero measure, and the* PSF *$K(\mathbf{x})$ is square integrable. Indeed, in the case of $\mathcal{D}$, by means of a change of variables, we get*

$$\|K\|^2 \le m(\mathcal{D}) \int_{\mathcal{D}'} |K(\mathbf{x})|^2 d\mathbf{x} \tag{7.66}$$

where $m(\mathcal{D})$ is the measure (area or volume) of $\mathcal{D}$. A similar inequality holds true when $m(\mathcal{D}') < \infty$, if we exchange $\mathcal{D}$ and $\mathcal{D}'$ in equation (7.66).

The adjoint A^* of the operator A is given by

$$(A^*g)(\mathbf{x}') = \int_{\mathcal{D}'} K^*(\mathbf{x},\mathbf{x}')\, g(\mathbf{x})\, d\mathbf{x}\ , \quad \mathbf{x}' \in \mathcal{D}\ . \tag{7.67}$$

Then, as in the case of a matrix, Section 7.2, we can introduce the operators $\bar{A} = A^*A$ and $\tilde{A} = AA^*$. Both are integral operators with integral kernels given by

$$\bar{K}(\mathbf{x},\mathbf{x}') = \int_{\mathcal{D}'} K^*(\mathbf{x}'',\mathbf{x})\, K(\mathbf{x}'',\mathbf{x}')\, d\mathbf{x}''\ , \quad \mathbf{x},\mathbf{x}' \in \mathcal{D}\ , \tag{7.68}$$

in the case of $\bar{A}$, and by

$$\tilde{K}(\mathbf{x},\mathbf{x}') = \int_{\mathcal{D}} K(\mathbf{x},\mathbf{x}'')\, K^*(\mathbf{x}',\mathbf{x}'')\, d\mathbf{x}''\ , \quad \mathbf{x},\mathbf{x}' \in \mathcal{D}'\ , \tag{7.69}$$

in the case of $\tilde{A}$. The integral operators $\bar{A}$ and $\tilde{A}$ have the following properties:

- both operators are self-adjoint, i.e. for any pair of functions f, h in $\mathcal{X}$ and any pair of functions g, w in $\mathcal{Y}$

$$\left(\bar{A}f, h\right)_{\mathcal{X}} = \left(f, \bar{A}h\right)_{\mathcal{X}}\ , \quad \left(\tilde{A}g, w\right)_{\mathcal{Y}} = \left(g, \tilde{A}w\right)_{\mathcal{Y}}\ ; \tag{7.70}$$

  this property is a consequence of the following relations

$$\bar{K}^*(\mathbf{x},\mathbf{x}') = \bar{K}(\mathbf{x}',\mathbf{x})\ , \quad \tilde{K}^*(\mathbf{x},\mathbf{x}') = \tilde{K}(\mathbf{x}',\mathbf{x}) \tag{7.71}$$

  which can be easily checked by means of equations (7.68) and (7.69);
- both operators are of the Hilbert-Schmidt class because their integral kernels are square integrable (the proof of this result is similar to the proof of the inequality (7.65); the starting point is the application of the Schwarz inequality to the r.h.s. of equations (7.68) and (7.69));
- both operators are positive semi-definite

$$\left(\bar{A}f, f\right)_{\mathcal{X}} \geq 0\ , \quad \left(\tilde{A}g, g\right)_{\mathcal{Y}} \geq 0\ . \tag{7.72}$$

REMARK 7.2   *We sketch the proof of the last property in the case of $\bar{A}$. Indeed, from equation (7.68), by means of an exchange of the integration order, we have*

$$
\begin{aligned}
\left(\bar{A}f, f\right)_{\mathcal{X}} &= \int_{\mathcal{D}} \left(\int_{\mathcal{D}} \bar{K}(\mathbf{x},\mathbf{x}')f(\mathbf{x}')d\mathbf{x}' \right) f^*(\mathbf{x})\, d\mathbf{x} \\
&= \int_{\mathcal{D}'} \left| \int_{\mathcal{D}} K(\mathbf{x}'',\mathbf{x})f(\mathbf{x})\, d\mathbf{x} \right|^2 d\mathbf{x}'' \geq 0\ .
\end{aligned} \tag{7.73}
$$

A similar proof applies to the case of $\tilde{A}$.

According to the Hilbert-Schmidt theory [209], a self-adjoint integral operator with a square-integrable kernel (hence also satisfying a condition like (7.71)) has real eigenvalues with finite multiplicity. Moreover the eigenfunctions associated with different eigenvalues are orthogonal. The eigenvalues form a countable set and accumulate to zero if the rank of the operator is not finite. On the other hand, a finite-rank integral operator, with rank M, has exactly M eigenvalues if each eigenvalue is counted as many times as its multiplicity, so that the null eigenvalue has infinite multiplicity.

These results apply, for instance, to the spherical convolution operators introduced in Section 6.4: they have a countable set of positive eigenvalues with finite multiplicity. Another example is the integral operator of equation (6.40) which is a self-adjoint integral operator with an infinite set of eigenvalues. Indeed, its kernel is symmetric (because $G^*(\mathbf{s}, \mathbf{s}') = G(\mathbf{s}', \mathbf{s})$, as follows from equation (6.41)), and is square-integrable (because $G(\mathbf{s}, \mathbf{s}')$ is a bounded function defined over a bounded domain). In agreement with the general result stated above, its eigenvalues, given in equation (6.42), have finite multiplicity (the multiplicity of λ_l is $2l+1$) and accumulate to zero, as follows from equation (6.43). Moreover the representation (6.41) is a particular case of the general result which we give now.

If $K(\mathbf{x}, \mathbf{x}')$ is a square-integrable kernel such that $K(\mathbf{x}', \mathbf{x}) = K^*(\mathbf{x}, \mathbf{x}')$, let $\lambda_1, \lambda_2, \lambda_3, \ldots$ be the sequence of the eigenvalues of the corresponding integral operator, ordered in such a way that $|\lambda_1| \geq |\lambda_2| \geq |\lambda_3| \geq \ldots$, each eigenvalue being counted as many times as its multiplicity. Moreover, let $v_1(\mathbf{x}), v_2(\mathbf{x}), v_3(\mathbf{x}), \ldots$ be the sequence of the eigenfuctions associated with these eigenvalues. They constitute an orthonormal set of square-integrable functions. Then the basic result of the Hilbert-Schmidt theory is the following *spectral representation* of the kernel $K(\mathbf{x}, \mathbf{x}')$

$$K(\mathbf{x}, \mathbf{x}') = \sum_{k=1}^{\infty} \lambda_k \, v_k(\mathbf{x}) \, v_k^*(\mathbf{x}') \tag{7.74}$$

the series being convergent in the L^2-norm. Accordingly, the eigenvalues λ_k satisfy the following condition

$$\sum_{k=1}^{\infty} \lambda_k^2 < \infty . \tag{7.75}$$

It is not difficult to understand that equation (7.74) implies a spectral representation of the corresponding operator A which is an extension of the spectral representation (7.23) of a symmetric matrix.

The results stated above apply to the operators $\bar{A}$ and $\tilde{A}$, whose non-zero eigenvalues are positive because they are positive semi-definite operators. Then, it is possible to generalize to the present case the method used in the case of a matrix and to show that the operators $\bar{A}$ and $\tilde{A}$ have the same positive eigenvalues with the same multiplicity.

Indeed, let us denote by σ_k^2 the positive eigenvalues of $\bar{A}$, with the ordering $\sigma_1^2 \geq \sigma_2^2 \geq \cdots \geq \sigma_k^2 \geq \ldots$ and $\sigma_k^2 \to 0$ for $k \to \infty$. Again each eigenvalue is counted as many times as its multiplicity, which is certainly finite as follows from the Hilbert-Schmidt theory. A normalized eigenfunction v_k is associated to each eigenvalue σ_k^2 and these eigenfunctions constitute an orthonormal system, i.e. $(v_k, v_j)_{\mathcal{X}} = \delta_{kj}$. Then, to each eigenfunction v_k of $\bar{A}$ we can associate a function u_k in $\mathcal{Y}$ defined by

$$u_k = \frac{1}{\sigma_k} A v_k . \tag{7.76}$$

Using the relation $\tilde{A}A = A\bar{A}$, which can be easily proved by an exchange of integration order in the definition of the integral kernels, we obtain

$$\tilde{A}u_k = \frac{1}{\sigma_k} A\bar{A}v_k = \sigma_k^2 \left(\frac{1}{\sigma_k} A v_k \right) = \sigma_k^2 u_k \tag{7.77}$$

and also

$$
\begin{aligned}
(u_k, u_j)_{\mathcal{Y}} &= \frac{1}{\sigma_k \sigma_j} (A v_k, A v_j)_{\mathcal{Y}} = \\
&= \frac{1}{\sigma_k \sigma_j} \left(v_k, \bar{A} v_j \right)_{\mathcal{X}} = \frac{\sigma_j}{\sigma_k} \left(v_k, v_j \right)_{\mathcal{X}} = \delta_{kj} .
\end{aligned}
\tag{7.78}
$$

Therefore all eigenvalues σ_k^2 of $\bar{A}$ are also eigenvalues of $\tilde{A}$ and the u_k are the corresponding eigenfunctions, which constitute an orthonormal system in $\mathcal{Y}$. In order to show that in this way we have obtained all eigenvalues and eigenfunctions of $\tilde{A}$, it is sufficient to repeat the same argument starting from $\tilde{A}$. If its eigenvalues and eigenfunctions are denoted by σ_k^2 and u_k respectively, then we can show, by means of the relation $\bar{A}A^* = A^*\tilde{A}$, that the functions of $\mathcal{X}$ defined by

$$v_k = \frac{1}{\sigma_k}A^*u_k \tag{7.79}$$

are orthonormal eigenfunctions of $\bar{A}$ associated with the eigenvalues σ_k^2.

The equations (7.76) and (7.79) define the usual shifted eigenvalue problem, which we write now explicitly in terms of the integral operators

$$\int_{\mathcal{D}} K(\mathbf{x},\mathbf{x}')\,v_k(\mathbf{x}')\,d\mathbf{x}' = \sigma_k\,u_k(\mathbf{x}) \tag{7.80}$$

$$\int_{\mathcal{D}'} K^*(\mathbf{x},\mathbf{x}')\,u_k(\mathbf{x})\,d\mathbf{x} = \sigma_k\,v_k(\mathbf{x}') \ .$$

The proof of the SVD of the operators A and A^* requires the completeness of the Hilbert spaces $\mathcal{X}$ and $\mathcal{Y}$ and therefore we do not report this proof here (see, for instance, [141]). We only give the result which takes the usual form except for the fact that now the sums are replaced by series. We have

$$Af = \sum_{k=1}^{\infty} \sigma_k\,(f,v_k)_{\mathcal{X}}\,u_k \ , \tag{7.81}$$

and also

$$A^*f = \sum_{k=1}^{\infty} \sigma_k\,(g,u_k)_{\mathcal{Y}}\,v_k \ , \tag{7.82}$$

the series being convergent in the norms of $\mathcal{Y}$ and $\mathcal{X}$ respectively.

We also remark that the SVD of A is equivalent to the following series expansion of the kernel $K(\mathbf{x},\mathbf{x}')$

$$K(\mathbf{x},\mathbf{x}') = \sum_{k=1}^{\infty} \sigma_k\,u_k(\mathbf{x})\,v_k^*(\mathbf{x}') \ , \tag{7.83}$$

the series being convergent with respect to the L^2-norm. From this expansion (which is a generalization of equation (7.74)), from the orthonormality of the singular functions and from the definition (7.63) of $\|K\|^2$ it follows that

$$\|K\|^2 = \sum_{k=1}^{\infty} \sigma_k^2 \tag{7.84}$$

and therefore *the sum of the squared singular values of a Hilbert-Schmidt integral operator is convergent.*

Using the SVD, equation (7.81), it is also possible to show that the norm of the integral operator A is given by

$$\|A\| = \sigma_1 \ . \tag{7.85}$$

From this equation and equation (7.84), it follows that the *the norm of the integral operator is never greater than the L^2-norm of the integral kernel; the two norms coincide if and only if the integral operator has rank one.*

We conclude this section with a few comments about the spherical convolution operator introduced in Section 6.4, equation (6.20). As already remarked, this operator is self-adjoint and of the Hilbert-Schmidt class. Therefore it has a countable set of real non-null eigenvalues accumulating to zero; these non-null eigenvalues are associated with even eigenfunctions and are alternately positive and negative. They have a finite multiplicity with the multiplicity increasing for decreasing eigenvalues. The null space of the operator is infinite-dimensional since it contains all the odd functions. Therefore, the spectral representation of a Hilbert-Schmidt operator can be applied to this case. To be pedantic, one can obtain the SVD of the operator by taking $\sigma_k = |\lambda_k|$ and $u_k = sgn(\lambda_k)v_k$, the v_k being given by the spherical harmonics $Y_{l,m}$.

7.5 SVD of the Radon transform

The SVD of the Radon transform, defined in Section 6.2, is derived in [83]. In this paper the general case of functions of n variables is considered and it is assumed that the object space is the Hilbert space of square-integrable functions on a bounded domain while the data space is a Hilbert space defined as a suitable weighted space of square-integrable functions. For simplicity, we consider the case of functions of two variables, which is also the important one for the applications, as discussed in Section 6.2. We first justify the choice of the Hilbert spaces and then we give the main results.

In practical applications, it is quite natural to assume that the objects are described by functions with support interior to a disc of radius a. It is not restrictive to take $a = 1$ because it is always possible to satisfy this condition by rescaling the space variables. Next we assume that the object functions f are square integrable, so that $\mathcal{X} = L^2(\mathcal{D})$, $\mathcal{D}$ being the disc of radius 1, and

$$\|f\|_{\mathcal{X}}^2 = \int_{\mathcal{D}} |f(\mathbf{x})|^2 d\mathbf{x} . \qquad (7.86)$$

We denote by $w(s)$ the half-length of the chord obtained by intersecting the disc with a straight line having signed distance s from the origin (see Figure 7.2), which is given by

$$w(s) = (1 - s^2)^{1/2} . \qquad (7.87)$$

Then the Radon transform of a function which is zero outside $\mathcal{D}$ is given by

$$(Rf)(s, \boldsymbol{\theta}) = \int_{-w(s)}^{w(s)} f(s\boldsymbol{\theta} + t\boldsymbol{\theta}^{\perp}) \, dt , \quad |s| \leq 1 . \qquad (7.88)$$

A simple property of the functions in the range of the operator R is obtained by applying the Schwarz inequality to the r.h.s. of this equation if we consider the integrand as the scalar product of the function f and of the function equal to one over the interval $[-w(s), w(s)]$. We get

$$|(Rf)(s, \boldsymbol{\theta})|^2 \leq 2 \, w(s) \int_{-w(s)}^{w(s)} |f(s\boldsymbol{\theta} + t\boldsymbol{\theta}^{\perp})|^2 dt \qquad (7.89)$$

so that, by means of a change of variables

$$\int_{-1}^{1} (w(s))^{-1} |(Rf)(s, \boldsymbol{\theta})|^2 \, ds \ \leq \ 2 \int_{-1}^{1} ds \int_{-w(s)}^{w(s)} |f(s\boldsymbol{\theta} + t\boldsymbol{\theta}^{\perp})|^2 dt$$

$$= 2 \int_{\mathcal{D}} |f(\mathbf{x})|^2 d\mathbf{x} . \qquad (7.90)$$

This remark suggests that a quite natural norm for the functions $g = g(s, \boldsymbol{\theta})$ of the data space $\mathcal{Y}$ is the following one

$$\|g\|_{\mathcal{Y}}^2 = \int_0^{2\pi} d\phi \int_{-1}^{1} \frac{ds}{w(s)} \, |g(s, \boldsymbol{\theta})|^2 . \qquad (7.91)$$

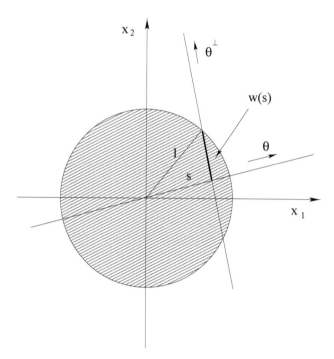

FIGURE 7.2 Geometry of the Radon transform in the case of functions with support interior to the disc of radius 1.

Then, from the inequality (7.90), by integrating with respect to the angle ϕ, it follows that R is a bounded operator from $\mathcal{X}$ into $\mathcal{Y}$

$$\|Rf\|_{\mathcal{Y}} \leq \sqrt{4\pi}\, \|f\|_{\mathcal{X}} \,. \qquad (7.92)$$

From the results reported in the following, it turns out that the largest singular value of R is $\sqrt{4\pi}$ so that $\|R\| = \sqrt{4\pi}$. In other words inequality (7.92) is precise and cannot be improved. Moreover, from the definition of the scalar products in $\mathcal{X}$ and $\mathcal{Y}$ and the definition of adjoint operator, one can easily derive that R^* is given by

$$(R^*g)(\mathbf{x}) = \int_0^{2\pi} g\left(\boldsymbol{\theta} \cdot \mathbf{x}, \boldsymbol{\theta}\right) w^{-1}\left((\boldsymbol{\theta} \cdot \mathbf{x})\right) d\phi \,. \qquad (7.93)$$

The starting point for the derivation of the singular system of R is the representation of a function $g(s, \boldsymbol{\theta})$ of $\mathcal{Y}$, for fixed $\boldsymbol{\theta}$, as a series of *Chebyshev polynomials of the second kind* [2], which are defined by

$$U_m(s) = \frac{\sin\left[(m+1)\arccos(s)\right]}{\sin\left(\arccos(s)\right)} \;;\quad m = 0, 1, 2, \ldots \qquad (7.94)$$

These polynomials are orthogonal with respect to the weight function $w(s)$; more precisely they satisfy the following orthogonality and normalization conditions

$$\int_{-1}^{1} w(s)\, U_m(s)\, U_{m'}(s)\, ds = \frac{\pi}{2}\, \delta_{m,m'} \qquad (7.95)$$

which can be easily obtained by remarking that they are related to trigonometric functions through the change of variable $s = \cos\xi$, $0 \leq \xi \leq \pi$.

Starting from these points and after some lengthy computations, one obtains the following results

- for $m = 0, 1, \dots$ the singular values are given by

$$\sigma_{m,k} = \left(\frac{4\pi}{m+1} \right)^{1/2} , \quad k = 0, 1, \dots, m \qquad (7.96)$$

so that, for a given m, the corresponding singular value has multiplicity $m + 1$; since the singular values decrease as m^{-1} and their multiplicity increases with m, this result implies that the inversion of the Radon transform is a mildly ill-posed problem, justifying the good reconstructions achievable in X-ray tomography;
- the singular functions in the image space, associated to the singular values indicated above, are given by

$$u_{m,k}(s, \boldsymbol{\theta}) = \sqrt{\frac{2}{\pi}} \, w(s) \, U_m(s) \, Y_{m-k}(\boldsymbol{\theta}) \qquad (7.97)$$

where the $U_m(s)$ are the Chebyshev polynomials and the angular functions $Y_l(\boldsymbol{\theta})$ are given by

$$Y_l(\boldsymbol{\theta}) = \frac{1}{\sqrt{2\pi}} \, e^{-il\phi} \, ; \qquad (7.98)$$

- the singular functions in the object space are given by [202]

$$v_{m,k}(\mathbf{x}) = (2m+2)^{1/2} Q_{m,|m-2k|}(|\mathbf{x}|) \, Y_{m-2k} \left(\frac{\mathbf{x}}{|\mathbf{x}|} \right) \qquad (7.99)$$

where

$$Q_{m,l}(r) = r^l P^{(0,l)}_{\frac{1}{2}(m+l)}(2r^2 - 1) , \qquad (7.100)$$

$P_n^{(\alpha,\beta)}(t)$ being the Jacobi polynomial of degree n [2].

These results can not be used for the numerical inversion of the Radon transform because too many singular values and singular functions should be computed. An efficient inversion algorithm, the so-called *filtered back-projection*, is discussed in the next chapter.

Before concluding this section, we spend a few words about some results concerning the case of a finite number of projections which is a semi-discrete version of the Radon transform because only discretization of the angular variable ϕ is performed. This problem is investigated in [84, 60] where explicit expressions of the singular values are given.

Using the previous notations the problem can be formulated as the inversion of the following operator

$$(Rf)(s, \boldsymbol{\theta}_j) = \int_{-w(s)}^{w(s)} f(s\boldsymbol{\theta}_j + t\boldsymbol{\theta}_j^{\perp}) \, dt , \quad \boldsymbol{\theta}_j \in \Theta_p , \qquad (7.101)$$

where Θ_p is the set of $2p$ directions $\boldsymbol{\theta}_j = \{\cos\phi_j, \sin\phi_j\}, j = 0, 1, \dots, 2p - 1$, satisfying the conditions $\phi_0 = 0, 0 < \phi_1 < \cdots < \phi_{2p-1}$ and also $\boldsymbol{\theta}_{p+j} = -\boldsymbol{\theta}_j$. Thanks to the last condition the set of the corresponding projections is redundant but this redundancy can simplify the analysis. Moreover the object and image space are given respectively by $\mathcal{X} = L^2(\mathcal{D})$ and $\mathcal{Y} = \{L^2([-1, 1], w^{-1})\}^{2p}$.

A detailed discussion of the results proved in the papers quoted above is beyond the scope of this book; the interested reader is referred to the original papers. We only point out that the singular values, denoted by $\sigma_{m,k}^{(p)}$, have a positive minimum and have a non-zero limit so that the generalized inverse exists and is continuous (see next chapter).

More precisely, in the case of equispaced projections, the singular values, with the first p values of m, coincide with the singular values of the complete Radon transform, with the same multiplicity and the same singular functions in object space. The total number of these singular values is $p(p+1)/2$. The minimum singular value in this group corresponds to $m = p-1$ and therefore it is given by $(4\pi/p)^{1/2}$, which is also the asymptotic limit of the singular values when $m \to \infty$. However, singular values smaller than the asymptotic limit exist in the other group. The Fourier transforms of the singular functions in object space, associated with these singular values, are characterized by an angular modulation around the observation directions $\boldsymbol{\theta}_j$ and therefore they are related to the aliasing effects due to insufficient sampling at high frequencies [176].

8

Inversion methods revisited

In Chapters 4 and 5, some of the classic inversion methods are presented and discussed in the case of image deconvolution. Here we use the singular value decomposition, presented in the previous chapter, for extending these methods to the more general linear inverse problems introduced in Chapter 6 and for understanding their main features. This chapter can also be used as a short summary of the content of the first part. At the end of the chapter we present two other well-known methods: a first one used for out-of-band extrapolation, a problem which is relevant in several instances as shown in Chapter 6, the second one used for X-ray tomography, namely the *filtered back-projection* (FBP), which is still relevant as a benchmark for the evaluation of more up-to-date inversion methods in the field.

8.1   The generalized solution

According to the mathematical description of a linear imaging system introduced in Section 7.1, the model of a noisy image of an object $f^{(0)}$ is given in equation (7.2) where A is a linear operator from the object space $\mathcal{X}$ into the image space $\mathcal{Y}$ and w is an element of $\mathcal{Y}$ describing the noise contribution to g. We assume that A is a matrix, i.e. an operator between vector spaces, or an operator between Hilbert spaces.

As in the case of image deconvolution, we no longer make explicit the noise term in equation (7.2) and we formulate the inverse problem as follows: given the noisy image g and the linear operator A describing the imaging system, solve the linear equation

$$Af = g \ . \tag{8.1}$$

In this chapter we assume that the operator A has a singular value decomposition so that we can write

$$Af = \sum_{j=1}^{p} \sigma_j \left(f, v_j\right)_{\mathcal{X}} u_j \tag{8.2}$$

DOI: 10.1201/9781003032755-8

where $p < \infty$ for a discrete or a semi-discrete problem (considered in Sections 7.2 and 7.3, respectively),while $p = \infty$ for the operators considered in Sections 7.4 and 7.5. Therefore, as far as possible, we treat simultaneously finite and infinite-dimensional problems. Only at the end of this section we consider separately the two cases.

We remark that, for simplicity, in equation (8.2) we use for the scalar product the notation used in the case of function spaces and not the one used in the case of vector spaces, otherwise in this equation we should write $\mathbf{f} \cdot \mathbf{v}_j$. Similarly we use the notation for an operator and not for a matrix.

The problem (8.1) is in general ill-posed in the sense that the solution is not unique, does not exist, or else does not depend continuously on the data.

Uniqueness does not hold when the null space of the operator A, $\mathcal{N}(A)$, is not trivial. For instance uniqueness never holds true in the case of the semi-discrete problems of Section 7.3, while it holds true in the case of a discrete problem if the rank of the matrix coincides with the number of unknowns. Uniqueness holds also true in the case of the inversion of the Radon transform, Section 7.5.

The most frequently used procedure for restoring uniqueness is the following. Any element f of the object space $\mathcal{X}$ can be represented by

$$f = \sum_{j=1}^{p} (f, v_j)_{\mathcal{X}} \, v_j + v \tag{8.3}$$

where v is the projection of f onto $\mathcal{N}(A)$ while the first term is the component of f orthogonal to $\mathcal{N}(A)$. The term v can be called the invisible component of the object f because it does not contribute to the image Af of f. Since this invisible component cannot be determined from equation (8.1), it may be natural to look for a solution of this equation whose invisible component is zero. Such a solution is unique because from equation (8.2) and equation (8.3), with $v = 0$, we easily deduce that $Af = 0$ implies $f = 0$. If this solution exists, it is denoted by $f^\dagger$ and is called the *minimal norm solution*. Indeed, any solution of equation (8.1) is given by

$$f = f^\dagger + v \tag{8.4}$$

where v is an arbitrary element of $\mathcal{N}(A)$. Since v is orthogonal to $f^\dagger$, we have

$$\|f\|_{\mathcal{X}}^2 = \|f^\dagger\|_{\mathcal{X}}^2 + \|v\|_{\mathcal{X}}^2 \tag{8.5}$$

and therefore the solution with $v = 0$, i.e. $f^\dagger$, is the solution of minimal norm.

As concerns the existence of a solution of equation (8.1) and, in particular, of $f^\dagger$, we first have to distinguish between the two following cases.

- The null space of A^*, $\mathcal{N}(A^*)$, contains only the zero element, $g = 0$. Then the singular functions (vectors) u_j constitute an orthonormal basis in $\mathcal{Y}$ and the noisy image g can be represented by

$$g = \sum_{j=1}^{p} (g, u_j)_{\mathcal{Y}} \, u_j \; . \tag{8.6}$$

  By comparing this representation with the SVD of A, equation (8.2), we see that a solution of equation (8.1) may exist. We reconsider this point in a moment.

- The null space of A^*, $\mathcal{N}(A^*)$, contains non-zero elements. In such a case the singular functions (vectors) u_j do not constitute an orthonormal basis in $\mathcal{Y}$ and the noisy image g can be represented as follows

$$g = \sum_{j=1}^{p} (g, u_j)_{\mathcal{Y}} \, u_j + u \tag{8.7}$$

where u is the component of g in $\mathcal{N}(A^*)$, i.e. the component of g orthogonal to the range of A. This term is analogous to the out-of-band noise which affects the noisy images formed by a bandlimited imaging system (see Section 3.5). Notice that, if the mathematical model of the imaging system is physically correct, the presence of this term is an effect of the noise. If $u \neq 0$, by comparing the representation (8.2) of Af with the representation (8.7) of g, we see that there does not exist any object f such that Af coincides with g. Then we can look for objects f such that Af is as close as possible to f, i.e. for objects which minimize the discrepancy functional

$$\|Af - g\|_{\mathcal{Y}}^2 = \text{minimum} . \tag{8.8}$$

Any solution of this variational problem is called a *least-squares solution.*

The concept of least-squares solution is more general than the concept of solution because a solution of equation (8.1) is also a least-squares solution but the converse is not true. Indeed, the set of the least-squares solutions coincide with the set of the solutions if and only if the minimum of the discrepancy functional (8.8) is zero. This remark shows that, without loss of generality, we can investigate the problem of existence in the case of the least-squares solutions.

Solving problem (8.8) is equivalent to solving the corresponding Euler-Lagrange equation which is given by

$$A^*Af = A^*g . \tag{8.9}$$

From the SVD of the operator A, equation (8.2), and of the operator A^*

$$A^*g = \sum_{j=1}^{p} \sigma_j \, (g, u_j)_{\mathcal{Y}} \, v_j \tag{8.10}$$

we obtain

$$A^*Af = \sum_{j=1}^{p} \sigma_j^2 \, (f, v_j)_{\mathcal{X}} \, v_j . \tag{8.11}$$

If we insert these representations into equation (8.9) and we compare the coefficients of v_j, we find that the components of any solution f of equation (8.9) are given by

$$\sigma_j^2 \, (f, v_j)_{\mathcal{X}} = \sigma_j \, (g, u_j)_{\mathcal{Y}} , \tag{8.12}$$

and therefore

$$(f, v_j)_{\mathcal{X}} = \frac{1}{\sigma_j} \, (g, u_j)_{\mathcal{Y}} . \tag{8.13}$$

In such a way the problem of the existence of least-squares solutions is reduced to the problem of the existence of elements of the object space $\mathcal{X}$ whose components with respect to the singular functions v_j are given by equation (8.13). To this purpose we consider separately the case $p < \infty$ and the case $p = +\infty$.

For the sake of simplicity, we first deal with the case $p < \infty$, even if this is not the natural order, because in the case of infinite-dimensional problems we must also consider convergence issues which may hamper the clarity of the treatment.

(A) *The case $p < \infty$*

According to equation (8.13) we introduce the following function (vector)

$$f^\dagger = \sum_{j=1}^{p} \frac{1}{\sigma_j} \, (g, u_j)_{\mathcal{Y}} \, v_j \tag{8.14}$$

which is a solution of equation (8.9) and therefore a least-squares solution.

As concerns uniqueness, let us first observe that the null space of the operator A and the null space of the operator A^*A coincide, $\mathcal{N}(A) = \mathcal{N}(A^*A)$, as already observed in Section 7.2. In any case the proof is easy and can be easily generalized to any linear operator. Thanks to this property, $f^\dagger$ is the unique least-squares solution if and only if $\mathcal{N}(A)$ contains only the zero element.

If uniqueness does not hold as a consequence of the existence of invisible objects, then any least-squares solution is given by equation (8.4), with $f^\dagger$ defined by equation (8.14) and therefore we can conclude that $f^\dagger$ is the *least-squares solution of minimal norm*. It is also called the *generalized solution* of the inverse problem and the results of the analysis outlined above can be summarized as follows: *in the case of a discrete or semi-discrete problem, for any image g (noise-free or noisy) there exists a unique generalized solution $f^\dagger$, whose singular function (vector) expansion is given by equation (8.14)*.

Equation (8.14) defines an operator from $\mathcal{Y}$ to $\mathcal{X}$, which is called the *generalized inverse operator* of A and denoted by $A^\dagger$

$$f^\dagger = A^\dagger g . \tag{8.15}$$

We observe that $A^\dagger g = 0$ if and only if $A^*g = 0$, so that $\mathcal{N}(A^\dagger) = \mathcal{N}(A^*)$.

In the case of a matrix $\mathbf{A}$, the generalized inverse matrix $\mathbf{A}^\dagger$ is also called the *Moore-Penrose inverse* of $\mathbf{A}$. If the SVD of $\mathbf{A}$ is given by equation (7.20), then the SVD of $\mathbf{A}^\dagger$ is given by

$$\mathbf{A}^\dagger = \mathbf{V}\mathbf{\Sigma}^{-1}\mathbf{U}^* , \tag{8.16}$$

as one can easily deduce from equation (8.14).

The generalized solution depends continuously on the image g and therefore $A^\dagger$ is a linear and continuous operator.

This property can be proved rather easily and, to this purpose as well as to the purpose of investigating the numerical stability of the solution, we first observe that only the component of g in the range of A, $\mathcal{R}(A)$, contributes to $f^\dagger$. This component is the first term in the r.h.s. of equation (8.7), and it can be denoted by $g^\dagger$ since it can be considered as the image of $f^\dagger$, i.e. $g^\dagger = Af^\dagger$. We can also write $f^\dagger = A^\dagger g^\dagger$. Let us write these relationships explicitly in terms of the singular system of the operator A

$$f^\dagger = \sum_{j=1}^{p} \frac{1}{\sigma_j} \left(g^\dagger, u_j\right)_{\mathcal{Y}} v_j \tag{8.17}$$

$$g^\dagger = \sum_{j=1}^{p} \sigma_j \left(f^\dagger, v_j\right)_{\mathcal{X}} u_j . \tag{8.18}$$

We also observe that the procedure which transforms g into $g^\dagger$ is analogous to the filtering operation considered in Section 3.5 for suppressing the out-of-band noise. Now, if $\delta g^\dagger$ is a variation of the image $g^\dagger$ and $\delta f^\dagger$ is the corresponding variation of $f^\dagger$, from equation (8.17), the linearity of the relationship and the fact that $\sigma_1 \geq \sigma_2 \geq \cdots \geq \sigma_p$, we obtain

$$\left\|\delta f^\dagger\right\|_{\mathcal{X}}^2 = \sum_{j=1}^{p} \frac{1}{\sigma_j^2} |\left(\delta g^\dagger, u_j\right)_{\mathcal{Y}}|^2 \leq \frac{1}{\sigma_p^2} \left\|\delta g^\dagger\right\|_{\mathcal{Y}}^2 . \tag{8.19}$$

This inequality proves, as already stated, that $f^\dagger$ depends continuously on $g^\dagger$ because, when $\delta g^\dagger$ tends to zero, also $\delta f^\dagger$ tends to zero.

Continuity, however, is not sufficient to ensure numerical stability, which can be investigated by estimating the relative error induced on $f^\dagger$ by the error on $g^\dagger$. To this purpose we can use the relation (8.18) to get the following inequality

$$\left\| g^\dagger \right\|_{\mathcal{Y}}^2 = \sum_{j=1}^{p} \sigma_j^2 \left| (f^\dagger, v_j)_{\mathcal{X}} \right|^2 \leq \sigma_1^2 \left\| f^\dagger \right\|_{\mathcal{X}}^2 , \tag{8.20}$$

which, combined with the inequality (8.19), provides the following estimate of the relative error

$$\frac{\| \delta f^\dagger \|_{\mathcal{X}}}{\| f^\dagger \|_{\mathcal{X}}} \leq \frac{\sigma_1}{\sigma_p} \frac{\| \delta g^\dagger \|_{\mathcal{Y}}}{\| g^\dagger \|_{\mathcal{X}}} . \tag{8.21}$$

The quantity

$$\alpha = \frac{\sigma_1}{\sigma_p} \tag{8.22}$$

is the *condition number* of the operator A, controlling the numerical stability of the inverse problem. It is the extension of the condition number introduced in Section 3.4, equation (3.30), and, as in that case, it is not difficult to prove that the inequality (8.21) cannot be improved even if it may be pessimistic.

According to the analysis by Twomey, already mentioned in Section 3.4, a better estimate of error propagation may be provided by the AREMF β which is now given by

$$\beta = \frac{1}{p} \left(\sum_{j=1}^{p} \sigma_j^2 \right)^{1/2} \left(\sum_{j=1}^{p} \sigma_j^{-2} \right)^{1/2} . \tag{8.23}$$

As in the case of the problems considered in Section 3.4, the condition number α or the AREMF β of a discrete or semi-discrete inverse problem can be quite large. Indeed, if the problem we are considering comes from the discretization of a problem with singular values tending to zero and if the discretization is fine enough, we expect some of its singular values to be small (even very small) so that the condition number can be extremely large. In such a case the generalized solution $f^\dagger$ is deprived of any physical meaning and cannot be accepted as a reasonable estimate of the unknown object $f^{(0)}$. We have to face the same difficulty as discussed in the case of the solution provided by the inverse filtering method for deconvolution problems.

(B) *The case $p = \infty$*

If we proceed as in the case $p < +\infty$, we can introduce the following formal solution

$$f^\dagger = \sum_{j=1}^{\infty} \frac{1}{\sigma_j} (g, u_j)_{\mathcal{Y}} v_j . \tag{8.24}$$

We say that this solution is formal because it is given by a series expansion so that the solution exists if and only if the series is convergent.

If we consider convergence in the sense of the norm of $\mathcal{X}$, then this convergence is ensured if and only if the sum of the squares of the coefficients of the singular functions v_j is convergent. We obtain the following condition

$$\sum_{j=1}^{\infty} \frac{1}{\sigma_j^2} \left| (g, u_j)_{\mathcal{Y}} \right|^2 < \infty \tag{8.25}$$

which is also called the *Picard criterion* for the existence of solutions or least-squares solutions of the linear inverse problem under consideration [141].

It is important to point out that, if the singular values σ_j accumulate to zero, as shown in several examples in Chapters 6 and 7, then condition (8.25) may not be satisfied by an arbitrary noisy image g. If it is not satisfied, then *no solution or least-squares solution of the inverse problem exists.*

The functions g satisfying the Picard criterion are the images in the range of A, $\mathcal{R}(A)$. For any of these functions, the series (8.24) defining $f^\dagger$ is convergent. Then by repeating the same arguments as in the case $p < \infty$ we can conclude that *for any image g satisfying the Picard criterion there exists a unique generalized solution $f^\dagger$, whose singular function expansion is given by equation (8.24).*

The generalized solution defines a generalized inverse operator $A^\dagger$ as in the case $p < \infty$. The main difference is that now this operator is not defined everywhere on $\mathcal{Y}$ but only on the set of the functions g satisfying the Picard criterion. This set is the domain of the operator $A^\dagger$, $\mathcal{D}(A^\dagger)$. Moreover the operator $A^\dagger$ is not continuous or, in other words, the generalized solution $f^\dagger$ does not depend continuously on the image g.

In order to prove this statement, let us assume that g is an image satisfying the Picard criterion and let us consider a sequence of images given by

$$g_j = g + \sqrt{\sigma_j}\, u_j \;. \tag{8.26}$$

It is obvious that

$$\|g_j - g\|_{\mathcal{Y}} = \sqrt{\sigma_j} \to 0 \tag{8.27}$$

because we have assumed that the singular values tend to zero. On the other hand, if we denote by $f^\dagger$ the generalized solution associated with g, and by $f_j^\dagger$ the generalized solution associated with g_j, we have

$$f_j^\dagger = f^\dagger + \frac{1}{\sqrt{\sigma_j}}\, v_j \tag{8.28}$$

so that

$$\|f_j^\dagger - f^\dagger\|_{\mathcal{X}} = \frac{1}{\sqrt{\sigma_j}} \to \infty \;. \tag{8.29}$$

In such a way we have found a sequence of images, converging to g, such that the sequence of the corresponding generalized solutions does not converge to $f^\dagger$. This pathology, related to the fact that the singular values tend to zero, generates the numerical instability, already discussed, of the discrete versions of the problem.

8.2   The Tikhonov regularization method

The generalized solution of an ill-posed or ill-conditioned problem is not physically meaningful because it is completely corrupted by noise propagation. This point has been discussed in detail in Chapter 3. For this reason we must look for approximate solutions satisfying additional constraints suggested by the physics of the problem and regularization is a way for obtaining such solutions. We reconsider here the Tikhonov regularization method introduced in Chapter 4 in the case of deconvolution problems.

The starting point is to define a family of *regularized solutions* f_μ, depending on the *regularization parameter* $\mu > 0$, as the family of the functions minimizing the functionals

$$\Phi_\mu(f; g) = \|Af - g\|_{\mathcal{Y}}^2 + \mu \|f\|_{\mathcal{X}}^2 \tag{8.30}$$

where g is the given image. The main formal difference with respect to the functional (4.6) is that the objects and the images may now belong to different spaces.

The Euler-Lagrange equation associated with the minimization of this functional is given by

$$(A^*A + \mu I)f = A^*g . \tag{8.31}$$

An object is a minimum point f_μ of the functional (8.30) if and only if it is a solution of this equation. In order to get the solution, let us represent an arbitrary element f of $\mathcal{X}$ in terms of the singular functions v_j of the operator A and of the elements v (orthogonal to all the v_j) of the null space of A, as in equation (8.3). Here p can be finite or infinite. If we insert this representation in equation (8.31) and if we take into account equations (8.10) and (8.11), we obtain

$$\sum_{j=1}^{p}(\sigma_j^2 + \mu)\,(f, v_j)_{\mathcal{X}}\,v_j + \mu v = \sum_{j=1}^{p} \sigma_j\,(g, u_j)_{\mathcal{Y}}\,v_j . \tag{8.32}$$

It follows that there exists a unique solution f_μ of equation (8.31), which can be obtained from equation (8.3) with $v = 0$ and with coefficients $(f, v_j)_{\mathcal{X}}$ given by

$$(\sigma_j^2 + \mu)\,(f, v_j)_{\mathcal{X}} = \sigma_j\,(g, u_j)_{\mathcal{Y}} . \tag{8.33}$$

In conclusion we get

$$f_\mu = \sum_{j=1}^{p} \frac{\sigma_j}{\sigma_j^2 + \mu}\,(g, u_j)_{\mathcal{Y}}\,v_j . \tag{8.34}$$

If $p = \infty$, the series at the r.h.s. of this equation is always convergent (thanks to the factors σ_j, the coefficients tend to zero more rapidly than the components $(g, u_j)_{\mathcal{Y}}$ of g) and therefore the regularized solution f_μ exists for any noisy image g and any μ.

The regularized solution f_μ can be rewritten in the form given by equations (4.25)-(4.26), i.e.

$$f_\mu = R_\mu\, g \tag{8.35}$$

$$R_\mu = (A^*A + \mu I)^{-1} A^* . \tag{8.36}$$

The operator R_μ is an approximation of the generalized inverse $A^\dagger$, equation (8.15), in the sense explained in Section 4.3. Indeed it is not difficult to prove that:

- for any $\mu > 0$, R_μ is a linear and continuous operator from $\mathcal{Y}$ into $\mathcal{X}$;
- for any image g, in the case $p < \infty$, and for any image g satisfying the Picard criterion in the case $p = \infty$, we have

$$\lim_{\mu \to 0} R_\mu g = A^\dagger g . \tag{8.37}$$

Therefore, the family of linear and bounded operators $\{R_\mu\}_{\mu>0}$ defines a family of *linear regularization operators for the approximation of the generalized inverse*. Moreover, in the case of noisy data, it is easy to extend the arguments used in Section 4.3 to show that this family is also defining a *regularization method* in the sense of Tikhonov.

REMARK 8.1   *In the case of a discrete problem, if its size is not excessive, then one can compute the regularized solution by computing R_μ, i.e. by inverting the matrix $\mathbf{A}_\mu = \mathbf{A}^*\mathbf{A} + \mu\mathbf{I}$. If this matrix must be inverted for several values of μ, a considerable saving in terms of operations can be obtained by a method due to Elden [112].*

By means of equation (8.34) we can compute all quantities which are relevant for analyzing the method.

- *Energy functional*

$$E^2(f_\mu) = \|f_\mu\|_\mathcal{X}^2 = \sum_{j=1}^{p} \frac{\sigma_j^2}{(\sigma_j^2 + \mu)^2} \left|(g, u_j)_\mathcal{Y}\right|^2 \; ; \tag{8.38}$$

- *Discrepancy functional*

$$\varepsilon^2(f_\mu; g) = \|Af_\mu - g\|_\mathcal{Y}^2 = \sum_{j=1}^{p} \frac{\mu^2}{(\sigma_j^2 + \mu)^2} \left|(g, u_j)_\mathcal{Y}\right|^2 + \|u\|_\mathcal{Y}^2 \; . \tag{8.39}$$

In equation (8.39) the representation (8.7) of the image g is used. We observe that, as in the case of image deconvolution, the energy is a decreasing function of the regularization parameter μ, while the discrepancy is an increasing function of μ.

If we assume that the noisy image g is given by equation (7.2), the difference between the regularized solution f_μ and the true object $f^{(0)}$ is given by

$$f_\mu - f^{(0)} = (R_\mu A f^{(0)} - f^{(0)}) + R_\mu w \; . \tag{8.40}$$

We have:

- *Approximation error*

$$\|R_\mu A f^{(0)} - f^{(0)}\|_\mathcal{X}^2 = \sum_{j=1}^{p} \frac{\mu^2}{(\sigma_j^2 + \mu)^2} \left|(f^{(0)}, v_j)_\mathcal{X}\right|^2 + \|v^{(0)}\|_\mathcal{X}^2 \tag{8.41}$$

where $v^{(0)}$ is the component of $f^{(0)}$ in the null space of A (i.e. the invisible component of $f^{(0)}$).

- *Noise-propagation error*

$$\|R_\mu w\|_\mathcal{X}^2 = \sum_{j=1}^{p} \frac{\sigma_j^2}{(\sigma_j^2 + \mu)^2} |(w, u_j)_\mathcal{Y}|^2 \; . \tag{8.42}$$

We observe that, as in the case of image deconvolution, the approximation error is an increasing function of μ while the noise-propagation error is a decreasing function of μ. As a consequence the total reconstruction error, given by

$$\varrho(\mu) = \|f_\mu - f^{(0)}\|_\mathcal{X} \tag{8.43}$$

is expected to have a minimum, i.e. there should exist an optimum value of the regularization parameter. We do not repeat here the discussion of this property, called semi-convergence, which was extensively discussed in Section 4.3.

As concerns the choice of the regularization parameter, the methods introduced in Section 4.6 can also be applied to the present case.

8.3   Truncated SVD

The representation (8.34) of the regularized solution can be recast in the following form

$$f_\mu = \sum_{j=1}^{p} \frac{W_{\mu,j}}{\sigma_j} (g, u_j)_\mathcal{Y} \, v_j \tag{8.44}$$

where

$$W_{\mu,j} = \frac{\sigma_j^2}{\sigma_j^2 + \mu} \; . \tag{8.45}$$

This expression shows that the regularized solution f_μ can be obtained by a filtering of the singular value decomposition of the generalized solution: the components of $f^\dagger$ corresponding to singular values much larger than μ are taken without any significant modification, whereas the components corresponding to singular values much smaller than μ are essentially removed.

This remark suggests to consider other kinds of filters in the vein of the filtering methods discussed in Section 4.4 for image deconvolution. Here we only discuss a method which has been widely used in the solution of linear inverse problem, the so-called *truncated* SVD. The idea is to replace the smooth filter given in equation (8.45) by a sharp one, i.e. to take in the singular function expansion of the generalized solution only the terms corresponding to singular values greater than a certain threshold value. Since the singular values are ordered to form a non-increasing sequence, those greater than the threshold value are those corresponding to values of the index less than a certain maximum integer.

Let us denote by J the number of singular values satisfying the condition

$$\sigma_j^2 \geq \mu \; , \tag{8.46}$$

then the approximate solution provided by the truncated SVD is defined as follows

$$f_J = \sum_{j=1}^{J} \frac{1}{\sigma_j} (g, u_j)_{\mathcal{Y}} \, v_j \; . \tag{8.47}$$

This equation can be obtained from equation (8.44) by taking $W_{\mu,j} = 1$ when $\sigma_j^2 \geq \mu$ and $W_{\mu,j} = 0$ when $\sigma_j^2 < \mu$. Moreover, it can be conveniently written in a form similar to that of equation (8.35) by introducing the operator $R^{(J)}$ from $\mathcal{Y}$ into $\mathcal{X}$ such that

$$f_J = R^{(J)} g \; . \tag{8.48}$$

If we observe that $J = J(\mu)$ increases when μ decreases and that $J(\mu) \to p$ when $\mu \to 0$, then it is not difficult to prove that $R^{(J)}$ has properties similar to R_μ, because, for any J, $R^{(J)}$ is a bounded operator from $\mathcal{Y}$ into $\mathcal{X}$; moreover $R^{(J)}$ converges to $A^\dagger$, in the sense of equation (8.37), when $J \to p$.

The computation of the relevant quantities for the analysis of the method is again very easy.

- *Energy functional*

$$E^2(f_J) = \|f_J\|_{\mathcal{X}}^2 = \sum_{j=1}^{J} \frac{1}{\sigma_j^2} \left| (g, u_j)_{\mathcal{Y}} \right|^2 \tag{8.49}$$

- *Discrepancy functional*

$$\varepsilon^2(f_J; g) = \|Af_J - g\|_{\mathcal{Y}}^2 = \sum_{j=J+1}^{p} \left| (g, u_j)_{\mathcal{Y}} \right|^2 + \|u\|_{\mathcal{Y}}^2 \; . \tag{8.50}$$

These expressions imply that the energy is an increasing function of J (and therefore a decreasing function of μ) while the discrepancy is a decreasing function of J (and therefore an increasing function of μ). These properties correspond to similar properties of the Tikhonov regularized solutions.

Let us now consider the difference between the truncated solution and the true object $f^{(0)}$. Using equation (8.48) and the model (7.2) for the noisy image g, we obtain

$$R^{(J)}g - f^{(0)} = (R^{(J)}Af^{(0)} - f^{(0)}) + R^{(J)}w . \qquad (8.51)$$

We have:

- *Approximation error*

$$\|R^{(J)}Af^{(0)} - f^{(0)}\|_{\mathcal{X}}^2 = \sum_{j=J+1}^{p} |(f^{(0)}, v_j)_{\mathcal{X}}|^2 + \|v^{(0)}\|_{\mathcal{X}}^2 \qquad (8.52)$$

  where $v^{(0)}$ is the component of $f^{(0)}$ in the null space of A.
- *Noise-propagation error*

$$\|R^{(J)}w\|_{\mathcal{X}}^2 = \sum_{j=1}^{J} \frac{1}{\sigma_j^2} \left|(w, u_j)_{\mathcal{Y}}\right|^2 . \qquad (8.53)$$

The approximation error is a decreasing function of J while the noise-propagation error is an increasing function of J. Therefore the total restoration error, $\rho_J = \|f_J - f^{(0)}\|_{\mathcal{X}}$, is expected to have a minimum for a certain value of J: when J increases the truncated SVD solution first approaches the true object $f^{(0)}$ and then goes away. It follows that also this method has the semi-convergence property which is typical of regularization methods.

The previous remark implies that there should exist an optimum number of terms in the truncated SVD. As usual the problem is to estimate this optimum number in the case of real data.

It is shown [95] that the Morozov discrepancy principle can provide such an estimate. Let ε^2 be an estimate of the norm of the noise term

$$\|w\|_{\mathcal{Y}}^2 \le \varepsilon^2 \qquad (8.54)$$

and let $b > 1$ be a fixed but arbitrary positive parameter. Then an estimate of the optimum number of terms is given by the integer $\tilde{J}$ such that

$$\|Af_{\tilde{J}} - g\|^2 \ge b\,\varepsilon^2 , \quad \|Af_{\tilde{J}+1} - g\|^2 < b\,\varepsilon^2 . \qquad (8.55)$$

The parameter b is introduced for theoretical (presumably technical) reasons because one needs the condition $b > 1$ to prove the convergence of the method (in the case $p = \infty$) for $\varepsilon \to 0$. In practice, we can take $b = 1$. The method provides a unique value $\tilde{J}$ because, as shown before, the discrepancy functional is a decreasing function of the number of terms in the truncated SVD .

If an estimate E^2 of the norm of the true object is also known

$$\|f^{(0)}\|_{\mathcal{X}}^2 \le E^2 \qquad (8.56)$$

then another method proposed by Miller [111] consists in taking as an estimate of the optimum number of terms the number of singular values satisfying the condition

$$\sigma_k^2 \ge \left(\frac{\varepsilon}{E}\right)^2 . \qquad (8.57)$$

Finally let us observe that in 2D problems the number of terms to be used in the truncated SVD can be rather large. For instance, in the case of tomography, the singular values are given by equation (7.96). They depend only on the index m and, for a given m, the multiplicity of the singular value is $m + 1$. Therefore, the total number of singular values with $m \leq m_0$ is $J_0 = (m_0 + 1)(m_0 + 2)/2$. If we consider, for instance, all the singular values greater than 10^{-1}, the maximum value of m is $m_0 = 1255$ and their total number is $J_0 = 788140$. If the threshold is 10^{-2}, then $m_0 = 125662$ and $J_0 \simeq 9 \, 10^9$. These large numbers occur because tomography is a mildly ill-posed problem. For other more severely ill-posed problems the number of terms can be much smaller.

8.4 Iterative regularization methods

In this section we discuss the extension of the iterative methods introduced in Chapter 5 to more general linear inverse problems. We consider the *Landweber, projected Landweber, steepest descent* and *conjugate gradient* methods. In the case of discrete or semi-discrete problems, the common feature of these methods is that the basic operation required for their implementation is a matrix-vector multiplication. Therefore these methods are particularly useful in practice when the matrix $\mathbf{A}$ is large and sparse.

(A) *The Landweber method*

As in the case of image deconvolution, the iterative scheme can be written as follows

$$f_{k+1} = f_k + \tau \left(A^* g - A^* A f_k \right) , \tag{8.58}$$

where f_k is the current iteration and τ is the relaxation parameter. For simplicity we consider only the most usual choice for the initial guess, i.e. $f_0 = 0$.

The first problem is to establish conditions on τ ensuring the convergence of the method. To this purpose we can insert in equation (8.58) the representations (8.10) and (8.11) of the operators A^* and $A^* A$. It follows that the components of the iterates with respect to the orthonormal system formed by the singular functions v_j satisfy the following recursive relation

$$(f_{k+1}, v_j)_\mathcal{X} = \tau \sigma_j \, (g, u_j)_\mathcal{Y} + (1 - \tau \sigma_j^2) \, (f_k, v_j)_\mathcal{X} . \tag{8.59}$$

On the other hand, the component of f_k in the null space of A is not modified by the iterations so that it must coincide with the corresponding component of f_0. Since we assume $f_0 = 0$, this component is also zero. From equation (8.59) we can easily derive by induction that

$$(f_k, v_j)_\mathcal{X} = \tau \left[1 + (1 - \tau \sigma_j^2) + \cdots + (1 - \tau \sigma_j^2)^{k-1} \right] \, \sigma_j \, (g, u_j)_\mathcal{Y} . \tag{8.60}$$

This equation is similar to equation (5.10), with σ_j^2 replacing $\hat{H}(\boldsymbol{\omega})$ and $\sigma_j \, (g, u_j)_\mathcal{Y}$ replacing $\hat{g}(\boldsymbol{\omega})$. Therefore, if we use equation (5.12), we get

$$(f_k, v_j)_\mathcal{X} = \left[1 - (1 - \tau \sigma_j^2)^k \right] \frac{1}{\sigma_j} \, (g, u_j)_\mathcal{Y} . \tag{8.61}$$

For any j, the sequence has a limit for $k \to \infty$ if and only if τ satisfies the condition

$$|1 - \tau \sigma_j^2| < 1 \text{ equivalent to } 0 < \tau < \frac{2}{\sigma_j^2} . \tag{8.62}$$

Since σ_1 is the largest singular value of the operator A, the conditions (8.62) are satisfied for any j if and only if

$$0 < \tau < \frac{2}{\sigma_1^2} . \tag{8.63}$$

This condition is analogous to condition (5.25) for the problem of image deconvolution. When it is satisfied, we get from equation (8.61) that

$$\lim_{k \to \infty} (f_k, v_j)_{\mathcal{X}} = \frac{1}{\sigma_j} (g, u_j)_{\mathcal{Y}} \; . \tag{8.64}$$

For a finite value of k, equation (8.61) implies the following representation of f_k in terms of the singular system of A

$$f_k = \sum_{j=1}^{p} \left[1 - (1 - \tau \sigma_j^2)^k\right] \frac{(g, u_j)_{\mathcal{Y}}}{\sigma_j} \; v_j \; . \tag{8.65}$$

Hence we find that f_k is a filtered version of the generalized solution $f^\dagger$. Moreover, we can deduce from equation (8.64) that, in the case $p = \infty$, the iterates f_k converge to $f^\dagger$ if the data satisfy the Picard criterion (8.25). In the discrete or semi-discrete case, the iterates f_k always converge to $f^\dagger$.

Since, in the case of an ill-conditioned or ill-posed problem, the limit is not physically meaningful we must investigate the regularization properties of the algorithm. We give again the expression of the most relevant quantities.

- *Energy functional*

$$E^2(f_k) = \|f_k\|_{\mathcal{X}}^2 = \sum_{j=1}^{p} \frac{1}{\sigma_j^2} \left|1 - (1 - \tau \sigma_j^2)^k\right|^2 \left|(g, u_j)_{\mathcal{Y}}\right|^2 \tag{8.66}$$

- *Discrepancy functional*

$$\varepsilon^2(f_k; g) = \|A f_k - g\|_{\mathcal{Y}}^2 = \sum_{j=1}^{p} (1 - \tau \sigma_j^2)^{2k} \left|(g, u_j)_{\mathcal{Y}}\right|^2 + \|u\|_{\mathcal{Y}}^2 \; . \tag{8.67}$$

We observe that the energy is an increasing function of k while the discrepancy is a decreasing function of k.

If we introduce the operator $R^{(k)}$ from $\mathcal{Y}$ into $\mathcal{X}$ associated with the mapping from g into f_k defined by equation (8.65)

$$f_k = R^{(k)} g \; , \tag{8.68}$$

then, by means of the usual model of the process of image formation, we can write the difference between f_k and the true object $f^{(0)}$ as follows

$$f_k - f^{(0)} = (R^{(k)} A f^{(0)} - f^{(0)}) + R^{(k)} w \; . \tag{8.69}$$

We have:

- *Approximation error*

$$\|R^{(k)} A f^{(0)} - f^{(0)}\|_{\mathcal{X}}^2 = \sum_{j=1}^{p} (1 - \tau \sigma_j^2)^{2k} |(f^{(0)}, v_j)_{\mathcal{X}}|^2 \tag{8.70}$$

- *Noise-propagation error*

$$\|R^{(k)} w\|_{\mathcal{X}}^2 = \sum_{j=1}^{p} \frac{1}{\sigma_j^2} |1 - (1 - \tau \sigma_j^2)^k|^2 \left|(g, u_j)_{\mathcal{Y}}\right|^2 \; . \tag{8.71}$$

It follows that the approximation error is a decreasing function of k while the noise-propagation error is an increasing function of k. As a result the restoration error, $\rho_k = \|f_k - f^{(0)}\|_{\mathcal{X}}$, is expected to have a minimum for a certain value of k, k_{opt}, so that the semi-convergence property holds true also in this case.

In the case of real data, an estimate $\tilde{k}$ of k_{opt} can be obtained by means of the Morozov discrepancy principle already discussed in Section 5.1. This method provides a unique value of $\tilde{k}$ because, as follows from equation (8.67) the discrepancy is a decreasing function of k.

Let us now comment on the implementation of the method for discrete or semi-discrete problems. If the singular system of the operator can be computed, then one can compute directly f_k by means of equation (8.65), i.e. one can consider the method as a filtering method and not as an iterative method. However, as we already observed, the computation of the singular system may not be practical for imaging problems. Then one has to implement the iterative scheme as given in equation (8.58).

In the case of a discrete problem, the basic operation in the implementation of equation (8.58) is a matrix-vector multiplication. This is not so obvious in the case of a semi-discrete problem but it is not difficult to modify equation (8.58) in such a way that the same property holds also in this case. Indeed, let us consider a semi-discrete mapping A_M, Section 7.3, defined by M linearly independent functions $\varphi_1, \varphi_2, \ldots, \varphi_M$. Then, as follows from equation (8.58), in the case $f_0 = 0$, f_k belongs to this subspace because it is in the range of the operator A_M^*. Since the null space of A_M^* contains only the null element, for any k there exists a unique vector $\mathbf{f}_k$ such that

$$f_k = A_M^* \mathbf{f}_k . \tag{8.72}$$

If we insert this representation into equation (8.58) and use again the property of the null space of A_M^*, we obtain the iteration

$$\mathbf{f}_{k+1} = \mathbf{f}_k + \tau(\mathbf{g} - A_M A_M^* \mathbf{f}_k) . \tag{8.73}$$

The operator $A_M A_M^*$ is the Gram matrix of the functions φ_m, so that we find that also in this case the implementation of equation (8.73) requires only a matrix-vector multiplication.

(B) *The projected Landweber method*

In the case of constrained least-squares solutions, the solutions being elements of a closed and convex set $\mathcal{C}$, the projected Landweber method can be used for obtaining approximations of these solutions. It is defined by the iterative scheme

$$f_{k+1}^{(\mathcal{C})} = P_{\mathcal{C}} \left[f_k^{(\mathcal{C})} + \tau(A^* g - A^* A f_k^{(\mathcal{C})}) \right] \tag{8.74}$$

where $P_{\mathcal{C}}$ is the projection operator onto the set $\mathcal{C}$. As in the case of the Landweber method one takes $f_0^{(\mathcal{C})} = 0$ and τ such that conditions (8.63) are satisfied.

In the case of a discrete or semi-discrete problem, the iterates $f_k^{(\mathcal{C})}$ converge, for $k \to \infty$, to a solution of the constrained least-squares problem

$$\|Af - g\|_{\mathcal{Y}} = \text{minimum} , \quad f \in \mathcal{C} . \tag{8.75}$$

When $p = \infty$, then we refer to the results of convergence discussed in Section 5.2.

We briefly consider the discrete and semi-discrete cases. In general, the solutions of problem (8.75) are unphysical; then, as follows from numerical practice, the method shows the semi-convergence property which is typical of the iterative regularization methods when applied to least-squares problems. Examples are given in Section 5.2.

From the computational point of view, the basic operations required at each step for the implementation of the method are a matrix-vector multiplication and a projection onto the set $\mathcal{C}$. In order to choose correctly the value of the relaxation parameter, an estimate of σ_1 is also required.

(C) *The steepest descent method*

The steepest descent method, shortly discussed at the beginning of Section 5.4, is defined by the following iterative scheme

$$f_{k+1} = f_k + \tau_k \, \bar{r}_k \tag{8.76}$$

where

$$\bar{r}_k = A^* g - A^* A f_k \ , \quad \tau_k = \frac{\|\bar{r}_k\|_{\mathcal{X}}^2}{\|A\bar{r}_k\|_{\mathcal{Y}}^2} \ . \tag{8.77}$$

It can be viewed as a Landweber method with a choice of the relaxation parameter depending on the iteration step.

The iterates f_k converge, for $k \to \infty$, to a least-squares solution in the case of a discrete or semi-discrete problem. If $p = \infty$ then it is proved [208] that the same result holds true if g is in the range of the operator A, or, in other words, if g satisfies the Picard criterion (8.25). Also the steepest descent method has the semi-convergence property and therefore it can be used as a regularization method.

(D) *The conjugate gradient method*

The method, discussed in Section 5.4, is defined by the following iterative scheme

$$\bar{r}_0 = \bar{p}_0 = A^* g \ , \quad f_0 = 0$$

$$\bar{\alpha}_k = \frac{\|\bar{r}_k\|_{\mathcal{X}}^2}{\|A\bar{p}_k\|_{\mathcal{Y}}^2} \ , \quad \bar{r}_{k+1} = \bar{r}_k - \bar{\alpha}_k A^* A \, \bar{p}_k \tag{8.78}$$

$$\bar{\beta}_k = \frac{\|\bar{r}_{k+1}\|_{\mathcal{X}}^2}{\|\bar{r}_k\|_{\mathcal{X}}^2} \ , \quad \bar{p}_{k+1} = \bar{r}_{k+1} + \bar{\beta}_k \, \bar{p}_k$$

$$f_{k+1} = f_k + \bar{\alpha}_k \, \bar{p}_k \ .$$

In the case of discrete or semi-discrete problems, the method converges to the generalized solution in M steps (M is the dimension of the data space) if $p = M$ and the Krylov subspace $\mathcal{K}^{(M)}(\bar{A}; \bar{g})$ has dimension M so that it coincides with the data space $\mathcal{Y}$. In the case $p = \infty$, it has been proved [132], that the iterates f_k converge, for $k \to \infty$, to the generalized solution $f^\dagger$, if g is in the range of the operator A and therefore satisfies the Picard criterion. It is also known that the method has the semi-convergence property, so that it is a powerful regularization method for the solution of linear inverse problems.

A glance at the iterative scheme (8.78) shows that also in this case the basic operation required at each step is a matrix-vector multiplication.

8.5 Super-resolution and out-of-band extrapolation

Nowadays, the term super-resolution is used with several different meaning; in particular, in microscopy, it is mainly used for indicating the improvement in resolution achievable with the new techniques such as STED microscopy (or nanoscopy). Historically, it was attributed to the possibility of improving the resolution of an optical instrument, such as a microscope or a telescope, by suitable data processing. In this section we refer to this last meaning and we show how, in such a case, the problem of super-resolution is strictly related to the problem of out-of-band extrapolation. A thorough presentation of the different facets and applications of resolution and super-resolution is the recent book by de Villers and Pike [93].

8.5.1 Super-resolution

In Chapters 2 and 6, several examples are given of imaging systems which provide bandlimited images of the objects to be analyzed, as e.g. optical or infrared telescopes, fluorescence microscopes or equipments for far-field acoustic holography. Other examples arise from inverse scattering problems in the Born approximation or in the physical-optics approximation. All of these systems provide a limited information in Fourier space about the imaged object.

Classical resolution limits induced by this limitation have been discussed briefly in Section 2.6. The most famous concept is the Rayleigh criterion for imaging by an ideal circular pupil with aperture Ω: two narrow peaks in the object $f^{(0)}$ cannot be distinguished in the image if their distance is smaller than the quantity $\delta = 1.22\pi/\Omega$, which is also of the order of $\lambda/2$, i.e. half the wavelength of the probing radiation (see equation (2.74)). Here the limit in resolution is strictly related to the loss of information at high frequencies.

Usually, one calls *super-resolution* the possibility of improving the classical resolving power of the instrument, i.e. to get a resolution better than the Rayleigh limit. To this purpose, one can think of deconvolving the image, but as we know, to overcome the ill-posedness of the problem, some regularization method has to be applied. As could be expected, we claim that limitations in resolution still apply to the reconstructed objects after deconvolution by means of a linear regularization method. Indeed, in the case of circular symmetry, the window functions $\hat{W}(\omega)$ characterizing these methods, as defined in Section 4.4, may be approximately one inside a disc with a certain radius $\tilde{\Omega}$ and negligible outside the same disc. The estimate of $\tilde{\Omega}$ depends mainly on the transfer function and on the noise affecting the data.

To support this claim and to fix the ideas, let us focus on the particular example of acoustic holography. In the case of planar surfaces (see Section 2.7), near-field acoustic holography is described by the PSF given in equation (2.94), with the corresponding TF given by equations (2.95) and (2.96). In this example the TF has a circular symmetry and is never zero. The unknown field amplitude on the boundary plane $x_3 = 0$ can be estimated by means of one of the basic regularization methods such as Tikhonov or Landweber. Since all linear methods provide, in general, quite similar results, we can take e.g. the Tikhonov method as a case example and discuss its application to the problem under consideration.

Assume that the regularization parameter has been determined using, for instance, the Miller method, equation (4.80) (this choice does also correspond to the Wiener filter method discussed in Section 12.2). If we denote by $\tilde{f}_0$ the corresponding regularized solution, then the relationship between $\tilde{f}_0$ and the unknown object $f^{(0)}(\mathbf{x})$ is given by (see Section 4.5)

$$\hat{\tilde{f}}_0(\omega) = \hat{W}_0(|\omega|)\,\hat{f}^{(0)}(\omega) + \hat{W}_0(|\omega|)\,\frac{\hat{w}(\omega)}{\hat{K}(|\omega|)} \tag{8.79}$$

where the second term is the noise contribution to $\tilde{f}_0$ and $\hat{W}_0(|\omega|)$ is the Tikhonov window function given by

$$\hat{W}_0(|\omega|) = \frac{|\hat{K}(|\omega|)|^2}{|\hat{K}(|\omega|)|^2 + (\varepsilon/E)^2}\,. \tag{8.80}$$

Now, let $\tilde{\Omega}$ be defined by

$$\hat{W}_0(\tilde{\Omega}) = \frac{1}{2} \ . \tag{8.81}$$

It follows that $\hat{W}_0(|\boldsymbol{\omega}|)$ is close to 1 for $|\boldsymbol{\omega}| < \tilde{\Omega}$ (remember that $|\hat{K}(\boldsymbol{\omega})|$ is exactly 1 for $|\boldsymbol{\omega}| < k = 2\pi/\lambda$) and very small for $|\boldsymbol{\omega}| > \tilde{\Omega}$. Therefore, if we neglect the noise term in equation (8.79) we see that $\widetilde{\hat{f}}_0(\boldsymbol{\omega})$ is approximately equal to $\hat{f}^{(0)}(\boldsymbol{\omega})$ for $|\boldsymbol{\omega}| < \tilde{\Omega}$ and approximately zero for $|\boldsymbol{\omega}| > \tilde{\Omega}$. We conclude that $\tilde{f}_0(\mathbf{x})$ provides, in practice, an $\tilde{\Omega}$-bandlimited approximation of $f_0(\mathbf{x})$. For the problem under consideration it is easy to compute $\tilde{\Omega}$. If we observe that condition (8.81) is equivalent to the following one

$$|\hat{K}(\tilde{\Omega})| = \frac{\varepsilon}{E} \ , \tag{8.82}$$

from equations (2.95) and (2.96) we get

$$\tilde{\Omega} = k \left[1 + \frac{1}{(ka)^2} \ln^2 \left(\frac{E}{\varepsilon} \right) \right]^{1/2} \ . \tag{8.83}$$

It is this frequency cut-off $\tilde{\Omega}$ which characterizes the resolving power of the imaging system after implementation of an inversion/regularization method to reconstruct the object. The resulting new resolution distance $\tilde{\delta} = 1.22\pi/\tilde{\Omega}$ should be compared to the Rayleigh limit $\delta = 1.22\pi/\Omega$. When $\tilde{\delta}$ is smaller than δ, we can say that *super-resolution* has been achieved. Let us observe that $\tilde{\Omega}$ and hence also $\tilde{\delta}$ depend on the noise through the quantity E/ε, this dependence being rather mild because of the logarithm in equation (8.83). More critically, in the present example of inverse diffraction, we see that $\tilde{\Omega}$ depends on the distance a between the two planes. When a is very large, i.e. in the far-field region, we see that $\tilde{\Omega} \cong k$, and therefore there is no gain in resolution with respect to the classical limit. On the other hand, if ka is small, i.e. $a < \lambda$, one can obtain a resolution much better than $\lambda/2$ so that a significant amount of super-resolution can be achieved in inverse diffraction from near-field data. This result is due to the combined effect of the existence of evanescent waves and of the use of regularization methods and explains why near-field acoustic holography (NFAH) and also scanning near-field optical microscopy (see Section 2.7) are considered as super-resolving techniques. A similar analysis applies to many other problems in image reconstruction.

REMARK 8.2   *When the* PSF *does not have a circular symmetry, the resolution depends on the direction in the image plane or image volume. Interesting examples can be found both in astronomy and in microscopy. For instance, in astronomy, the use of Fizeau interferometry, as in the instrument LBTI (Large Binocular Telescope Interferometer) of the Large Binocular Telescope (LBT) allows to observe images with a resolution in the direction of the baseline of the interferometer which is three times better than the resolution in the orthogonal direction. By recording images with different orientations and combining them with suitable regularization methods, one can obtain a resolution approximately uniform in all directions. This is not always possible because of possible observational restrictions due to the Earth rotation; however, even if the resolution is not uniform, one can get more important information than with a single image of one mirror of the telescope.*

The second important example is confocal microscopy. As discussed in Section 2.6.1, by means of the technique known as optical sectioning, confocal microscopes can produce 3D images; however, the resolution in the direction of the optical axis of the instrument is worse than that in the lateral directions by about a factor of 3. In this case, by means of regularization methods focused on the axial direction, it is possible to reduce the difference between the two resolutions.

Anyway, as already argued, the achieved limits in resolution are not expected to critically depend on the choice of a particular linear regularization method. The dependence may be more significant, however, on the available additional prior knowledge, such as non-negativity constraints, which could be implemented in the data inversion process by means of nonlinear methods.

In the next subsection, we will exploit the a priori information that the object has a bounded known support, a property which can still be implemented in linear methods. It implies that the Fourier transform of the object is an analytic function which can in principle be extrapolated outside the available frequency band. We will investigate this out-of-band extrapolation problem, its ill-posedness and its regularization, as well as the achievable gain in resolution. We will show that to obtain a significant gain, the diameter of the support of the object must be of the order of the resolution distance. In a wave propagation problem with far-field data, this means that the linear dimensions of the object must be of the order of the wavelength λ of the radiation.

8.5.2 Out-of-band extrapolation

In this section we discuss the problem of out-of-band extrapolation which is interesting because it can be analyzed in terms of singular systems and solved by very simple and classical iterative methods. We restrict the analysis to the 2D case, which is the most frequent one in imaging. However the extension to the 3D case is straightforward.

We consider an object $f^{(0)}(\mathbf{x})$ which is a function whose support is a bounded domain $\mathcal{D}$. We assume that this domain is sufficiently regular and has a finite measure (area) $m(\mathcal{D})$. As a consequence of this assumption, its Fourier transform $\hat{f}^{(0)}(\boldsymbol{\omega})$ is a so-called entire analytic function. Moreover, we assume that $\hat{f}^{(0)}(\boldsymbol{\omega})$ is known over a bounded domain $\mathcal{B}$ in frequency space and that this band $\mathcal{B}$ has also a finite non-zero measure (area) $m(\mathcal{B})$. The problem of out-of-band extrapolation is the problem of determining $f^{(0)}(\mathbf{x})$ from the limited knowledge of its Fourier transform.

In principle, this problem has a unique solution thanks to existence of a unique analytic continuation of the Fourier transform of $f^{(0)}(\mathbf{x})$. Indeed, if we have two solutions f_1, f_2, the Fourier transform of their difference, $f_1 - f_2$, is an entire analytic function which is zero over $\mathcal{B}$. Since an analytic function which is zero over a set with non-zero measure is zero everywhere, it follows that $f_1 = f_2$. Unfortunately this problem is ill-posed and no analytic continuation exists if the data are noisy. As usual in similar cases, the problem is to find an estimate of the object which provides an improvement with respect to the bandlimited approximation obtained by applying the inverse Fourier transform to the available data.

We remark that, if we can estimate the FT of $f^{(0)}$ over a broader domain, then we reach a knowledge of $f^{(0)}$ at higher frequencies and therefore we get information on finer details of $f^{(0)}$. Hence the problem of out-of-band extrapolation is related to the problem of super-resolution intended as resolution improvement by data processing.

An equivalent formulation of the problem of out-of-band extrapolation is the following: estimate a function $f^{(0)}(\mathbf{x})$, given a function $g^{(0)}(\mathbf{x})$ whose Fourier transform coincides with the Fourier transform of $f^{(0)}(\mathbf{x})$ on $\mathcal{B}$ and is zero outside $\mathcal{B}$, i.e.

$$\hat{g}^{(0)}(\boldsymbol{\omega}) = \chi_{\mathcal{B}}(\boldsymbol{\omega}) \, \hat{f}^{(0)}(\boldsymbol{\omega}) \tag{8.84}$$

where $\chi_{\mathcal{B}}(\boldsymbol{\omega})$ is the characteristic function of $\mathcal{B}$. If we denote by $H_{\mathcal{B}}(\mathbf{x})$ the inverse Fourier transform of $\chi_{\mathcal{B}}(\boldsymbol{\omega})$

$$H_{\mathcal{B}}(\mathbf{x}) = \frac{1}{(2\pi)^2} \int \chi_{\mathcal{B}}(\boldsymbol{\omega}) \, e^{i\mathbf{x}\cdot\boldsymbol{\omega}} \, d\boldsymbol{\omega} \tag{8.85}$$

then, by means of the convolution theorem, from equation (8.84) we obtain

$$g^{(0)}(\mathbf{x}) = \int_{\mathcal{D}} H_{\mathcal{B}}(\mathbf{x} - \mathbf{x}') \, f^{(0)}(\mathbf{x}') \, d\mathbf{x}' \qquad (8.86)$$

where we have used the *a priori* knowledge of the support of $f^{(0)}(\mathbf{x})$ by restricting the integral to the domain $\mathcal{D}$. This is a Fredholm integral equation of the first kind, corresponding to the integral operator

$$(Af)(\mathbf{x}) = \int_{\mathcal{D}} H_{\mathcal{B}}(\mathbf{x} - \mathbf{x}') \, f(\mathbf{x}') \, d\mathbf{x}' \qquad (8.87)$$

which transforms functions defined on $\mathcal{D}$ into functions defined everywhere. If we consider as usual square-integrable functions, then A is an operator from $L^2(\mathcal{D})$ into $L^2(\mathbb{R}^2)$. Its adjoint A^* is an operator from $L^2(\mathbb{R}^2)$ into $L^2(\mathcal{D})$ and is given by

$$(A^*g)(\mathbf{x}') = \int_{\mathbb{R}^2} H_{\mathcal{B}}^*(\mathbf{x} - \mathbf{x}') \, g(\mathbf{x}) \, d\mathbf{x} \, , \quad \mathbf{x}' \in \mathcal{D} \, . \qquad (8.88)$$

The uniqueness of the solution of the problem of out-of-band extrapolation implies that the null space of the operator A contains only the zero element. On the other hand, the functions in the range of the operator A are $\mathcal{B}$-bandlimited, as follows from equation (8.84). As a consequence, the null space of the operator A^* is the orthogonal complement of the subspace of the $\mathcal{B}$-bandlimited functions.

It is important to remark that the operator A is related to the bandlimiting projection operator $P^{(\mathcal{B})}$ defined in Section 3.5, equation (3.37). Equation (8.87) suggests that we can write $A = P^{(\mathcal{B})}P^{(\mathcal{D})}$, where $P^{(\mathcal{D})}$ is the projection onto the subspace of the square-integrable functions which are zero outside $\mathcal{D}$. To be pedantic, this form is not completely equivalent to equation (8.87). Indeed, this form implies that A is considered as an operator in $L^2(\mathbb{R}^2)$ and therefore has a null space which contains all square-integrable functions which are zero on $\mathcal{D}$. However, this form is very useful; by taking into account properties of orthogonal projection operators, namely that $P^* = P$ and $P^2 = P$, we immediately obtain that $A^* = P^{(\mathcal{D})}P^{(\mathcal{B})}$ and $A^*A = P^{(\mathcal{D})}P^{(\mathcal{B})}P^{(\mathcal{D})}$. The last relation is equivalent to the following representation of A^*A

$$(A^*Af)(\mathbf{x}) = \int_{\mathcal{D}} H_{\mathcal{B}}(\mathbf{x} - \mathbf{x}') \, f(\mathbf{x}') \, d\mathbf{x}' \, ; \quad \mathbf{x} \in \mathcal{D} \, . \qquad (8.89)$$

This self-adjoint and positive-definite operator is an integral operator whose properties have been extensively investigated by Slepian [256]. It is an operator of the Hilbert-Schmidt class since the operator A also belongs to this class. Indeed, if we define the L^2-norm of its kernel as in equation (7.63) (with $K(\mathbf{x}, \mathbf{x}') = H_{\mathcal{B}}(\mathbf{x} - \mathbf{x}')$) then, by means of Parseval's equality, we get

$$\|K\|^2 = m(\mathcal{D}) \int |H_{\mathcal{B}}(\mathbf{x})|^2 d\mathbf{x} = \frac{1}{(2\pi)^2} \, m(\mathcal{D}) \, m(\mathcal{B}) \, . \qquad (8.90)$$

It follows that A has a singular value decomposition with the properties described in Section 7.4. We denote by $\{\sigma_j; u_j, v_j\}$ its singular system. The singular values are the square roots of the eigenvalues of the operator (8.89), $\lambda_1 \geq \lambda_2 \geq \lambda_3 \geq \ldots$, i. e. $\sigma_j = \sqrt{\lambda_j}$, while the singular functions are related to the *generalized prolate spheroidal functions* introduced by Slepian [256] and defined as follows. If we denote the eigenfunctions associated to the eigenvalues $\lambda_1 \geq \lambda_2 \geq \lambda_3 \geq \ldots$ by $\psi_1, \psi_2, \psi_3, \ldots$, they form an orthogonal basis in $L^2(\mathcal{D})$ since the null space of A contains only the null element. These functions, defined on $\mathcal{D}$, can be extended to functions defined everywhere by means of the eigenvalue equation

$$\psi_j(\mathbf{x}) = \frac{1}{\lambda_j} \int_{\mathcal{D}} H_{\mathcal{B}}(\mathbf{x} - \mathbf{x}') \, \psi_j(\mathbf{x}') \, d\mathbf{x}' \qquad (8.91)$$

using the fact that the kernel is defined everywhere. The extended eigenfunctions, denoted also by ψ_j, are precisely the *generalized prolate spheroidal functions* introduced by Slepian in the paper mentioned above. These functions have *a double orthogonality* property. They are bandlimited, as follows from their definition, and are normalized to 1 with respect to the norm of $L^2(\mathbb{R}^2)$; moreover the eigenfunctions of A^*A are given by $P^{(\mathcal{D})}\psi_j$ and form an orthogonal basis in $L^2(\mathcal{D})$, but are not normalized to 1.

Indeed, let us compute the scalar product of two functions ψ_j. Expressing ψ_j and ψ_k in terms of projection operators (as follows from equation (8.91)) as well as the operator A^*A and taking into account the properties of the projection operators, we get

$$(\psi_j, \psi_k)_{L^2(\mathbb{R}^2)} = \frac{1}{\lambda_j \lambda_k}(P^{(\mathcal{B})}P^{(\mathcal{D})}\psi_j, P^{(\mathcal{B})}P^{(\mathcal{D})}\psi_k)_{L^2(\mathbb{R}^2)} \tag{8.92}$$

$$= \frac{1}{\lambda_j \lambda_k}(A^*AP^{(\mathcal{D})}\psi_j, P^{(\mathcal{D})}\psi_k)_{L^2(\mathcal{D})} = \frac{1}{\lambda_k}(P^{(\mathcal{D})}\psi_j, P^{(\mathcal{D})}\psi_k)_{L^2(\mathcal{D})} .$$

It follows that the scalar product is zero if $k \neq j$ and that the norm of the functions $P^{(\mathcal{D})}\psi_j$ is given by $\|P^{(\mathcal{D})}\psi_j\|^2 = \lambda_j$.

To summarize we have the following properties for the singular system of the operator A, equation (8.87):

- the singular values σ_j of the operator A are the square roots of the eigenvalues λ_j of the Slepian operator (8.89);
- the singular functions u_j coincide with the generalized prolate spheroidal functions

$$u_j(\mathbf{x}) = \psi_j(\mathbf{x}) ; \tag{8.93}$$

- the singular functions v_j are proportional to the generalized prolate spheroidal functions restricted to the domain $\mathcal{D}$ and precisely

$$v_j(\mathbf{x}) = \frac{1}{\sigma_j}P^{(\mathcal{D})}\psi_j(\mathbf{x}) . \tag{8.94}$$

It is possible to give now the solution of the problem of out-of-band extrapolation by solving the integral equation (8.86) in terms of the singular system of the operator A, namely

$$f(\mathbf{x}) = \sum_{j=1}^{\infty} \frac{1}{\sqrt{\lambda_j}}(g, u_j) \, v_j(\mathbf{x}) \tag{8.95}$$

where

$$(g, u_j) = \int g(\mathbf{x}) \, \psi_j^*(\mathbf{x}) \, d\mathbf{x} = \frac{1}{(2\pi)^2}\int_{\mathcal{B}} \hat{f}(\boldsymbol{\omega}) \, \hat{\psi}_j^*(\boldsymbol{\omega}) \, d\boldsymbol{\omega} . \tag{8.96}$$

This expression shows that $f(\mathbf{x})$ can be uniquely determined from the values of its Fourier transform on the set $\mathcal{B}$. In other words, it is the solution of the problem of analytic continuation expressed as a series of generalized prolate spheroidal functions.

The series (8.95) is convergent if and only if we know the exact values of $\hat{f}(\boldsymbol{\omega})$ on $\mathcal{B}$. This is equivalent to require that the coefficients (8.96) satisfy the Picard criterion (8.25). If we only have approximate values of $\hat{f}(\boldsymbol{\omega})$, then the series (8.95) in general is not convergent and even if it converges, it can give completely unreliable values of $f(\mathbf{x})$. As already remarked, the problem of out-of-band extrapolation is ill-posed and therefore one can only get estimates of $f(\mathbf{x})$ by the use of regularization methods. A consequence of this fact is that it is not possible to extrapolate $\hat{f}(\boldsymbol{\omega})$ everywhere but only over a bounded set $\mathcal{B}' \supset \mathcal{B}$. If the set $\mathcal{B}'$ is not significantly broader than $\mathcal{B}$, then no significant improvement in resolution can be obtained by means of out-of-band extrapolation. The improvement depends essentially on the rate of decay of the singular values but the discussion of this point is beyond the scope of this book.

From the previous remark it is obvious that, in order to estimate the feasibility of out-of-band extrapolation, one would have to compute the eigenvalues and the eigenfunctions of the Slepian operator (8.89). This is not easy in the general case of arbitrary sets $\mathcal{B}$ and $\mathcal{D}$ so that methods requiring this computation are not useful in practice. Nevertheless, results are available in the rather frequent case where both $\mathcal{B}$ and $\mathcal{D}$ are discs, with radii respectively Ω and R [256, 124]. In such a case the operator A is given by

$$(Af)(\mathbf{x}) = \frac{\Omega}{2\pi} \int_{|\mathbf{x}'| \leq R} \frac{J_1(\Omega|\mathbf{x} - \mathbf{x}'|)}{|\mathbf{x} - \mathbf{x}'|} f(\mathbf{x}') \, d\mathbf{x}' \ . \tag{8.97}$$

A simple change of variables, transforming the disc of radius R into the disc of radius 1, shows that its eigenfunctions, called *circular prolate functions* depend only on the parameter

$$c = R \, \Omega \tag{8.98}$$

which is the so-called *space-bandwidth product*. We will see in a moment that this parameter is of critical importance for evaluating the amount of achievable out-of-band extrapolation. An even simpler situation is the case where both $\mathcal{B}$ and $\mathcal{D}$ are squares of sides 2Ω and $2R$, respectively. Using again a change of variables which transforms the square of side $2R$ into the square of side 2 and introducing the parameter c, as defined in equation (8.98), we find the corresponding operator

$$(Af)(\mathbf{x}) = \int_{-1}^{1} dx_1' \int_{-1}^{1} dx_2' \, \frac{\sin c(x_1 - x_1')}{\pi(x_1 - x_1')} \, \frac{\sin c(x_2 - x_2')}{\pi(x_2 - x_2')} f(x_1', x_2') \ . \tag{8.99}$$

The singular system of this operator can be given in terms of the eigenvalues and eigenfunctions of the Slepian-Pollack operator [257]

$$(A_{SP}f)(x) = \int_{-1}^{1} \frac{\sin c(x - x')}{\pi(x - x')} f(x') \, dx' \ , \quad |x| \leq 1 \ . \tag{8.100}$$

Its eigenfunctions are the well-known *prolate spheroidal wave functions* (PSWF) denoted by $\psi_k(c, x)$, $k = 0, 1, 2, \ldots$. If we denote by λ_k the eigenvalue of A_{SP} associated with $\psi_k(c, x)$, then the singular system of the operator (8.99) is given by

$$\begin{aligned}
\sigma_{j,k} &= \sqrt{\lambda_j \lambda_k} \ ; \quad j, k = 0, 1, 2, \ldots \\
u_{j,k}(\mathbf{x}) &= \psi_j(c, x_1) \, \psi_k(c, x_2) \\
v_{j,k}(\mathbf{x}) &= \frac{1}{\sqrt{\lambda_j \lambda_k}} \, \psi_j(c, x_1) \, \psi_k(c, x_2) \ .
\end{aligned} \tag{8.101}$$

$$\tag{8.102}$$

Using the properties of the eigenvalues of the Slepian-Pollack operator, it is possible to understand to what extent out-of-band extrapolation is feasible and, in particular, how this depends on the parameter c. If the space-bandwidth product is sufficiently large, the prolate eigenvalues $\lambda_k = \lambda_k(c)$ have a rather typical behavior as a function of k: they are approximately equal to 1 for $k < 2c/\pi$ and tend to zero very rapidly for $k > 2c/\pi$. The quantity

$$S = \frac{2c}{\pi} = \frac{2\Omega R}{\pi} \tag{8.103}$$

is called the *Shannon number* [285] and has a very simple meaning: it is the number of sampling points, spaced by the Nyquist distance π/Ω, interior to the interval $[-R, R]$. According to such behavior, if we consider a regularized solution provided by a truncated SVD, for any reasonable threshold value the number of terms will be of the order of S^2, i.e. of the order of the number of sampling points interior to the square of side $2R$. Since in this case the sampling distance coincides with the resolution distance, it is clear that no improvement of resolution has been obtained. The result is interpreted by saying that no significant amount of out-of-band extrapolation can be achieved when c is large, i.e. when the size of the support of f is large with respect to the resolution distance π/Ω. The situation is different when c is not much larger than 1, i.e. when the linear dimensions of the support of the object f are comparable with the resolution distance. Clearly this condition can be satisfied in microscopy or astronomy. In such a case, using tables of the prolate eigenvalues, it is possible to show [35] that the number of terms in a truncated SVD solution can be greater than the Shannon number S, even by factors of 2 or 3. Hence super-resolution is achieved as well as a significant amount of out-of-band extrapolation.

8.5.3 The Gerchberg-Papoulis method

In the case of arbitrary regions $\mathcal{B}$ and $\mathcal{D}$, a regularized approximation of the SVD expansion (8.95)-(8.96) can be easily obtained by means of a generalization of the very simple iterative scheme proposed by Gerchberg [130] and Papoulis [227] in the 1D case, and usually referred to as the Gerchberg-Papoulis method. In fact it turns out that this method is a very clever implementation of the Landweber method for this particular problem.

In order to derive the Gerchberg-Papoulis method from the Landweber method for out-of-band extrapolation, it is convenient to write the integral operator (8.87) in terms of projection operators, as remarked above. We first observe that, since A is the product of two non-commuting projection operators, its norm is certainly smaller than 1, i.e. $\|A\| < 1$. It follows that the relaxation parameter τ can take values between 0 and 2 and that we can choose $\tau = 1$, the choice made in the original method.

The iterative scheme of the Landweber method, combined with the representation of the operators A, A^* and A^*A in terms of projection operators, provides the following iteration

$$f_{k+1} = f_k + P^{(\mathcal{D})}[P^{(\mathcal{B})}g - P^{(\mathcal{B})}P^{(\mathcal{D})}f_k] \qquad (8.104)$$

where g is the $\mathcal{B}$-bandlimited data function yielding the approximate Fourier transform of the object on the band $\mathcal{B}$ (see equation (8.84)). If we take as initial guess $f_0 = 0$, we find

$$f_1 = P^{(\mathcal{D})}P^{(\mathcal{B})}g = P^{(\mathcal{D})}g \qquad (8.105)$$

where we have used the property $P^{(\mathcal{B})}g = g$. Since this relation implies that $P^{(\mathcal{D})}f_1 = f_1$, by induction, we can derive that

$$P^{(\mathcal{D})}f_k = f_k \qquad (8.106)$$

for any k. Equation (8.106) and the linearity of the projection operators show that the iterative scheme equation (8.104) is equivalent to the following one

$$f_{k+1} = P^{(\mathcal{D})}[g + (I - P^{(\mathcal{B})})f_k] . \qquad (8.107)$$

The implementation of this method is precisely the Gerchberg-Papoulis algorithm, generalized here to the case of an arbitrary number of variables and of arbitrary domains:

- compute $f_1(\mathbf{x}) = (P^{(\mathcal{D})}g)(\mathbf{x})$

- from $f_k(\mathbf{x})$ compute $\hat{f}_k(\boldsymbol{\omega})$
- compute $\hat{g}_{k+1}(\boldsymbol{\omega})$ given by

$$\hat{g}_{k+1}(\boldsymbol{\omega}) = \hat{g}(\boldsymbol{\omega}) + (1 - \chi_{\mathcal{B}}(\boldsymbol{\omega}))\,\hat{f}_k(\boldsymbol{\omega}) \tag{8.108}$$

- from $\hat{g}_{k+1}(\boldsymbol{\omega})$ compute $g_{k+1}(\mathbf{x})$
- compute $f_{k+1}(\mathbf{x}) = (P^{(\mathcal{D})}g_{k+1})(\mathbf{x})$.

A few remarks are in order. The projection operator $P^{(\mathcal{D})}$ is easily implemented because it amounts to zeroing a function outside $\mathcal{D}$. Analogously the third step, equation (8.108), amounts to replacing $\hat{f}_k(\boldsymbol{\omega})$ on $\mathcal{B}$ by the data values $\hat{g}(\boldsymbol{\omega})$, without modifying $\hat{f}_k(\boldsymbol{\omega})$ outside $\mathcal{B}$ (since $\hat{g}(\boldsymbol{\omega})$ is zero outside $\mathcal{B}$). It follows that the implementation of the method requires essentially the computation of one direct and one inverse Fourier transform.

Since the Gerchberg-Papoulis method is a special version of the Landweber method, its convergence and regularization properties are a direct consequence of the same properties of the Landweber method, already discussed in the previous section. Moreover, it is obvious that, at step k, it provides, in a very cheap way from the computational point of view, a filtered version of the singular function expansion (8.95)-(8.96), since we have

$$f_k(\mathbf{x}) = \sum_{j=1}^{\infty} \frac{1 - (1 - \lambda_j)^k}{\sqrt{\lambda_j}}\,(g, u_j)\, v_j(\mathbf{x}) \ . \tag{8.109}$$

It is interesting to remark that the Gerchberg-Papoulis method can be further generalized to the inversion of the operator (7.62), i.e. the inversion of a convolution operator with the additional information that the object to be estimated has a support interior to a domain $\mathcal{D}$. If we set $Af = K * f$, where K is the PSF of the imaging system, then the basic equation to solve can be written in the following form

$$g = A\,P^{(\mathcal{D})}f \ . \tag{8.110}$$

If we write the Landweber method for the operator $A_{\mathcal{D}} = AP^{(\mathcal{D})}$, we obtain

$$f_{k+1} = f_k + \tau P^{(\mathcal{D})}\left[A^*g - A^*AP^{(\mathcal{D})}f_k\right] \ . \tag{8.111}$$

In the case $f_0 = 0$, we can prove again by induction that $P^{(\mathcal{D})}f_k = f_k$, for any k, so that equation (8.111) is equivalent to the following one

$$f_{k+1} = P^{(\mathcal{D})}[f_k + \tau\,(A^*g - A^*Af_k)] \ . \tag{8.112}$$

When written in this form the method looks like a projected Landweber method. This form is important for its implementation which is the same as for the projected Landweber method and which is repeated here for the reader's convenience:

- compute $f_1(\mathbf{x}) = \tau\,(P^{(\mathcal{D})}A^*g)\,(\mathbf{x})$
- from $f_k(\mathbf{x})$ compute $\hat{f}_k(\boldsymbol{\omega})$
- compute $\hat{h}_{k+1}(\boldsymbol{\omega})$ given by

$$\hat{h}_{k+1}(\boldsymbol{\omega}) = \tau\hat{K}^*(\boldsymbol{\omega})\hat{g}(\boldsymbol{\omega}) + (1 - \tau|\hat{K}(\boldsymbol{\omega})|^2)\hat{f}_k(\boldsymbol{\omega}) \tag{8.113}$$

- from $\hat{h}_{k+1}(\boldsymbol{\omega})$ compute $h_{k+1}(\mathbf{x})$
- compute $f_{k+1}(\mathbf{x}) = \left(P^{(\mathcal{D})}h_{k+1}\right)(\mathbf{x})$.

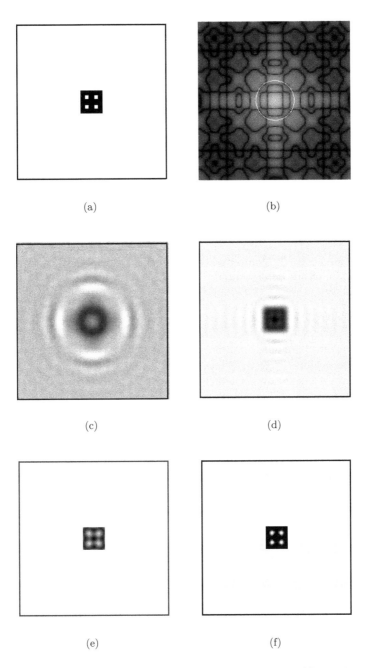

FIGURE 8.1   Example of super-resolution in far-field acoustic holography. (a) The object: a grid $2.5\lambda \times$ 2.5λ with bars 0.5λ wide. (b) The modulus of the FT of the object in (a); the white circle indicates the band of the far-field data. (c) The modulus of the noisy image at the distance 5λ from the object plane. (d) The reconstruction obtained by means of the inverse filtering method. (e) The reconstruction obtained by means of the iterative method with the constraint of bounded support. (f) The reconstruction obtained by means of the iterative method with the constraints of bounded support and non-negativity.

We remark that, even if the method looks like a projected Landweber method for the operator A, it is the Landweber method for the operator $A_{\mathcal{D}} = AP^{(\mathcal{D})}$ and therefore it has the same convergence and regularization properties as the Landweber method. Moreover, it provides a filtered version of the solution of the problem as given in terms of the singular system of the operator $A_{\mathcal{D}}$. Indeed, while the operator A is a convolution operator so that it does not have a singular value decomposition, the operator $A_{\mathcal{D}}$ has a singular value decomposition, as discussed in Section 7.4 (Remark 7.1).

It is also important to observe that the projected Landweber method can be used for introducing the constraint of non-negativity in addition to the constraint on the support of the object. Since the projection operator P_+ onto the convex set of all nonnegative functions commutes with the orthogonal projection $P^{(\mathcal{D})}$, as one can easily verify,

$$P_+ P^{(\mathcal{D})} = P^{(\mathcal{D})} P_+ \, , \qquad (8.114)$$

it follows that the projection operator onto the convex set of the nonnegative functions with support interior to $\mathcal{D}$ is given by

$$P_{\mathcal{C}} = P_+ P^{(\mathcal{D})} \, . \qquad (8.115)$$

This operator can be easily computed, so that the corresponding projected Landweber method can also be easily implemented.

We illustrate the method by means of an application to the problem of far-field acoustic holography, considered in Section 3.5. In such a case we know that we can use the inverse filtering method because the problem of determining the generalized solution is well posed. The resolution achievable is of the order of $\lambda/2$, λ being the wavelength of the monochromatic acoustic radiation.

We consider again a binary object which consists of a grid $2.5\lambda \times 2.5\lambda$ with vertical and horizontal bars 0.5λ wide. Therefore the size of this object, which is shown in Figure 8.1(a), is not much larger than the resolution distance. If we apply the inverse filtering method to the noisy image of the object at a distance 5λ from the boundary plane (see Figure 8.1(c)), the structure of the grid is not clearly resolved, as shown in Figure 8.1(d). The basic reason is illustrated in Figure 8.1(b): the bandlimiting due to wave propagation annihilates an important part of the Fourier spectrum of the object. The result obtained by means of the iterative method (8.112) where the iterates are constrained to be zero outside the square $2.5\lambda \times 2.5\lambda$ is shown in Figure 8.1(e). The improvement with respect to the inverse filtering solution is visible, even if the size of the object is larger than the resolution distance $\lambda/2$ by a factor 5. Finally, in Figure 8.1(f), we give the result obtained when also the non-negativity constraint on the support of the object is implemented in the iterative method. The restoration obtained is now quite satisfactory.

8.6   The filtered back-projection (FBP) method in tomography

In Section 6.2, it is shown that the basic problem in X-ray tomography is the inversion of the Radon transform R defined in equations (6.11)-(6.12). If we denote by $g(s, \boldsymbol{\theta})$ the measured values of the Radon transform of an unknown function f, then the problem is to solve the linear equation $Rf = g$.

In Section 7.5, the singular system of R is given, assuming that the object f is a square-integrable function with support interior to the disc of radius 1, while g belongs to a suitable weighted space of square-integrable functions. From this result we deduce that Radon inversion is ill-posed because the singular values of R accumulate to zero. The singular system, however, is not very convenient in practice for computing the regularized solutions because,

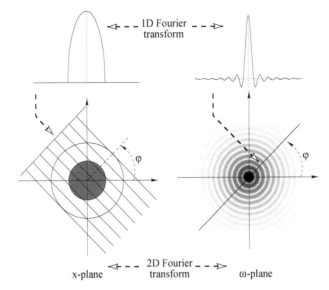

FIGURE 8.2 Illustrating the Fourier slice theorem in the case of a function f which is the characteristic function of a disc.

as observed at the end of Section 8.3, for a given noise level the number of singular values which must be used is exceedingly large. This fact, which is negative from the computational point of view, is positive from the point of view of the accuracy of suitable regularized solutions, because it implies that the problem is mildly ill-posed. In other words, it is very important to know that it is possible to design methods which provide a high accuracy for the reconstructed objects. One of these methods is the *filtered back-projection* presented in the present section. Since it is based on the Fourier transform, it also leads to a fast algorithm for the inversion of the Radon transform.

The first step in the derivation of the method is the so-called *Fourier slice theorem*, which can be formulated as follows: *let* $(R_{\boldsymbol{\theta}}f)\,(s) = (Rf)\,(s, \boldsymbol{\theta})$ *be the projection of* f *in the direction* $\boldsymbol{\theta}$, *then the Fourier transform of* $R_{\boldsymbol{\theta}}f$ *is the Fourier transform of* f *on the straight line passing through the origin and having direction* $\boldsymbol{\theta}$, *i.e. the following relation holds true*

$$(\widehat{R_{\boldsymbol{\theta}}f})(\omega) = \hat{f}(\omega\boldsymbol{\theta}) \; . \tag{8.116}$$

We point out that the FT in the l.h.s. of equation (8.116) is one-dimensional while the FT in the r.h.s. is two-dimensional. Figure 8.2 provides an illustration of this result.

In order to prove equation (8.116) we insert the definition of $R_{\boldsymbol{\theta}}f$ into the expression of its Fourier transform. We get

$$
\begin{aligned}
(\widehat{R_{\boldsymbol{\theta}}f})(\omega) &= \int_{-\infty}^{+\infty} e^{-i\omega s}(R_{\boldsymbol{\theta}}f)(s)\,ds \\
&= \int_{-\infty}^{+\infty} e^{-i\omega s}\left(\int_{-\infty}^{+\infty} f(s\boldsymbol{\theta} + t\boldsymbol{\theta}^{\perp})\,dt\right)ds \; .
\end{aligned}
\tag{8.117}
$$

The variables $\{s, t\}$ are the coordinates of a point $\mathbf{x} = \{x_1, x_2\}$ with respect to the Cartesian system defined by the unit vectors $\boldsymbol{\theta}, \boldsymbol{\theta}^{\perp}$, so that the relation between $\{s, t\}$ and $\{x_1, x_2\}$ is given by $\mathbf{x} = s\boldsymbol{\theta} + t\boldsymbol{\theta}^{\perp}$. Since this change of variables is a rotation, its Jacobian is one. If we further remark that $s = (\boldsymbol{\theta} \cdot \mathbf{x})$, we see that equation (8.117) can be written as follows

$$(\widehat{R_{\boldsymbol{\theta}}f})(\omega) = \int f(\mathbf{x})\exp\left(-i\omega(\boldsymbol{\theta} \cdot \mathbf{x})\right)d\mathbf{x} \tag{8.118}$$

and this is precisely the content of equation (8.116).

This very simple result has several important consequences:

- *The solution of the equation $Rf = g$, when it exists, is unique.* Indeed, if a function f is such that $Rf = 0$, this means that all projections of f are zero as well as their Fourier transforms. The Fourier slice theorem implies that $\hat{f} = 0$, so that $f = 0$.

- *The singular functions $v_{m,k}$ of R, given in Section 7.5, equations (7.99)-(7.100), form an orthonormal set in the space of all square-integrable functions with support in the disc of radius 1.* Indeed, the singular functions in the object space form, in general, an orthonormal basis in the orthogonal complement of the null space of R. Since this null space contains only the null element they are a basis in the object space.

- *The knowledge of the Radon transform of f implies the knowledge of the Fourier transform of f.* More precisely, since the observed projections g are affected by noise, by computing their Fourier transform, we obtain a noisy version of the Fourier transform of f.

The last point implies that the inversion of the Radon transform can be obtained by means of a Fourier transform inversion. This inversion, however, is not completely trivial because the Fourier transform of f, as given by the Fourier transforms of the projections, is represented in polar coordinates.

In order to prepare the tools required to derive the inversion formula, we introduce the *back-projection operator* $R^{\#}$, defined as follows: let $g(s, \boldsymbol{\theta})$ be a function defined for $-\infty < s < \infty$, $0 \leq \phi < 2\pi$, then

$$\left(R^{\#}g\right)(\mathbf{x}) = \int_0^{2\pi} g(\mathbf{x} \cdot \boldsymbol{\theta}, \boldsymbol{\theta}) \, d\phi . \tag{8.119}$$

Therefore $R^{\#}$ transforms the data function (represented by the sinogram) into a function of the space variables $\mathbf{x}$, i.e. an element of the object space.

REMARK 8.3   *The notation $R^{\#}$ derives from the fact that this operator can be formally viewed as the adjoint of the operator R when we consider the usual L^2 scalar product for both the data functions and the objects. Indeed, we have*

$$\begin{aligned}
(Rf, g)_{\mathcal{Y}} &= \int_0^{2\pi} \left(\int_{-\infty}^{+\infty} (Rf)(s, \boldsymbol{\theta}) \, g(s, \boldsymbol{\theta}) \, ds \right) d\phi \tag{8.120} \\
&= \int_0^{2\pi} \left[\int_{-\infty}^{+\infty} \left(\int_{-\infty}^{+\infty} f(s\boldsymbol{\theta} + t\boldsymbol{\theta}^{\perp}) \, dt \right) g(s, \boldsymbol{\theta}) \, ds \right] d\phi ,
\end{aligned}$$

and if we replace, for a given $\boldsymbol{\theta}$, the variables $\{s, t\}$ with the variables $\{x_1, x_2\}$ defined by $\mathbf{x} = s\boldsymbol{\theta} + t\boldsymbol{\theta}^{\perp}$, we obtain

$$\begin{aligned}
(Rf, g)_{\mathcal{Y}} &= \int_0^{2\pi} \left(\int f(\mathbf{x}) \, g(\mathbf{x} \cdot \boldsymbol{\theta}, \boldsymbol{\theta}) \, d\mathbf{x} \right) d\phi \tag{8.121} \\
&= \int f(\mathbf{x}) \left(\int_0^{2\pi} g(\mathbf{x} \cdot \boldsymbol{\theta}, \boldsymbol{\theta}) \, d\phi \right) d\mathbf{x} = \left(f, R^{\#}g \right)_{\mathcal{X}} .
\end{aligned}$$

In the last step an exchange in the order of integration is performed.

It is interesting to point out that, while the Radon transform integrates over all points in a straight line, the back-projection operator integrates over all straight lines through a point. Indeed, the integral of equation (8.119) is the integral of $g(s, \boldsymbol{\theta})$ along the curve $s = (\mathbf{x} \cdot \boldsymbol{\theta}) = x_1 \cos \phi + x_2 \sin \phi$. For a given $\mathbf{x}$, the points of this curve correspond to the straight lines through the point $\mathbf{x}$.

Let $g(s, \boldsymbol{\theta})$ be the Radon transform of a function $f(\mathbf{x})$, $g = Rf$. If $f(\mathbf{x}_1)$ is larger (smaller) than $f(\mathbf{x}_2)$, with $\mathbf{x}_1$ close to $\mathbf{x}_2$, then the integrals over the straight lines through $\mathbf{x}_1$ are, in general, larger (smaller) than the integrals over the straight lines through $\mathbf{x}_2$. It follows that $\left(R^{\#} g\right)(\mathbf{x}_1)$ is also larger (smaller) than $\left(R^{\#} g\right)(\mathbf{x}_2)$. This remark suggest that $R^{\#} g = R^{\#} Rf$ provides a blurred image of f. Indeed, it is possible to prove ([218], chapter 2) that

$$\left(R^{\#} Rf\right)(\mathbf{x}) = 2 \int \frac{f(\mathbf{x}')}{|\mathbf{x} - \mathbf{x}'|} \, d\mathbf{x}' . \tag{8.122}$$

This result implies the possibility of using the methods of image deconvolution for estimating the unknown object $f^{(0)}$. Filtered back-projection, however, provides a more elegant and practical solution of this problem.

In order to derive this algorithm, since the Fourier slice theorem provides the FT of f in polar coordinates, we must obtain a suitable expression of the inverse Fourier transform, not in the usual form

$$f(\mathbf{x}) = \frac{1}{(2\pi)^2} \int_{-\infty}^{+\infty} \int_{-\infty}^{+\infty} \hat{f}(\boldsymbol{\omega}) \, \exp\left(i(\boldsymbol{\omega} \cdot \mathbf{x}) \, d\omega_x d\omega_y \right. \tag{8.123}$$

but in terms of the polar coordinates, $\omega = |\boldsymbol{\omega}|$, $\phi = \arctan(\omega_y / \omega_x)$, $\boldsymbol{\theta} = \{\cos \phi, \sin \phi\}$; it is given by

$$f(\mathbf{x}) = \frac{1}{(2\pi)^2} \int_0^{2\pi} \left(\int_0^{+\infty} \omega \, \hat{f}(\omega\boldsymbol{\theta}) \, \exp\left(i\omega(\boldsymbol{\theta} \cdot \mathbf{x})\right) \, d\omega \right) d\phi . \tag{8.124}$$

Now we split the integral with respect to ϕ into two integrals, respectively over $[0, \pi]$ and $[\pi, 2\pi]$ obtaining the following equation

$$\begin{aligned} f(\mathbf{x}) &= \frac{1}{(2\pi)^2} \int_0^{\pi} \left(\int_0^{+\infty} \omega \, \hat{f}(\omega\boldsymbol{\theta}) \, \exp\left(i\omega(\boldsymbol{\theta} \cdot \mathbf{x})\right) \, d\omega \right) d\phi \\ &+ \frac{1}{(2\pi)^2} \int_{\pi}^{2\pi} \left(\int_0^{+\infty} \omega \, \hat{f}(\omega\boldsymbol{\theta}) \, \exp\left(i\omega(\boldsymbol{\theta} \cdot \mathbf{x})\right) \, d\omega \right) d\phi . \end{aligned} \tag{8.125}$$

If we observe that $\hat{f}(\omega\boldsymbol{\theta})$ is a periodic function of ϕ, with period 2π, by replacing ϕ with $\phi + \pi$, i.e. $\boldsymbol{\theta}$ with $-\boldsymbol{\theta}$, in the second integral we obtain

$$\begin{aligned} f(\mathbf{x}) &= \frac{1}{(2\pi)^2} \int_0^{\pi} \left(\int_0^{+\infty} \omega \, \hat{f}(\omega\boldsymbol{\theta}) \, \exp\left(i\omega(\boldsymbol{\theta} \cdot \mathbf{x})\right) \, d\omega \right) d\phi \\ &+ \frac{1}{(2\pi)^2} \int_0^{\pi} \left(\int_0^{+\infty} \omega \, \hat{f}(-\omega\boldsymbol{\theta}) \, \exp\left(-i\omega(\boldsymbol{\theta} \cdot \mathbf{x})\right) \, d\omega \right) d\phi \end{aligned} \tag{8.126}$$

From this expression, by means of a further change of variable in the second integral, which consists in replacing ω with $-\omega$, we get

$$\begin{aligned} f(\mathbf{x}) &= \frac{1}{(2\pi)^2} \int_0^{\pi} \left(\int_0^{+\infty} \omega \, \hat{f}(\omega\boldsymbol{\theta}) \, \exp\left(i\omega(\boldsymbol{\theta} \cdot \mathbf{x})\right) \, d\omega \right) d\phi \\ &+ \frac{1}{(2\pi)^2} \int_0^{\pi} \left(\int_{-\infty}^{0} (-\omega) \, \hat{f}(\omega\boldsymbol{\theta}) \, \exp\left(i\omega(\boldsymbol{\theta} \cdot \mathbf{x})\right) \, d\omega \right) d\phi . \end{aligned} \tag{8.127}$$

Finally, by observing that in the second integral $-\omega = |\omega|$ and recombining the two integrals over ω we obtain the desired representation of $f(\mathbf{x})$ in terms of its FT in polar coordinates

$$f(\mathbf{x}) = \frac{1}{(2\pi)^2} \int_0^{\pi} \left(\int_{-\infty}^{+\infty} |\omega| \, \hat{f}(\omega\boldsymbol{\theta}) \, \exp\left(i\omega(\boldsymbol{\theta} \cdot \mathbf{x})\right) \, d\omega \right) d\phi . \tag{8.128}$$

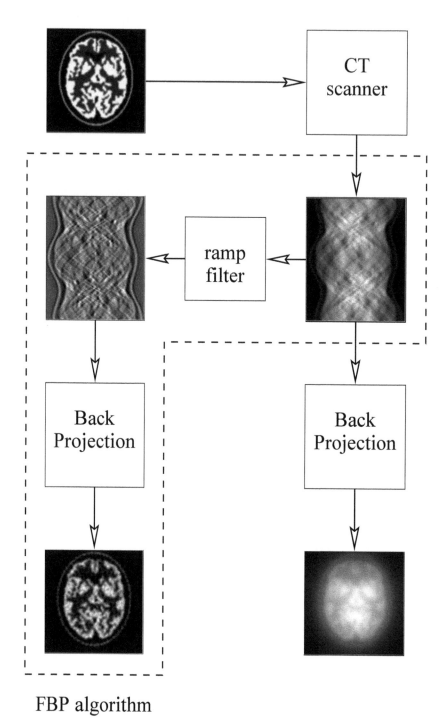

FBP algorithm

FIGURE 8.3 Scheme of the FBP algorithm and comparison of the results obtained by applying the back-projection operator to the original sinogram (to the right) and to the filtered one (to the left). In this figure we represent the sinogram with values of ϕ between 0 and 2π.

Since this FT of f is related to the Fourier transform of its projections by the Fourier slice theorem, if we denote by $\hat{g}(\omega, \boldsymbol{\theta})$ the FT of the measured projections, then $\hat{f}(\omega\boldsymbol{\theta}) = \hat{g}(\omega, \boldsymbol{\theta})$. By inserting this relation in equation (8.128) we get

$$f(\mathbf{x}) = \frac{1}{(2\pi)^2} \int_0^\pi \left(\int_{-\infty}^{+\infty} |\omega|\, \hat{g}(\omega, \boldsymbol{\theta})\, \exp\left(i\omega(\boldsymbol{\theta} \cdot \mathbf{x})\right) d\omega \right) d\phi \qquad (8.129)$$

and this formula is the key of the *filtered back-projection* (FBP) algorithm. Indeed, if we define the *filtered projections* $G(s, \boldsymbol{\theta})$ as follows

$$G(s, \boldsymbol{\theta}) = \frac{1}{2\pi} \int_{-\infty}^{+\infty} |\omega|\, \hat{g}(\omega, \boldsymbol{\theta})\, \exp\left(i\omega s\right) d\omega \;, \qquad (8.130)$$

then equation (8.129) can be written in the following form

$$f = \frac{1}{4\pi} R^\# G \;, \qquad (8.131)$$

i.e. f is obtained by applying the back-projection operator to the filtered projections. In conclusion, the inversion procedure can be decomposed into the following steps:

- for each value of $\boldsymbol{\theta}$ compute the Fourier transform $\hat{g}(\omega, \boldsymbol{\theta})$ of $g(s, \boldsymbol{\theta})$;
- multiply $\hat{g}(\omega, \boldsymbol{\theta})$ by the so-called ramp filter $|\omega|$;
- compute the inverse Fourier transform of $|\omega|\, \hat{g}(\omega, \boldsymbol{\theta})$ to obtain the filtered projections $G(s, \boldsymbol{\theta})$;
- apply the back-projection operator to $G(s, \boldsymbol{\theta})$.

Figure 8.3 illustrates this scheme. In this figure we also show the difference between the original sinogram and the filtered one as well as the difference between the reconstruction provided by FBP and the one obtained by applying the back-projection operator to the original sinogram.

The filtered back-projection algorithm provides further insight into the numerical instability of tomography. The ill-posedness of the problem is already established by means of the computation of the singular values in Section 7.5, because it is found that the singular values of the Radon transform tend to zero. As a consequence, the Picard criterion, equation (8.25), is not satisfied by arbitrary data functions. This criterion is now replaced by the condition that

$$\int_{-\infty}^{+\infty} |\omega|^2 |\hat{g}(\omega, \boldsymbol{\theta})|^2 d\omega < \infty \qquad (8.132)$$

which is also not satisfied by arbitrary square-integrable functions. In other words the ramp filter can amplify the high frequency noise in such a way that the restored function is not acceptable from the physical point of view.

In order to avoid this effect it is necessary to regularize the second step of the FBP algorithm or else to introduce a low-pass filter $\hat{W}_\Omega(|\omega|)$ characterized by a cut-off frequency Ω. Then the second step must be modified as follows:

- multiply $\hat{g}(\omega, \boldsymbol{\theta})$ by the ramp filter $|\omega|$ and a suitable low-pass filter $W_\Omega(|\omega|)$.

It follows that the global filter $F_\Omega(|\omega|) = |\omega|\, W_\Omega(|\omega|)$ has in general a behavior similar to the one plotted in Figure 8.4. The filtered projections, including the low-pass filter, are now given by

$$G_\Omega(s, \boldsymbol{\theta}) = \frac{1}{2\pi} \int_{-\infty}^{+\infty} |\omega|\, W_\Omega(|\omega|)\, \hat{g}(\omega, \boldsymbol{\theta}) e^{-i\omega s}\, d\omega \qquad (8.133)$$

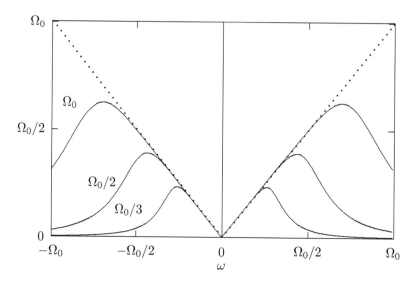

FIGURE 8.4   Picture of the filter $F_\Omega(|\omega|) = |\omega| W_\Omega(|\omega|)$ on an interval $[-\Omega_0, \Omega_0]$, when $W_\Omega(|\omega|)$ is a Butterworth filter – equation (4.64) – of order $n = 10$, for various values of Ω (given in terms of Ω_0). The dotted line represents the pure ramp filter.

and the back-projection operator must be applied to these filtered projections.
If we denote by f_Ω the approximation of the unknown object $f^{(0)}$ which can be obtained in this way

$$f_\Omega = \frac{1}{4\pi} R^{\#} G_\Omega \tag{8.134}$$

and if we assume the usual model for the noisy data, i.e. $g = Rf^{(0)} + w$, then it is easy to show that

$$f_\Omega(\mathbf{x}) = \frac{1}{(2\pi)^2} \int W_\Omega(|\boldsymbol{\omega}|)\, \hat{f}^{(0)}(\boldsymbol{\omega})\, e^{i\mathbf{x}\cdot\boldsymbol{\omega}}\, d\boldsymbol{\omega} + \frac{1}{4\pi}\left(R^{\#} w_\Omega\right)(\mathbf{x}) , \tag{8.135}$$

where $w_\Omega(s, \boldsymbol{\theta})$ is the noise contribution to $G_\Omega(s, \boldsymbol{\theta})$. Therefore the result is a noisy and bandlimited approximation of $f^{(0)}(\mathbf{x})$ and some of the considerations of the two previous subsections also apply to this case.

9

Edge-preserving regularization

In Section 4.5 we discuss one of the artifacts produced by quadratic Tikhonov regularization, namely the appearance of Gibbs oscillations in the neighborhood of sharp intensity variations as those due to the boundary of an object over a background. A regularization approach able to suppress these artifacts was introduced in the nineties as an extension of a denoising method proposed by Rudin, Osher and Fatemi. The approach is formulated in a Banach space and precisely the space of functions of bounded variation or BV space in the terminology of functional analysis. In this chapter we first provide a short, non-technical account of this method starting from its original denoising formulation; next we discuss its extension to inverse problems showing both the improvements and the new artifacts introduced by the method.

9.1 Total variation (TV) based image denoising and deblurring

The problem of image denoising is quite important in practice since a strong perturbation of a signal due to noise may render its interpretation and analysis difficult. For this reason, methods for 'cleaning' a signal or an image were proposed since the early developments in the field of signal and image processing.

The problem of denoising is not an inverse problem. If g is the noisy signal (image) and $g^{(0)}$ the noise-free one, then a least-squares approach for estimating a denoised image f leads to a trivial problem: indeed the minimization of the L^2-norm $\|f - g\|$ has the trivial solution $f = g$. However, as stated in [211], 'image denoising is among the most fundamental problems in image processing, not only for the sake of improving the image quality, but also as the first proof-of-concept for the development of virtually any new regularization term for inverse problems in imaging'. This remark applies in particular to the variational methods which have been proposed for denoising.

We first point out that, also in denoising, a rough model of a noisy image is very frequently used, similar to the one used for linear inverse problems, equation (7.2), i.e.

$$g(\mathbf{x}) = g^{(0)}(\mathbf{x}) + w(\mathbf{x}) , \qquad (9.1)$$

DOI: 10.1201/9781003032755-9

in the continuous case and

$$g_m = g_m^{(0)} + w_m , \quad m = 1, 2, \ldots, M , \tag{9.2}$$

in the discrete one. In both cases g is the observed image and $g^{(0)}$ the true one.

The literature on denoising is extremely vast and many researchers are still attracted by this problem in spite of the excellent results achievable by means of a variety of methods. An account is provided in a 2005 review paper by Buades et al. [55] where eight different denoising methods [197, 229, 6, 244, 310, 108, 69] are described and compared, including one proposed by the authors. Two of the methods mentioned in [55] can be extended to linear inverse problems and provide important generalizations of the quadratic regularizations discussed so far in the previous chapters. In this chapter we focus on the method proposed in [244] and in the next chapter on the method proposed in [108].

The *total variation* (TV) method is a variational approach proposed for denoising in 1992 by Rudin, Osher and Fatemi [244]. It is a deep method which inspired a large number of beautiful and interesting mathematical papers, first on denoising and slightly later on data inversion (this application being already mentioned in the original paper). It is out of the scope of this book to give a complete account of the wide mathematical literature on this subject. We only provide an account of the main ideas and properties of the method, with examples of practical applications.

In [244], thanks to previous experience of two of the authors on shock related images, it is pointed out that the appropriate space for solutions with discontinuities is a space of functions with bounded variation, the so-called BV-space; therefore the proposed method for the estimation of the denoised image f consists in the minimization of the total variation (TV) of f, defined on a bounded domain $\mathcal{D}$, with a constraint on the noise:

$$\min \int_{\mathcal{D}} |\nabla f(\mathbf{x})| \, d\mathbf{x} , \quad \text{subject to} \quad \int_{\mathcal{D}} |f(\mathbf{x}) - g(\mathbf{x})|^2 \, d\mathbf{x} = \varepsilon^2 . \tag{9.3}$$

In this formulation we omit a constraint on the integral of f, introduced in the original formulation and derived from the assumption that the image is perturbed by additive and white Gaussian noise (see Section 11.1) with zero mean; indeed, this constraint is automatically satisfied by the solution of the problem.

We point out that this problem has a structure similar to that of the problem considered in Section 4.2, even if the mathematics required for its solution is much more refined in the present case. Indeed, it consists in the minimization of a functional (a norm in Section 4.2, a seminorm here) with a constraint on the noise which derives from *a priori* information on it. For instance, in the case of a discrete image with M pixels/voxels, if we know that it is perturbed by additive and white Gaussian noise with zero mean and known variance σ^2, then one can choose $\varepsilon^2 = M\sigma^2$.

The problem (9.3) is usually called the ROF problem from the initials of the three authors. The solution method they propose is based on the solution of a nonlinear parabolic equation related to the Euler-Lagrange equation, which in $\mathcal{D}$ is given by

$$0 = \frac{\partial}{\partial x} \left(\frac{f_x}{\sqrt{f_x^2 + f_y^2}} \right) + \frac{\partial}{\partial y} \left(\frac{f_y}{\sqrt{f_x^2 + f_y^2}} \right) - \lambda(f - g) \tag{9.4}$$

with the boundary condition $\partial f/\partial n = 0$ on $\partial \mathcal{D}$. As it has been remarked [274], the proposed solution method can actually be viewed as a model case of the denoising method of *anisotropic diffusion* introduced by Perona and Malik [229].

Let us now give the abstract definition of the total variation of a function, a definition not used in [244] for computational reasons. For a discussion of this point and of the use of TV in imaging we refer to the lecture notes by Chambolle et al [62] while for an analysis of the properties of the functions of bounded variation we refer to the book by Giusti [133]. For a function f, defined and integrable on a bounded domain $\mathcal{D} \subset \mathbb{R}^d$, its total variation is defined as follows: let φ be a vector-valued function whose components $\{\varphi_i\}_{i=1}^d$ are infinitely differentiable with bounded support interior to $\mathcal{D}$, i.e $\varphi_i \in C_0^\infty(\mathcal{D})$, and, in addition, satisfy the condition $\sum_{i=1}^d \varphi_i^2(\mathbf{x}) \leq 1$, then the total variation of f on $\mathcal{D}$ is defined by

$$V_\mathcal{D}(f) = \sup_\varphi \left\{ -\int_\mathcal{D} f(\mathbf{x}) \sum_{i=1}^d \frac{\partial \varphi_i}{\partial x_i}(\mathbf{x}) \, d\mathbf{x} \right\} \tag{9.5}$$

and it is easy to prove that it is a *convex function* of f. A function of bounded variation in $\mathcal{D}$ is a function such that $V_\mathcal{D}(f) < \infty$. It can be shown that the linear space of the functions of bounded variation equipped with the norm

$$\|f\|_{BV(\mathcal{D})} = \|f\|_{L^1(\mathcal{D})} + V_\mathcal{D}(f) \tag{9.6}$$

is a complete normed space and therefore a Banach space. Hence a Tikhonov theory for the reconstruction of functions of bounded variation requires that we leave the familiar framework of Hilbert spaces for working into the framework of Banach spaces.

REMARK 9.1   *If f is differentiable, then, by partial integration and by taking into account that the functions φ_i are zero on the boundary of $\mathcal{D}$, it is easy to recognize that the following equality holds true*

$$-\int_\mathcal{D} f(\mathbf{x}) \sum_{i=1}^d \frac{\partial \varphi_i}{\partial x_i}(\mathbf{x}) \, d\mathbf{x} = \int_\mathcal{D} \sum_{i=1}^d \varphi_i(\mathbf{x}) \frac{\partial f}{\partial x_i}(\mathbf{x}) \, d\mathbf{x} \,. \tag{9.7}$$

Next, by applying the Schwarz inequality to the integrand of the r.h.s. and taking into account the bound on the functions φ_i, we obtain the following bound for the integral in the r.h.s.

$$\int_\mathcal{D} \sqrt{\sum_{i=1}^d \left(\frac{\partial f}{\partial x_i}\right)^2} \, d\mathbf{x} \tag{9.8}$$

which coincides with the functional of equation (9.3). Finally one can prove that the bound is reached by observing that, for the computation of the sup of the r.h.s. of equation (9.7), one can replace φ_i with the functions

$$\frac{\partial f}{\partial x_i} \frac{1}{\sqrt{\sum_{i=1}^d \left(\frac{\partial f}{\partial x_i}\right)^2}} \tag{9.9}$$

as it is allowed by suitable continuity properties of the functional.

Another interesting functional to be considered in the context of functions of bounded variation is that introduced in [3] to overcome the computational difficulties due to the non-differentiability of the TV functional (see also [299], chapter 8); it is given by

$$S_{\delta,\mathcal{D}}(f) = \int_\mathcal{D} \sqrt{\delta^2 + |\nabla f(\mathbf{x})|^2} \, d\mathbf{x} \tag{9.10}$$

where δ has to be intended as a small parameter such that $S_{\delta,\mathcal{D}}(f)$ provides a suitable approximation of the TV functional as written in equation (9.3).

The definition (9.10) applies to the case of a differentiable function but it can be extended to functions of bounded variation in a way similar to that used for the TV functional. To this purpose, in addition to the functions $\varphi_i \in C_0^\infty(\mathcal{D})$ introduced above, we introduce another function $\varphi_0 \in C_0^\infty(\mathcal{D})$ such that $\varphi_0^2(\mathbf{x}) + \sum_{i=1}^d \varphi_i^2(\mathbf{x}) \le 1$ for any $\mathbf{x}$ and we define the functional as follows (φ now includes also φ_0)

$$S_{\delta,\mathcal{D}}(f) = \sup_\varphi \left\{ \int_\mathcal{D} \left(\delta\,\varphi_0(\mathbf{x}) - f(\mathbf{x}) \sum_{i=1}^d \frac{\partial \varphi_i}{\partial x_i}(\mathbf{x}) \right) d\mathbf{x} \right\} . \tag{9.11}$$

The functional $S_{\delta,\mathcal{D}}(f)$ is also convex but it is not a semi-norm as the TV functional $V_\mathcal{D}(f)$.

REMARK 9.2   *In order to show that the new definition encompasses the previous one in the case of a differentiable function, by partial integration we obtain that the integral appearing in equation (9.11) is equal to*

$$\int_\mathcal{D} \left(\delta\,\varphi_0(\mathbf{x}) + \sum_{i=1}^d \varphi_i(\mathbf{x}) \frac{\partial f}{\partial x_i}(\mathbf{x}) \right) d\mathbf{x} \le \int_\mathcal{D} \sqrt{ \delta^2 + \sum_{i=1}^d \left(\frac{\partial f}{\partial x_i}(\mathbf{x}) \right)^2 } \, d\mathbf{x} \tag{9.12}$$

the upper bound resulting from applying again the Schwarz inequality to the integrand and taking into account the bound on φ. Moreover, one can prove that the upper bound coincides with the sup thanks to the continuity properties of the functional.

REMARK 9.3   *In the case $\delta = 1$, equation (9.10) becomes the well-known formula of the surface area of the graph of a differentiable function f*

$$S_\mathcal{D}(f) = \int_\mathcal{D} \sqrt{1 + |\nabla f(\mathbf{x})|^2} \, d\mathbf{x} . \tag{9.13}$$

However it must be remarked that this formula is valid if the quantity $|\nabla f(\mathbf{x})|$ is dimensionless. In practical applications, however, the function f may be a physical quantity (a pressure, a density of some quantity, etc.), hence has a dimension; similarly, the variables $\mathbf{x}$ may have a dimension (length, angle, etc.). In such a case, $|\nabla f(\mathbf{x})|$ in general has a dimension and the formula (9.13) is not valid.
To obtain a valid alternative one can introduce an (arbitrary) quantity δ with the dimension of $|\nabla f(\mathbf{x})|$. Then, one can apply the formula (9.13) to $\delta^{-1}|\nabla f(\mathbf{x})|$ and obtain

$$\int_\mathcal{D} \sqrt{1 + \left(\frac{1}{\delta}|\nabla f(\mathbf{x})| \right)^2} \, d\mathbf{x} = \frac{1}{\delta} \int_\mathcal{D} \sqrt{\delta^2 + |\nabla f(\mathbf{x})|^2} \, d\mathbf{x} . \tag{9.14}$$

We re-obtain equation (9.10), except for the factor $1/\delta$, but now δ has a physical dimension and can be a large number even if it should be smaller than possible jumps in the values of $f(\mathbf{x})$. This point must be clear if the functional (9.10) is used in applications.

A lower and upper bound of $S_{\delta,\mathcal{D}}(f)$ in terms of $V_\mathcal{D}(f)$ can be easily obtained. By comparing the definition (9.5) of the total variation and the definition (9.11) of $S_{\delta,\mathcal{D}}(f)$ it is easy to show that, for any δ,

$$V_\mathcal{D}(f) \le S_{\delta,\mathcal{D}}(f) \le V_\mathcal{D}(f) + m(\mathcal{D})\,\delta \tag{9.15}$$

where $m(\mathcal{D})$ is the measure of the bounded domain $\mathcal{D}$. The first inequality is obvious since the function φ_0 realizing the sup is certainly nonnegative; the second inequality follows from the remark that the first term in the integrand of equation (9.11) is bounded by δ. We point out that, in the dimensionless case, if δ is small then the inequality can be sharp.

The two inequalities imply that the effective domain of the two functionals is the space of the functions of bounded variation and, therefore, also the domain of $S_{\delta,\mathcal{D}}(f)$ contains functions with jumps, discontinuities, etc. It also follows that, for any function of bounded variation f, $S_{\delta,\mathcal{D}}(f)$ converges to $V_{\mathcal{D}}(f)$ when δ tends to zero.

The ROF problem is naturally linked, via the method of Lagrange multipliers, to the unconstrained minimization of the following functional of the Tikhonov type

$$\Phi_\mu(f) = \|f - g\|^2_{L^2(\mathcal{D})} + \mu\, V_{\mathcal{D}}(f) \tag{9.16}$$

where $\mu > 0$, a regularization parameter, is the inverse of a Lagrange multiplier. Let f_μ be a minimum point of the functional (9.16). Its existence can be proved for any $\mu > 0$ (for a proof, see for instance [3, 64, 62]); moreover, the functional is strictly convex because the quadratic term is strictly convex and therefore the minimizer is also unique. This unique solution depends continuously on the data g [3] so that, in conclusion, the minimization problem is well posed.

If μ is the inverse of a Lagrange multiplier, then the solution of the ROF problem should be obtained by finding a value of μ such that the following condition is satisfied

$$\|f_\mu - g\|^2_{L^2(\mathcal{D})} = \varepsilon^2\ ; \tag{9.17}$$

it is interesting to remark that we re-obtain the criterion, discussed in Sections 4.2 and 4.6, for the determination of the regularization parameter in the case of Tikhonov theory and known, in general, as Morozov's discrepancy principle [212]. An efficient method for the computation of μ as prescribed by this criterion in the case of TV regularization, is proposed in [305].

The solution of equation (9.17) requires some comments.

- As proved, for instance in [56], the discrepancy function $\varepsilon^2(f_\mu; g)$, defined by the l.h.s. of equation (9.17), is a non-decreasing function of μ, which is zero for $\mu = 0$.
- The function $\varepsilon^2(f_\mu; g)$ is bounded by $\|g - g_{mean}\|^2_{L^2(\mathcal{D})}$ where g_{mean} is the mean-value of g, i.e. by the square of the distance of g from the space of the constant functions. The proof of this inequality is easy. Indeed, if f_μ is the minimum point of the functional (9.16), then its value, computed in f_μ is smaller than its value computed, for instance, in a constant function; it follows that

$$\|f_\mu - g\|^2_{L^2(\mathcal{D})} \le \|c - g\|^2_{L^2(\mathcal{D})} - \mu\, V_{\mathcal{D}}(f_\mu) \le \|c - g\|^2_{L^2(\mathcal{D})}\ . \tag{9.18}$$

  The inequality holds also true when the distance of g from c is replaced by the minimum distance of g from the space of the constant functions, which is reached when $c = g_{mean}$. Moreover, this upper bound is exact since it is reached when $\mu \to 0$.

The previous results imply that equation (9.17) provides a unique value of μ when ε satisfies the following conditions

$$0 \le \varepsilon \le \|g - g_{mean}\|_{L^2(\mathcal{D})} \tag{9.19}$$

which are more restrictive than the conditions given in equation (4.22) for the case of Tikhonov's regularization. In conclusion, as proved in [64], the solution of the ROF problem is equivalent to the unconstrained minimization of the functional (9.16), combined with the selection rule (9.17), if the previous conditions on ε are satisfied.

It is important to understand how TV regularization affects image reconstruction. Interesting results are proved in [275] where exact analytic solutions of the ROF problem are given in simple cases. In that paper radially symmetric piecewise constant functions are considered, assuming that they are contaminated by a noise which is also radially symmetric. This assumption is not realistic but numerical simulations show that the analytic results obtained in that case are preserved also when the noise is not radially symmetric. The assumption of a radially symmetric noisy image implies that the minimization of the functional (9.16) in $\mathbb{R}^d$, $d = 1, 2, 3$ can be reduced to a one-dimensional problem.

As stated by the authors, the main results deduced in [275] from their analytic solutions can be summarized as follows.

- TV regularization tends to preserve edge location in the solution, and in some cases it is exactly preserved.
- TV regularization produces a change in the intensity of a feature which is inversely proportional to the scale of the feature (for instance the size of the region where the function is constant) and directly proportional to the value of the regularization parameter μ. Therefore the change is larger for the smaller features. However, the change does not depend on the value of the intensity of the feature.

The second property explains why TV is able to remove noise since noise can be considered as consisting of features with very small scales; on the other hand, the intensity of features with larger scales is changed according to their scale, with a result consisting in a contrast reduction. This effect can be explained by a simple argument in the case of noise-free data. Indeed, if f_μ is the unique minimizer of the functional (9.16) with g replaced by $g^{(0)}$, then it is obvious that $\Phi_\mu(f_\mu) \leq \Phi_\mu(g^{(0)})$, i.e.

$$V_{\mathcal{D}}(f_\mu) + \frac{1}{\mu}\|f_\mu - g^{(0)}\|^2_{L^2(\mathcal{D})} \leq V_{\mathcal{D}}(g^{(0)}) \tag{9.20}$$

or also

$$V_{\mathcal{D}}(f_\mu) \leq V_{\mathcal{D}}(g^{(0)}) - \frac{1}{\mu}\|f_\mu - g^{(0)}\|^2_{L^2(\mathcal{D})} < V_{\mathcal{D}}(g^{(0)}) \,. \tag{9.21}$$

Therefore the minimizer of the functional (9.16) has a total variation smaller than that of the true image. In the next section we visualize this effect by means of numerical examples. An interesting approach for reducing this loss of contrast is proposed in [224]. It is an iterative method referred to as *Bregman iteration*. The method, at each iteration, requires the solution of a variational problem of a similar structure as the total variation regularized least-squares problem. Therefore, even if it provides an improvement of the solution, it requires an increase of the computational cost. For more details as well as for a way to improve efficiency we refer again to [56].

If we take now as regularizing penalty the functional (9.11), then the problem is the minimization of the functional

$$\Phi_\mu(f) = \|f - g\|^2_{L^2(\mathcal{D})} + \mu\, S_{\delta,\mathcal{D}}(f) \tag{9.22}$$

which is also strictly convex. Moreover, the existence of a solution can be proved as in the case of the TV-penalization. This unconstrained minimization amounts to a variational problem similar to ROF with the TV-penalty replaced by the functional (9.11). Therefore it is quite natural to investigate whether the parameter μ can be selected by means of the criterion (9.17). The answer is positive, with exactly the same conditions on ε. Indeed, if the discrepancy function $\varepsilon^2(f_\mu; g)$ is computed in the unique minimum point of the functional (9.22), we obtain again a non-decreasing function of μ. Moreover, it is a bounded function with the same bound as obtained in the TV case. Indeed the inequality (9.18) is now replaced by

$$\|f_\mu - g\|^2_{L^2(\mathcal{D})} \leq \|c - g\|^2_{L^2(\mathcal{D})} + \mu\,[m(\mathcal{D})\delta - S_{\delta,\mathcal{D}}(f_\mu)] \tag{9.23}$$

where the second term in the r.h.s. is negative, so that we get again the same upper bound. As far as we know no analytic solution of this problem is available. Therefore a first step in understanding the pros and cons of this approach is to apply the method to the cases where an exact analytic solution of the TV method is available. This is one of the topics of the next section. As concerns the properties of the minimizers of the functional (9.22), we point out that the argument leading to the bound (9.21) can be used also in this case. Therefore, for exact data $g^{(0)}$, if we denote again by f_μ the minimizer for any μ, it satisfies the inequality

$$S_{\delta,\mathcal{D}}(f_\mu) < S_{\delta,\mathcal{D}}(g^{(0)}) . \tag{9.24}$$

This inequality implies again a loss of contrast in the reconstructed images. The Bregman-iteration method can be applied also to this problem.

In the case of the function (9.22), the main problem is the choice of the two parameters μ and δ. For a given δ, the parameter μ can be selected again by means of the Morozov discrepancy principle. However, as far as we know, no criterion has been proposed for the selection of δ. In practice, it should be sufficiently small to obtain reconstructions close to those provided by the TV method but also sufficiently large to lead to more efficient reconstruction algorithms.

The extension of the previous denoising framework to the case of linear inverse problems is quite natural. Indeed, if A is the usual linear and continuous imaging operator, then the two functionals introduced above in the case of denoising are replaced by the following ones

$$\Phi_\mu(f) = \|Af - g\|_{L^2(\mathcal{D})}^2 + \mu V_{\mathcal{D}}(f) , \tag{9.25}$$

$$\Phi_\mu(f) = \|Af - g\|_{L^2(\mathcal{D})}^2 + \mu S_{\delta,\mathcal{D}}(f) . \tag{9.26}$$

The presence of the operator A in the least-squares term has no incidence on the proof of existence of a solution; the arguments used in the case of denoising can also be used in this case. However it has an incidence on the uniqueness. Indeed, if the null space of the operator A is trivial, i.e. the equation $Af = 0$ has only the solution $f = 0$, then the first term is strictly convex and therefore the solution is unique. If the null space is not trivial, however, then the solution is not necessarily unique.

As concerns other properties, let us remark that the inequalities (9.21) and (9.24) hold true also in the case of the functionals (9.25) and (9.26). Therefore one can conclude that also in the applications to linear inverse problems a regularization based on total variation or on the functional (9.11) implies a loss of contrast. As a remedy, the Bregman-iteration method can be applied to these problems as well.

9.2 Discrete problems: image denoising

The results outlined in the previous section provide a well-founded approach to the reconstruction of images with edges. However, for applying the methods to the reconstruction of real images, a discretization of the problems is required. We first consider the denoising problem.

In general, a quite straightforward discretization of the total variation in 2D is utilized. Consider the discretization of a function f, defined on a square. First, by means of a grid of equispaced points, we subdivide the square into $N \times N$ pixels with area s^2 and we denote by $\mathbf{f}$ the $N \times N$ array with elements $f_{m,n}$, where $f_{m,n}$ is, for instance, the mean value of f on the pixel m, n. Next, we consider the following substitutions

$$\int \ldots dx \;\rightarrow\; \sum \ldots s^2 \,, \quad |\nabla f| \;\rightarrow\; \frac{D\mathbf{f}}{s} \tag{9.27}$$

where $(D\mathbf{f})_{m,n}$ is the modulus of the vector whose components are the increments at pixel m, n of $\mathbf{f}$ in the two directions

$$(D\mathbf{f})_{m,n} = \sqrt{|f_{m+1,n} - f_{m,n}|^2 + |f_{m,n+1} - f_{m,n}|^2} \ . \tag{9.28}$$

Then, except for a factor s (which can be absorbed by μ), the discrete total variation is given by

$$V_N(\mathbf{f}) = \sum_{m,n}(D\mathbf{f})_{m,n} \ , \tag{9.29}$$

and the discrete version of the functional (9.10) by

$$S_N(\mathbf{f}) = \sum_{m,n}\sqrt{\delta^2 + |(D\mathbf{f})_{m,n}|^2} \ . \tag{9.30}$$

We recall that, in the case of an application, δ should have the same dimension as $|D\mathbf{f}|$. This quantity, as remarked above, is a free parameter but, in any case, it must be smaller than the maximum value of $|D\mathbf{f}|$.

In the literature on edge-preserving methods the functional (9.30) is usually called the TV-*smooth* regularizer and hence we will use this denomination in the following. Notice that in the book [33], this functional is also referred to as *hypersurface regularizer*, a denomination introduced in [67] and related to Remark 9.3.

In the discrete version an important difference between the two regularizers is evident: while $V_N(\mathbf{f})$ is not differentiable, $S_N(\mathbf{f})$ is differentiable and this difference implies important differences in the design of minimization algorithms. Indeed, the design of algorithms in the non-differentiable case needs a deep knowledge of *nonsmooth numerical optimization*, a knowledge which is not expected from the readers of our book. On the other hand, the minimization of the differentiable case can be obtained by means of standard methods and one can also use some of the iterative methods described in the previous chapters (even though they may not be very efficient). For the reader who intends to study the methods applicable to the TV case there are many lecture notes written on the subject, of which we just single out [62] and [56]. We also mention that one of the most cited and used method is the primal-dual method proposed by Chambolle and Pock [65].

In the present section we compare results obtained with two regularizers, one nonsmooth (TV) and one smooth (TV-smooth). To this purpose we use different algorithms, both developed in our group and not necessarily the most efficient ones. More precisely, we use the *alternating extragradient method* (AEM) for TV, a method proposed in [47] in the case of Poisson data inversion and easily adapted to the least-squares case, and the *scaled gradient projection* method (SGP) for TV-smooth, a method proposed in [48] for Poisson data, and also adapted to the least-squares case [30].

9.2.1 Images with radial symmetry

In this subsection as well as in the rest of this chapter we use images which are rescaled to take values between 0 and 1. The reason is that, for perturbing these images with additive Gaussian noise, we use the MATLAB© function *imnoise* for noise generation.

One of the purposes of the subsequent sections is to compare the two denoising methods mentioned above in the case of objects for which an exact solution of the ROF problem exists, as shown in [275]. Since these are essentially toy objects we only consider two examples. The first one is shown in the left panel of Figure 9.1. It is a 256×256 array obtained by sampling a function defined on the square $\{|x_1| \leq 1, |x_2| \leq 1\}$ and consisting of a central disc with radius 0.1, surrounded by annuli with maximum radius 0.3 and 0.8, respectively,

FIGURE 9.1 Left panel: the original object. The intensity values for the different regions from center to boundary are 0.867, 0.467, 0.233 and 0.0833 (the background). Right panel: the noisy version obtained by adding a white Gaussian noise with zero mean and variance $\sigma^2 = 10^{-3}$. The SNR for the different regions, as defined in equation (2.6), are 14.6 dB, 11.6 dB, 8.7 dB and 4.2 dB.

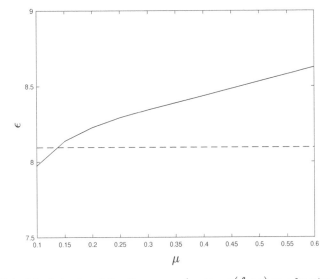

FIGURE 9.2 Plot of the behavior of the discrepancy function $\varepsilon(f_\mu; g)$ as a function of μ in the case of the image of Figure 9.1.

over a background. The intensity values of the inner circle, of the two annuli and of the background are given in the caption of the figure. The image is perturbed by additive Gaussian noise and the noisy image is shown in the right panel of the same figure. In the caption we give in dB the SNR values for the different levels. The noise is clearly visible in all different parts of the image.

In order to find the solution of the ROF problem we need an estimate of the value of the parameter ε appearing in equation (9.17). Since the exact image is available it can be exactly computed. However, in practice, this information is not necessary if the variance σ^2 of the noise is known. Indeed, it is easy to verify that it is then very well approximated by $\varepsilon = N\sigma$; in the case of the image of Figure 9.1 the value of ε is 8.09.

In order to apply the Morozov discrepancy principle which, as already remarked, provides the value of the regularization parameter μ corresponding to the solution of the ROF problem, we compute, as a function of μ, the discrepancy function $\varepsilon^2(f_\mu; g) = \|f_\mu - g\|^2$, where f_μ is the minimizer of the functional (9.16). In Figure 9.2, we plot the behavior of $\varepsilon(f_\mu; g)$ as a function of μ for the denoising of the image of Figure 9.1. The intersection with the value

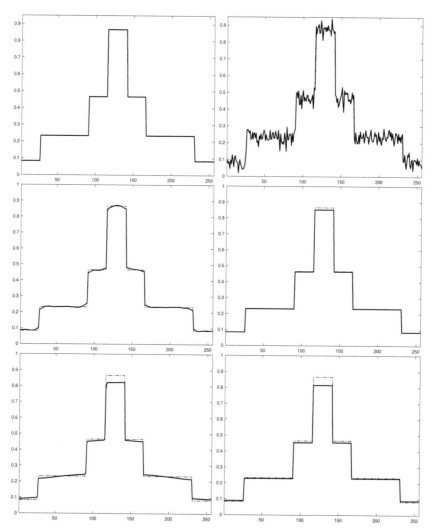

FIGURE 9.3 Cuts of the different images along the horizontal radial line. First row: original object (left panel) and noisy image (right panel). Second row: reconstruction obtained with $\mu = 0.15$ (left panel) and the image obtained by replacing the reconstructed values of the different regions by their average value (right panel). Third row: reconstruction obtained with $\mu = 0.6$ (left panel) and the result obtained with the procedure indicated above (right panel).

8.09 occurs when $\mu = 0.15$. Therefore this is the value we take for a first reconstruction of our image.

In Figure 9.3, we represent cuts along the horizontal radial line. In the first row we show, in the left panel, the cut of the original object and, in the right panel, that of the noisy image to show the level of the noise. In the second row, in the left panel, we show the cut of the reconstruction obtained with $\mu = 0.15$. We find that the positions of the discontinuities are perfectly reconstructed but it is also evident that the noise is not completely removed, especially in the annuli. Indeed some small undulations appear which can be seen as a residual effect of the noise (which is not radially symmetric here, contrary to what is assumed in [275]). We also observe some curvature at the edges whereas, in the image, the edges are sharp because of the discontinuity. Nevertheless the improvement of the SNR is remarkable: for the different regions of the image it is now ranging from 17.1 dB (central disc) to 11.9 dB (background), a very important gain, especially remarkable in the case of the background. Obviously the small effects due to noise are different on the different radial directions. Therefore, for the central disc as well as for the two annuli and the background we compute the mean value of the reconstruction and for each region of the image we replace the reconstructed values with this constant average value. The result is shown in the right panel of the second row. The reconstruction obtained with this trick is perfect, without residual noise effects (and this is obvious) and also without loss of contrast. If we increase the value of μ beyond the value provided by the Morozov discrepancy principle, we obtain an attenuation of the noise effect but also a loss in contrast. The cut of the reconstruction obtained with $\mu = 0.6$ is shown in the left panel of the third row. The noise is completely smoothed even if the reconstruction does not provide constant values in the different regions of the image, but the loss of contrast is now evident especially in the central disc. If we apply to this reconstruction the same trick used in the previous case, we obtain the result shown in the right panel of the third row. Differences on all the constant values are now visible and these differences are inversely proportional to the scale of the corresponding region as indicated in [275].

For a first comparison of TV-smooth with TV, we select two values of δ, 10^{-2} and 10^{-3}, i.e. values which are smaller than the smallest jump in the image, approximately 0.1. In the second case, the values of the TV-smooth regularizer are close to those of the TV regularizer but not excessively so that algorithms for differentiable functions can still converge sufficiently fast in such a case. As concerns μ we use the values provided by the discrepancy principle; we obtain $\mu = 0.2$ for $\delta = 10^{-2}$ and $\mu = 0.1$ for $\delta = 10^{-3}$, a value close to the one used in the TV case. In the left panels of Figure 9.4 we show the results obtained with these values of δ and μ for a comparison with the result provided by TV (with μ also given by the discrepancy principle) which is shown in the left panel of the second row of Figure 9.3. In both cases the positions of the discontinuities are correctly reproduced, coherently with the remark made above that the effective domains of TV and TV-smooth coincide and contain BV functions. On the other hand, in both the TV-smooth cases residual noise effects are slightly more significant than in the TV case, as well as the curvature at the edges. Finally, we used the same averaging trick as in the TV case; the results are shown in the right panels of Figure 9.4. We observe a very small loss of contrast. In spite of these differences, a visual inspection of the two reconstructions (TV and TV-smooth) does not reveal significant variations.

In a second experiment, we insert in the object a domain where the intensity is varying smoothly. The object has again a radial symmetry and is shown in Figure 9.5. In the first row we display the noise-free object (left panel) and the noisy one (right panel), perturbed with the same Gaussian noise as used in the first experiment. In the panel of the second row we show the superposition of their cuts along the horizontal radial direction. In the case of TV-smooth regularization, we choose $\delta = 0.01$ and for μ, in both cases, we use the value

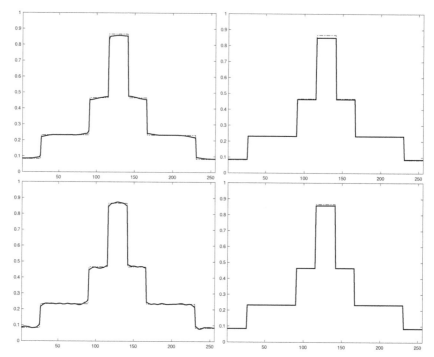

FIGURE 9.4   TV-smooth reconstructions. First row: reconstruction with $\delta = 0.01$ and $\mu = 0.2$ (left panel) and the result obtained with the trick indicated above (right panel). Second row: reconstruction with $\delta = 0.001$ and $\mu = 0.1$ (left panel) and the result obtained with the trick indicated above (right panel)

provided by the discrepancy principle. The reconstructions are shown in Figure 9.6, by TV in the left panel and by TV-smooth in the right one. We observe that in both cases the staircase effect, observed in [272] in the reconstruction of a similar smooth region, does not appear. Both reconstructions look quite satisfactory. Again the positions of the discontinuities are correctly reproduced.

As a conclusion of this subsection devoted to objects with a radial symmetry, i.e. to objects such that in [275] an exact solution of the ROF problem is available, we summarize the results obtained by means of a couple of examples and showing that also in the discrete case and for non-symmetric noise the following statements hold true.

- Both TV and TV-smooth regularization suppress noise and provide the correct positions of the discontinuities.
- The value of the regularization parameter provided by the Morozov discrepancy principle, which corresponds to the solution of the ROF problem, leads to reconstructions without a significant loss of contrast between regions with approximately the same scale.
- The same criterion applied to both regularizers in the case of an object with a smoothly varying region provides a satisfactory reconstruction.

Therefore the discrepancy principle seems to provide a value of μ which is a sensible starting point for the application of edge-preserving image denoising. It is important to remark that this principle is easily applicable also in the case of real images perturbed by white Gaussian noise. Indeed, the variance σ^2 of the noise can be estimated by analyzing regions of the image where the intensity is essentially constant. Then, as already remarked, the value of ε^2 to be used is given by $\varepsilon^2 = N^2 \sigma^2$.

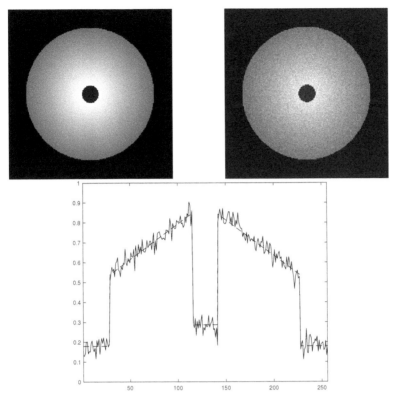

FIGURE 9.5 First row: the noise-free object (left panel) and the noisy one (right panel). Second row: the superposition of their cuts along the horizontal radial direction.

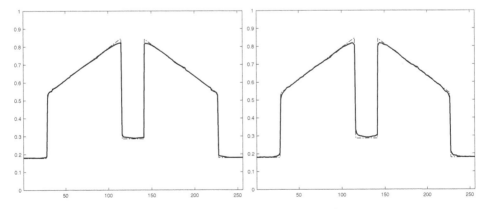

FIGURE 9.6 Left panel: TV reconstruction. Right panel: TV-smooth reconstruction.

FIGURE 9.7   The 'Barbara' image used for testing the performance of TV and TV-smooth.

9.2.2   Natural images

In the previous subsection, we did not find any significant loss of contrast, a possible artifact of TV and also of TV-smooth method as discussed in Section 9.1. The reason is that we considered regions with different intensity level but with roughly the same scale. Examples of such loss are given in [275]. The situation may change in the case of natural images because in such images it is frequent to find regions with very different scales and different intensities.

To clarify this point it is sufficient to assess the reconstruction of the noisy version of a standard image certainly known to our readers because it appears in hundreds, presumably thousands of papers on image processing. We mean the so-called *'Barbara' image* which is shown in Figure 9.7. This image, whose dimension is 512×512 pixels, is interesting for our purposes because it contains features, such as the checkered tablecloth or the striped scarf, showing very small details. Accordingly, two interesting regions of these objects are delimited by the white squares drawn on the image and will be the object of particular attention in the following. Also in this case the image is scaled in such a way as to take values between 0 and 1 and it is perturbed with additive Gaussian noise with zero mean and two different values of σ^2: 0.001, i.e. the noise considered in the previous subsection (denoted as *noise1*), and 0.005, i.e. a higher noise level (denoted as *noise2*). In this way we can test the effect of the noise level on the quality and contrast of the denoised image.

We reconstruct the image using TV and TV-smooth (with $\delta = 0.001$) and in both cases we set the value of μ by means of the discrepancy principle, the value of ε being assessed from the standard deviation of the noise. In the case of *noise1*, we find $\mu = 0.05$ for TV and the same value for TV-smooth while, in the case of *noise2*, we find $\mu = 0.12$ for TV and $\mu = 0.13$ for TV-smooth. The noisy images are shown in the first row of Figure 9.8; the corresponding reconstructions achieved with TV in the second row and those achieved with TV-smooth in the third row.

First, we point out the excellent noise reduction obtained with these reconstruction. We quantify this effect by computing the PSNR as defined in equation (2.7). Indeed, in the case of *noise1* its value on the noisy image is 30 dB while it is 32.2 dB on both reconstructions (remember that the scale is logarithmic so that a gain of about 2 dB is significant); in the case of *noise2*, its value is 23 dB on the noisy image and 28 dB on both reconstructions, with a gain of 5 dB. As concerns the loss of contrast, to quantify this effect we use one

FIGURE 9.8   First row: the noisy images, with *noise1* to the left and *noise2* to the right. Second row: the corresponding reconstructions obtained with TV. Third row: the corresponding reconstructions obtained with TV-smooth ($\delta = 0.001$).

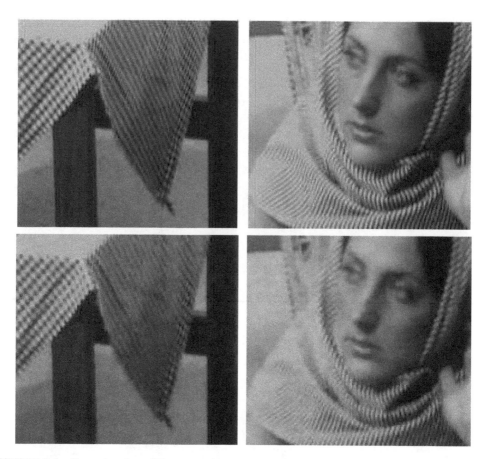

FIGURE 9.9   Reconstruction of the two regions corresponding to the white squares of Figure 9.7 in the case of *noise1* with $\mu = 0.05$. First row: the original images without noise. Second row: the reconstructed ones.

of the possible definitions of the so-called *contrast ratio* or *dynamic range* which takes into account the fact that the human visual system is more sensitive to differences in image intensities rather than to absolute values; accordingly and roughly speaking, the contrast ratio is given by the ratio between difference in intensity and average intensity of the image. For instance, the so-called *Michelson contrast* is given by

$$\text{contrast} = \frac{f_{max} - f_{min}}{f_{max} + f_{min}} \tag{9.31}$$

where f_{max} and f_{min} are, respectively, the highest and lowest value of the image intensity. Using this definition we find that for the original image the contrast is 0.914 while, in the case of *noise1*, it is 0.879 for both reconstructed images and in the case of *noise2*, it is 0.846 for TV and 0.851 for TV-smooth. Hence the loss of contrast does not appear to strongly depend on the noise level. In order to highlight the visual loss of contrast and also a related loss of details in the smooth regions, an effect increasing with noise, in Figures 9.9 and 9.10 we show, for the two noise levels, the reconstructions of the two subregions, corresponding to the two white squares drawn in Figure 9.7. In the first row we show the regions as they appear in the original image while in the second one we show the corresponding reconstructions as produced by TV regularization with the values of μ indicated above. We do not show those obtained with TV-smooth regularization because they are practically identical.

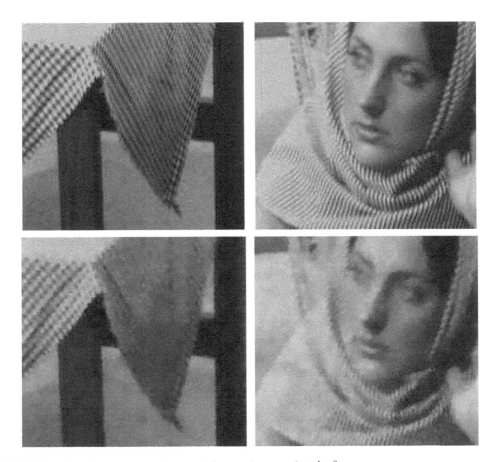

FIGURE 9.10 The same as in Figure 9.9 but in the case of *noise2*.

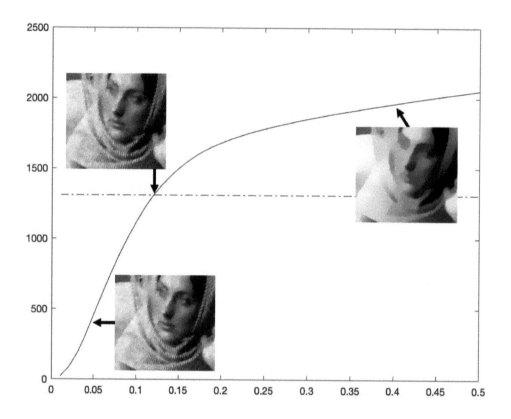

FIGURE 9.11 Plot of the discrepancy function as a function of μ in the case of *noise1*. We indicate the point where the function crosses the discrepancy value corresponding to the noise. We attach to this point a miniature of the corresponding reconstruction of the face of 'Barbara'; moreover we also show such a miniature for a greater and a smaller value of μ.

If we look first at Figure 9.9, corresponding to *noise1*, we notice a sharp reconstruction of the details of the tablecloth and of the scarf, even in the shadowed regions, but, besides the loss of contrast, we also remark a loss of details in the face of 'Barbara', a smooth region of this image. This effect is presumably also due to the loss of contrast. Next, if we look at Figure 9.10 we notice that the small details of the tablecloth and of the scarf are almost lost in their shadowed parts, as well as in the most illuminated ones, and we notice an increased loss of details in the face of 'Barbara'. Hence the increase in noise, even if it does not increase significantly the loss of contrast, has a significant effect on the quality of the reconstructed image.

One could hope to obtain better results by changing the value of μ, i.e. by taking a value of μ different from that given by the discrepancy principle. In Figure 9.11, we give a picture showing what happens. Indeed, if we decrease the value of μ, then an increased noise contamination appears while, if we increase μ, then cartoon effects affect the reconstruction. This could be however an interesting effect but only for other purposes such as image segmentation. Hence it is possible that very little improvements can be obtained by means of small variations of μ, but it seems that, at least in the case of denoising, the discrepancy principle provides an optimal or almost optimal value of the regularization parameter. In the next subsection we will find that, in the case of deblurring, the situation looks much more complex.

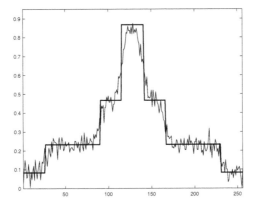

FIGURE 9.12 Radial cut of the object superimposed to that of the blurred and noisy image.

9.3   Discrete problems: image deblurring

In order to investigate the application of TV and TV-smooth regularization to the solution of linear inverse problems we consider the most simple case, namely image deconvolution as in Part I. To this purpose we consider a Gaussian blur, which can be seen as a difficult example because convolution with a Gaussian PSF practically acts as a low pass filter (with a band determined by the variance of the Gaussian).

9.3.1   Objects with radial symmetry

We first consider the objects with radial symmetry used above for illustrating the denoising effect of TV and TV-smooth. The first example is the piecewise constant function consisting of a central disc surrounded by annuli with decreasing intensity described at the beginning of Section 9.2.1. We use a Gaussian PSF with a standard deviation of about two pixels; its effect extends over a disc with a radius which is 5 times broader than the standard deviation, i.e with a radius of about 10 pixels. The blurred image of the object is perturbed with additive Gaussian noise with variance $\sigma^2 = 0.001$, i.e. the lowest noise level considered in the previous subsections. The result is shown in Figure 9.12, where the horizontal radial cut of the original object is drawn together with that of the blurred and noisy image, where the discontinuities are no longer visible. For TV regularization the functional to be minimized is given by equation (9.25) and for TV-smooth by equation (9.26). Again, for TV we must estimate a suitable value of μ and for TV-smooth we must also choose a value for δ. In view of its good performances in the case of denoising, we use the discrepancy principle also for deblurring. For TV deconvolution we get $\mu = 0.048$. The usual cut of the corresponding reconstruction is shown in the left panel of Figure 9.13, while the reconstructed image is shown in the middle panel of Figure 9.14. Visually, the reconstructed image looks fine. In the cut, one can observe that the locations of the discontinuities are correctly reconstructed even if in a couple of cases the discontinuity is replaced by a very steep ramp; moreover a small residual effect of the noise produces small undulations as already observed in the case of denoising. In an attempt to obtain a better reconstruction we changed the value of μ around the value of 0.048. Even if we discovered a minimum of the reconstruction error, defined as the MSE, for a value of μ slightly smaller than the one provided by the discrepancy principle, the corresponding reconstruction did not exhibit some improvement with respect to the reconstructions shown in the figures.

For TV-smooth, if we use $\delta = 0.01$ as in the case of denoising, the corresponding reconstruction is exceedingly smooth. Therefore we reduced δ to $\delta = 10^{-4}$, a value providing a

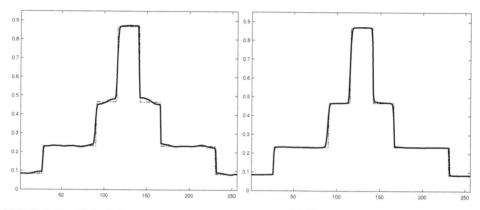

FIGURE 9.13   Cut of the TV reconstruction with $\mu = 0.048$ (left panel) and of the TV-smooth reconstruction with $\delta = 0.0001$ and $\mu = 0.01$ (right panel).

FIGURE 9.14   Blurred and noisy image (left panel), TV reconstruction (middle panel) and TV-smooth reconstruction (right panel).

better approximation of the TV regularizer; the corresponding value of μ coming from the discrepancy principle is $\mu = 0.01$ (for this value of μ the TV reconstruction is too noisy). The usual cut is shown in the right panel of Figure 9.13 and the complete reconstructed image is shown in the right panel of Figure 9.14. A visual inspection of the two reconstructed images does not reveal significant differences.

As a second example we consider the radially symmetric object with a circular 'hole' in the central region, surrounded by a linearly decreasing smooth region, as shown in Figure 9.5. We use the same PSF as in the previous example and we add the same Gaussian noise to the blurred image. A radial cut of the result is shown in Figure 9.15. We use again the discrepancy principle for estimating the value of μ both for TV and for TV-smooth with $\delta = 10^{-4}$ and we have $\mu = 0.1$ in the first case and $\mu = 0.07$ in the second case.

The usual horizontal radial cuts of the reconstructed images with TV and TV-smooth are shown in Figure 9.16, in the left and right panel, respectively. The corresponding reconstructed images are shown in Figure 9.17, with TV in the middle panel and TV-smooth in the right panel. By looking at Figure 9.16 we remark that, in the case of deconvolution, a strong staircase effect appears in the smooth region and in both reconstructions. On the other hand, the discontinuities are correctly detected and again only in a couple of cases one observes a slight deviation from the vertical line. Moreover, in both reconstructions, the top angles are not correctly reproduced and are somehow 'snipped'. Also in the present case a visual inspection of the two reconstructed images does not reveal significant differences; in both cases the circular artifacts around the central hole are a consequence of the staircase effect observed in the previous figure.

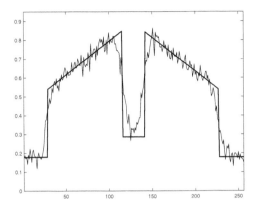

FIGURE 9.15 Radial cut of the object superimposed to that of the blurred and noisy image.

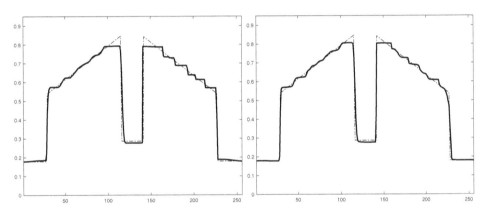

FIGURE 9.16 Cut of the TV reconstruction with $\mu = 0.01$ (left panel) and of the TV-smooth reconstruction with $\delta = 0.0001$ and $\mu = 0.07$ (right panel).

FIGURE 9.17 Blurred and noisy image (left panel), TV reconstruction (middle panel) and TV-smooth reconstruction (right panel). The values of the parameters are indicated in the previous figure.

FIGURE 9.18 In the left column we show the blurred and noisy images. Upper panel: PSF with a standard deviation of 2 pixels and noise with a variance of 10^{-3}; second panel: PSF with a standard deviation of 5 pixels and the same noise. Right column: the corresponding reconstructions; the values of μ are given in the text. Remark that, for each image we use a range of gray levels defined by the upper and lower bound of its values so that a possible loss of contrast may not be visible.

9.3.2 Natural images

We use again the challenging image of 'Barbara' for testing the performance of TV and TV-smooth and, for simplicity, we only show the results obtained with TV since those obtained with TV-smooth are pretty similar.

The first simulation consists in perturbing the object as in the case of the objects with radial symmetry, i.e. we use a Gaussian PSF with a standard deviation of 2 pixels and additive Gaussian noise with a variance of 10^{-3}. The blurred and noisy image is shown in the left panel of the first row of Figure 9.18. With $\mu = 0.02$, as prescribed by the discrepancy principle, we get the reconstruction displayed in the right panel of the same row. The blurring is evident in the original image and it is essentially removed in the deconvolved one where we recognize details of the tablecloth which are evident in the original one. While the boundaries of the table and in particular, the table legs are very well reconstructed, the details of the scarf are almost completely lost. Presumably they are too small and comparable with noise fluctuations.

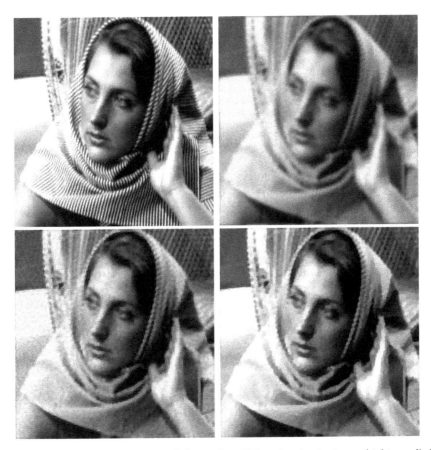

FIGURE 9.19 First row: original image (left panel) and blurred and noisy image (right panel). Here the standard deviation of the PSF is 2 pixels while the noise variance is 10^{-4}. Second row: TV reconstruction with $\mu = 0.004$ (left panel) and TV-smooth reconstruction with $\delta = 10^{-4}$ and $\mu = 0.003$ (right panel).

To verify possible improvement due to noise reduction we considered the same image with the same blurring but perturbed by a noise with $\sigma^2 = 10^{-4}$. We show only the detail of the face of 'Barbara' in Figure 9.19. Moreover, for at least one comparison, we also show the reconstruction obtained with TV-smooth. Effectively, a few details of the scarf, not visible in the blurred image, are now recovered by both reconstructions methods. However, the details of the stripes appear only to the right of the face of 'Barbara' while they are not reconstructed in the other parts of the scarf.

Finally, we tested the limits of the method, by increasing the standard deviation of the PSF from 2 to 5 pixels with the same noise. In such a case we found a rapid degradation of the quality of the reconstruction. In the second row of Figure 9.18 we show the reconstruction obtained with a standard deviation of 5 pixels. The value of μ obtained from the discrepancy principle is 0.01. At first glance, the reconstruction is affected by strong cartoon effects. However, the boundaries of the objects, dresses and face present in this scene are reconstructed correctly; this remark could be the starting point for an attempt of segmentation of a scene which is blurred and noisy, two perturbations which certainly increase the difficulty of this operation.

As a conclusion of the previous numerical experiments in the case of deblurring, we point out that, especially in the case of natural images, we used a case-example whose reconstruction is very difficult because it contains very small and complex details. Anyway we found that, even in the case of a very critical blurring such as the Gaussian one, it is possible to obtain satisfactory reconstructions, even if with some restrictions. Obviously, in the case of objects without the complexity of the 'Barbara' image it is possible to obtain excellent reconstructions which can be found everywhere in the literature on TV regularization. Our purpose was to detect the limitations of this celebrated method according to the general philosophy that *no regularization method is always working well* in the case of ill-posed problems.

We did not investigate in detail the problem of removing contrast reduction nor the problem of reducing the staircase effect. Further investigations of these problem are still in progress. We recall the already mentioned method of *Bregman iterations* as well as the method of Total Generalized Variation (TGV) [52] which deserves further investigation and testing.

10

Sparsity-enforcing regularization

In regularization methods, constraints or penalties enforcing some kind of sparsity on the recovered objects have emerged in recent years as an alternative to the usual quadratic constraints or penalties considered in the first part of the book. In the previous chapter, edge-preserving denoising or inversion methods are derived through total-variation penalties which enforce the sparsity of the gradient.

In the present chapter, this idea is extended to enforcing sparsity in any specified basis, by means of an ℓ^1-type penalty on the object coefficients on this basis, a penalty which favors the presence of many zeroes among these coefficients. For example, natural images appear to be typically sparse in so-called wavelet bases. We build on denoising techniques in wavelet bases to derive sparsity-enforcing inversion methods in arbitrary bases. We also present a simple iterative thresholding algorithm generalizing the Landweber scheme and allowing to compute the regularized solutions. Moreover, we show how the theory generalizes to the case of objects which are a superposition of a sparse and a smooth component. Finally, we provide examples of sparse recovery and some illustrative numerical simulations.

10.1 Sparse expansions and wavelet denoising

In the years 1980's, the construction of wavelet bases had a deep impact on the field of imaging. Wavelet analysis offers an alternative to traditional Fourier analysis. For the basics of wavelet theory we refer the reader to [79, 206]. Wavelets are basis functions which, contrary to trigonometric functions, are well localized. Therefore they can provide representations of images which are adaptive to their local features such as discontinuities. This led to the observation that images and in particular, natural images admit sparse representations on wavelet bases. This means that they are represented by linear combinations of only a few basis functions or in other words that many coefficients of the expansion are zero. Notice that this concept of sparse representation can be extended to other types of signals or images, as well as to other bases.

DOI: 10.1201/9781003032755-10

The fact that an image or signal is expected to admit a sparse representation on a wavelet basis naturally led to denoising methods as proposed in [108, 104]. Indeed, noise will typically be spread more or less evenly over all wavelet coefficients whereas the image will have a few large coefficients *spiking* above the uniform background noise. The idea is to apply *thresholding* to the coefficients in order to select the largest coefficients corresponding to the significant features of the image while disregarding the smallest ones corresponding to noise. The simplest way to do this, referred to as *hard thresholding*, consists in selecting all coefficients with absolute value larger than some prescribed threshold, without modifying them, and in setting to zero all other coefficients.

As an alternative to hard thresholding, one can use a variational framework and introduce a penalty on the sum of the absolute values of the wavelet coefficients (i.e. on the ℓ^1-norm of the sequence of these coefficients); this will lead to so-called *soft thresholding*, as defined below. In [63], more general penalties were proposed, which involve Besov-space norms of f. Besov spaces are function spaces which are naturally linked to wavelets in the sense that membership in a space of this type can be revealed by a test on the decay of the wavelet coefficients of the candidate functions. Roughly speaking, the Besov space $B^s_{p,p}(\mathbb{R}^d)$ consists of functions which have derivatives up to order s in the Lebesgue space L^p. Notice that the full family of Besov spaces $B^s_{p,q}(\mathbb{R}^d)$ is labeled by three indices, but we take here for simplicity $q = p$ which is sufficient for our purposes. The norm $\| \cdot \|_{s,p}$ in $B^s_{p,p}(\mathbb{R}^d)$ is equivalent to a weighted ℓ^p-norm of the sequence of the corresponding wavelet coefficients (f, φ_γ), namely

$$\|f\|_{s,p} \sim \left[\sum_{\gamma \in \Gamma} w_\gamma \, |(f, \varphi_\gamma)|^p \right]^{1/p} \tag{10.1}$$

where we denote by $(\varphi_\gamma)_{\gamma \in \Gamma}$ an orthonormal basis of wavelets in $L^2(\mathbb{R}^d)$ (with sufficient regularity, i.e. a bit smoother than typical members of the Besov space) and where the weights are given by

$$w_\gamma = 2^{\sigma p |\gamma|} \quad \text{with} \quad \sigma = s + d\left(\frac{1}{2} - \frac{1}{p}\right) . \tag{10.2}$$

By $|\gamma|$ we denote the wavelet scale parameter (j in a more common notation) and σ is assumed to be nonnegative to ensure that the Besov space $B^s_{p,p}(\mathbb{R}^d)$ is a subspace of L^2. We recall that by the norm equivalence $\|f\|_{\mathcal{X}} \sim \|f\|_{\mathcal{Y}}$ it is meant that there exist strictly positive constants C_1 and C_2 such that

$$C_1 \|f\|_{\mathcal{X}} \leq \|f\|_{\mathcal{Y}} \leq C_2 \|f\|_{\mathcal{X}} . \tag{10.3}$$

In the special case $p = 2$ and $s = 0$, the equivalence (10.1) is simply the well-known Parseval equality expressing the unitarity of the wavelet transform in L^2. Due to this norm equivalence, we can say that the properties of functions belonging to a Besov space can be directly translated in a diagonal form on the wavelet coefficients.

With these notations, we can formulate the variational denoising problem as the minimization of the functional

$$\Phi_\mu(f) = \sum_{\gamma \in \Gamma} \left[|f_\gamma - g_\gamma|^2 + \mu \, w_\gamma |f_\gamma|^p \right] \tag{10.4}$$

where $f_\gamma \equiv (f, \varphi_\gamma)$ and $g_\gamma \equiv (g, \varphi_\gamma)$ are the coefficients of f, respectively g, with respect to the chosen orthonormal wavelet basis and $\mu > 0$ is the regularization parameter. For $p = 2$, i.e. with a quadratic penalty corresponding to the usual Tikhonov regularization, the minimizer of (10.4) is

$$f_\gamma = (1 + \mu \, w_\gamma)^{-1} g_\gamma \tag{10.5}$$

and we see that regularization implies a *linear shrinkage*, by a factor $1 + \mu w_\gamma$, of the noisy coefficients g_γ, and this independently of their actual value. Moreover, no sparsity is enforced: all non-zero coefficients will remain so after denoising.

On the other hand, for $p = 1$, as easily checked, the minimizer is given by

$$f_\gamma = S_{\mu w_\gamma}(g_\gamma) \tag{10.6}$$

where S_τ is the *soft thresholding* function

$$S_\tau(x) = \begin{cases} x + \tau/2 & \text{if} \quad x \le -\tau/2 \\ 0 & \text{if} \quad |x| < \tau/2 \\ x - \tau/2 & \text{if} \quad x \ge \tau/2 \end{cases} \tag{10.7}$$

for the case of real-valued x, which we will consider here for the sake of simplicity. Nevertheless, the extension to the complex case is straightforward provided (10.7) is replaced by $S_\tau(r e^{i\phi}) = S_\tau(r)e^{i\phi}$, for $x = r\,e^{i\phi}$. Notice that for $p = 1$ the coefficients g_γ are still *shrunk*, that is replaced by coefficients having a smaller absolute value. However, the operation is no longer linear – whence the appellation of *nonlinear shrinkage*. Notice also that in this case the objective function (10.4) is convex but not continuously differentiable at $f_\gamma = 0$, so that we have to deal with a so-called nonsmooth optimization problem.

Intermediate values of p, $1 < p < 2$, can be considered as well (see e.g. [80]). For such values of $p \ne 1, 2$, however, the corresponding shrinkage function $S_\tau^{(p)}$ has no explicit expression, but can be determined implicitly and numerically. For $p < 1$, the optimization problem is no longer convex. We can also refer to [9] for more general shrinkage functions for wavelet denoising, in particular, for the nonconvex case.

10.2 Regularization through sparsity-enforcing penalties

The previous considerations naturally lead to the question of their possible extension from denoising to inverse problems. In a variational framework, we would like to introduce a linear imaging operator A different from the identity in the functional (10.4). Moreover, a generalization to objects which have sparse expansions on bases which are not necessarily wavelets would also be desirable. Indeed, there are cases in which an object is sparse directly in pixel space, namely on a basis of characteristic functions of pixels, e.g. with a few bright spots on a zero background such as stars in astronomical imaging. In other problems, one expects the object to be sparse in a Fourier basis, i.e. to be made of few harmonic components oscillating at well-defined frequencies. This was the motivation for the sparsity-enforcing regularization theory first developed in [86, 80].

We will now consider the following functional

$$\Phi_\mu(f) = \|Af - g\|^2 + \mu \|f\|_{\mathbf{w},p}^p \tag{10.8}$$

where $\|f\|_{\mathbf{w},p}^p$ is defined by

$$\|f\|_{\mathbf{w},p}^p = \sum_{\gamma \in \Gamma} w_\gamma |f_\gamma|^p \,. \tag{10.9}$$

We specify that $(\varphi_\gamma)_{\gamma \in \Gamma}$ is a fixed – but otherwise arbitrary – orthonormal basis of the object space $\mathcal{X}$, assumed to be a Hilbert space, typically L^2, to which f is assumed to belong. The discrepancy $\|Af - g\|$ is measured by means of the norm of the image/data space $\mathcal{Y}$, also assumed to be a Hilbert space. The operator A is defined as a bounded operator from $\mathcal{X}$ to $\mathcal{Y}$. The expression (10.9) is nothing else than a weighted ℓ^p-norm – to the power p – of the sequence of coefficients $f_\gamma = (f, \varphi_\gamma)$ on the chosen basis. The (fixed) weights w_γ are only assumed to be uniformly bounded below by some constant $c > 0$. It means that the weight sequence does not need to be of the special form (10.2), which is associated with the equivalent norm for Besov spaces.

The choice $p = 2$ corresponds to the usual quadratic Tikhonov regularization, where typically, in the presence of noise, all recovered coefficients f_γ of the object f on the chosen basis $(\varphi_\gamma)_{\gamma \in \Gamma}$ will turn out to be nonzero, albeit many can be very small. The choice $p = 1$ instead enforces sparsity namely favors the recovery of objects that are sparse in the chosen basis, i.e. have only a few nonzero coefficients f_γ. For $1 < p < 2$, we get an intermediate type of shrinkage. The case $p < 1$ also enforces sparsity but leads to a nonconvex optimization problem; therefore we will not consider it here. Notice that when p decreases from 2 to 1 while keeping the weights and μ fixed, if the coefficient is small, i.e. if $|(f, \varphi_\gamma)| < 1$, the penalization increases. On the other hand, if the coefficient is large, i.e. if $|(f, \varphi_\gamma)| > 1$, the penalization decreases. Hence choosing values of $p < 2$ allows to favor the recovery of objects having few large components rather than many small components. Let us stress the fact that the choice to enforce sparse recovery in a predefined basis should reflect our expectations about the nature of the objects we are looking for. In other words it should be viewed as an assumption or prior knowledge about the class of objects to be recovered. However, one is agnostic about which components are non-zero (active) or not.

An important question arising when doing so is whether the sparsity-enforcing penalties can provide proper regularization of the inverse problem as for the quadratic case. Indeed, as shown in Section 4.3, for $p = 2$ one can prove that with some appropriate choice of the regularization parameter μ as a function of the noise level on the data, one can guarantee that the regularized solution minimizing (10.8) tends, when the noise level tends to zero, to the true exact solution $f^{(0)}$ – or to the generalized solution in the presence of a nontrivial null space $\mathcal{N}(A)$ of A (see Section 4.3).

A positive answer to this question was given in [80] (with some preliminary results already contained in [86]), where the following regularization theorem is proved.

- Let $\{g_\varepsilon\}$ be a sequence (family) depending on the parameter $\varepsilon > 0$ such that, when ε tends to zero, $\|g_\varepsilon - g^{(0)}\| \leq \varepsilon$, i.e. g_ε tends to the noise-free data $g^{(0)} = Af^{(0)}$, where $f^{(0)}$ is the true solution.
- Let us assume that either $\mathcal{N}(A) = \{0\}$ or $p > 1$ so that there exists a unique minimizer $f_{\mu(\varepsilon)}$ of the functional (10.8) with g replaced by g_ε.
- Let us adjust the dependence of the regularization parameter $\mu = \mu(\varepsilon)$ on the noise level ε in such a fashion that

$$\lim_{\varepsilon \to 0} \mu(\varepsilon) = 0 \quad \text{and} \quad \lim_{\varepsilon \to 0} \varepsilon^{-2} \mu(\varepsilon) = \infty , \qquad (10.10)$$

- then the following regularization result holds true:

$$\lim_{\varepsilon \to 0} \|f_{\mu(\varepsilon)} - f^\dagger\| = 0 \qquad (10.11)$$

where $f^\dagger$ is the unique element of minimum $\| \cdot \|_{\mathbf{w},p}$–norm in $\mathcal{S} = \mathcal{N}(A) + f^{(0)} = \{f; Af = Af^{(0)}\} \subset \mathcal{X}$. If $\mathcal{N}(A) = \{0\}$ then $f^\dagger = f^{(0)}$. When $p = 2$, $f^\dagger$ coincides with the generalized solution defined in Chapter 8.

This theorem encompasses the well-known result mentioned above for quadratic Tikhonov regularization (see also the book [114]).

In the special case of weights growing to infinity, or more precisely, if, for all $C > 0$, the number of weights such that $w_\gamma \leq C$ is finite, then the above regularization theorem simply results from a general compactness argument. Indeed, the following set

$\mathcal{B}_{\mathbf{w},p} = \{f \in L^2 \mid \sum_{\gamma \in \Gamma} w_\gamma |f_\gamma|^p < \infty\}$ is then compact. This is the case e.g. for the exponentially growing weights corresponding to wavelets and Besov spaces (for $\sigma > 0$) considered in Section 10.1.

It has been shown later in [46] that the choice of the regularization parameter μ provided by the Morozov discrepancy principle also yields a proper regularization method, generalizing for penalties of the type (10.9) the result described in Section 4.6 for the classical quadratic case and in Chapter 9 for the TV penalty.

In addition to guaranteeing the regularization property, there has been a lot of work aiming at deriving stability estimates and convergence rates, namely at assessing the rate at which the regularized solution $f_{\mu(\varepsilon)}$ approaches the exact solution $f^{(0)}$ when $\varepsilon \to 0$ (assuming $\mathcal{N}(A) = \{0\}$). Convergence rates of this type were derived in [86, 80]. They generalize those well known for $p = 2$ (see e.g. [114]).

In the next section we will discuss the algorithmic aspects of the minimization of the cost function (10.8). Indeed, having defined sparse regularized solutions is not enough, we must also be able to compute them. Before doing so, let us stress the fact that while the papers [86, 80] first established the method as a proper regularization method in an infinite-dimensional Hilbert/Besov space setting, the idea of using ℓ^1-norm constraints or penalties to enforce sparsity in the solution of inverse problems is quite old and can be traced back to some forerunner papers including [277, 246, 109] in a finite-dimensional setting and in special instances. In statistics, the same idea has been promoted as a variable selection method for linear regression under the acronym *lasso* by Tibshirani [281] and as an alternative to the (Tikhonov) quadratic case named *ridge regression* after [159].

10.3 Iterative Soft Thresholding as a MM algorithm

Although algorithms for computing regularized solutions of linear inverse problems through the minimization of (10.8)-(10.9) with quadratic Tikhonov penalties ($p = 2$) are easily derived and implemented, the question is no longer straightforward for sparsity-enforcing penalties with $1 \leq p < 2$. The resulting optimization problem is still a convex problem. Indeed, the penalty is still convex, strictly when $p > 1$, non strictly for $p = 1$. On the other hand, the least-squares discrepancy is strictly convex if $\mathcal{N}(A) \neq \{0\}$, but not strictly in the presence of a nontrivial null space. Hence, the minimizer f_μ of (10.8)-(10.9) is unique for $1 < p \leq 2$, and also for $p = 1$ provided that $\mathcal{N}(A) = \{0\}$. Uniqueness, however, is not guaranteed for $p = 1$ and $\mathcal{N}(A) \neq \{0\}$.

Let us first consider the special case of an operator A which can be expressed in diagonal form in the φ_γ–basis, i.e. $A\varphi_\gamma = \alpha_\gamma \varphi_\gamma$. Then the problem is reduced to the minimization of

$$\Phi_\mu(f) = \sum_{\gamma \in \Gamma} \left[|\alpha_\gamma f_\gamma - g_\gamma|^2 + \mu w_\gamma |f_\gamma|^p \right] \tag{10.12}$$

which can be done for each γ separately. As for denoising, we get a family of uncoupled one-dimensional minimizations and the explicit solution we derived above for the case of denoising ($A = I$) can be adapted for this case in a straightforward manner.

When the operator A is not diagonal, however, it will couple the components of the object in the φ_γ–basis, so that we have to solve a coupled system of nonlinear equations for the f_γ's. In the case $p > 1$, the Euler-Lagrange equations become

$$(A^*Af, \varphi_\gamma) - (A^*g, \varphi_\gamma) + \frac{\mu w_\gamma p}{2} |f_\gamma|^{p-1} \mathrm{sign}(f_\gamma) = 0 \tag{10.13}$$

where, since $p > 1$, the last term is assigned the value zero when $f_\gamma = 0$ ($0 = \lim_{t \to 0} \left[|t|^{p-1} \mathrm{sign}(t) \right]$), so that this equation is well defined for all γ in the real case. In the complex case, when $f_\gamma = r_\gamma e^{i\alpha_\gamma}$, the expression $|f_\gamma|^{p-1} \mathrm{sign}(f_\gamma)$ must be replaced by

$r_\gamma^{p-1}e^{i\alpha_\gamma}$ when $r_\gamma \neq 0$ and by 0 when $r_\gamma = 0$. When $p = 1$, the penalty term is no longer differentiable at the origin and the usual way of writing the corresponding Euler equations makes use of the so-called subdifferential of the penalty term. For simplicity, we omit this discussion here.

As known from Chapter 7, in some cases, the operator A can be diagonalized through the SVD. The penalty, however, is expressed in diagonal form in the φ_γ–basis, e.g. wavelets. Unfortunately, we cannot in general diagonalize both in the same basis. Let us mention early attempts to go around this difficulty and to avoid tackling the coupling problem using a 'quasi-diagonalization' in terms of *wavelet/vaguelettes* bases, such as the *Vaguelette–Wavelet Decomposition* (VWD) [1] and the *Wavelet–Vaguelette Decomposition* (WVD) [105]. Unfortunately, these approaches are limited to special types of operators.

In [86], the coupling problem was solved in another way and in full generality, through an *iterative denoising* process, combining denoising with the Landweber gradient-descent algorithm considered in Sections 5.1 and 8.4. In [80], the resulting algorithm or *Iterative Soft Thresholding Algorithm* (later referred to in the literature through the acronym ISTA) was further studied and its convergence properties were established, both for $p = 1$ and for the more general shrinkage corresponding to $1 < p \leq 2$. The algorithm has been derived using a methodology called *optimization transfer* or *Majorization-Minimization* (MM), which consists in successive minimizations of majorizing *surrogate functionals* (we refer to [192] for a review). For example, using a MM method, one can simply derive (for a different cost function) the so-called ML-EM (*Maximum Likelihood-Expectation Maximization*) algorithm, widely used in medical imaging and astronomy [92]; this will be discussed in Chapter 13.

If one has a complicated cost function to minimize, say $\Phi_\mu(f)$, the idea is to majorize it by means of a family of auxiliary functionals, $\Phi_\mu^{SUR}(f;a)$ – depending on an auxiliary function a – which lie above $\Phi_\mu(f)$ and are strictly convex as well as easy to minimize. The precise requirements on the surrogates are the majorization property

$$\Phi_\mu(f) \leq \Phi_\mu^{SUR}(f;a), \quad \text{for all } f \text{ and for all } a, \tag{10.14}$$

and the anchoring property

$$\Phi_\mu(f) = \Phi_\mu^{SUR}(f;f), \quad \text{for all } f. \tag{10.15}$$

The minimizer of $\Phi_\mu(f)$ is then approached by means of the iterative algorithm ($k = 0, 1, \dots$)

$$f_0 \text{ arbitrary}; \quad f_{k+1} = \arg\min_f \Phi_\mu^{SUR}(f;f_k) \tag{10.16}$$

(see Figure 10.1). Let us remark that this ensures a monotonic decrease of the functional $\Phi_\mu(f)$ at each iteration, i.e. $\Phi_\mu(f^{k+1}) \leq \Phi_\mu(f^k)$. Indeed, we have

$$\Phi_\mu(f_{k+1}) \leq \Phi_\mu^{SUR}(f_{k+1};f_k) \leq \Phi_\mu^{SUR}(f_k;f_k) = \Phi_\mu(f_k), \tag{10.17}$$

where the first inequality results from the majorization property and the second one from the minimizing property of f_{k+1}.

To tackle our problem, we will construct *surrogate functionals* in which the coupling term A^*Af in (10.13) is no longer present. This can be done as follows. Define the following functional depending on an auxiliary $a \in \mathcal{X}$

$$\Xi(f;a) = C\|f - a\|^2 - \|Af - Aa\|^2 \geq 0 \tag{10.18}$$

where C is a constant such that $\|A^*A\| < C$. For the sake of simplicity and without loss of generality, we will take $C = 1$ and renormalize the operator to ensure that $\|A\| < 1$. The operator $I - A^*A$ is then strictly positive definite. Hence $\Xi(f;a)$ is strictly convex in f, for

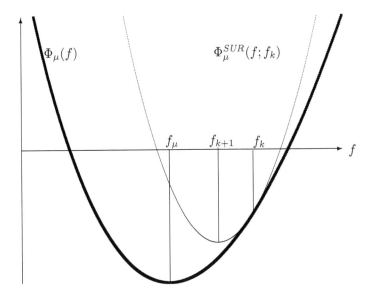

FIGURE 10.1 Majorization-Minimization method through surrogates.

all a, and vanishes for $f = a$. We will add $\Xi(f; a)$ to $\Phi_\mu(f)$ to form the following surrogate functional (we work in the real case only; the complex one is an easy variant)

$$\Phi_\mu^{SUR}(f; a) = \|Af - g\|^2 + \mu \sum_{\gamma \in \Gamma} w_\gamma |f_\gamma|^p - \|Af - Aa\|^2 + \|f - a\|^2 \quad (10.19)$$

$$= \sum_{\gamma \in \Gamma} \left[f_\gamma^2 - 2f_\gamma (a + A^*g - A^*Aa)_\gamma + \mu w_\gamma |f_\gamma|^p \right] + \|g\|^2 + \|a\|^2 - \|Aa\|^2$$

which coincides with $\Phi_\mu(f)$ for $a = f$ and clearly *majorizes* it: $\Phi_\mu(f) \leq \Phi_\mu^{SUR}(f; a)$, for all f and for all a. Moreover, for all a, this functional is strictly convex in f, hence it has a unique minimizer. Since the Euler equations for the f_γ's decouple, the minimizer can be computed easily. Defining the iterative algorithm as (10.16), we see that the iterate f_{k+1} minimizing the surrogate functional anchored at the previous iterate f_k is simply given by

$$f_{k+1} = S_{\mu,\mathbf{w}}^{(p)}(f_k + A^*(g - Af_k)) \ , \quad k = 0, 1, \ldots , \quad (10.20)$$

where $S_{\mu,\mathbf{w}}^{(p)}$ denotes the (nonlinear) *shrinkage operator*, acting component-wise. This means that its action on $h = \sum_\gamma h_\gamma \varphi_\gamma$ is defined by $S_{\mu,\mathbf{w}}^{(p)}(h) = \sum_\gamma S_{\mu w_\gamma}^{(p)}(h_\gamma)\varphi_\gamma$, where $S_\tau^{(p)}(x)$ is the appropriate shrinkage function described in Section 10.1 and reducing to soft thresholding as in equation (10.7) for $p = 1$. In the absence of regularization, i.e. for $\mu = 0$, the shrinkage operator reduces to the identity and the algorithm reduces to the Landweber iterative algorithm. When initialized by $f_0 = 0$ and when g belongs to the range of A, this algorithm is known to converge to the generalized solution of $Af = g$ (see Section 8.4). For $\mu > 0$, we see that the algorithm (10.20) consists in applying 'denoising by shrinkage' to each Landweber iterate.

Let us rewrite (10.20) by means of a mapping $T_{\mu,\mathbf{w}}^{(p)}$ as follows: $f_{k+1} = T_{\mu,\mathbf{w}}^{(p)} f_k$. The convergence of the sequence of iterates f_k to a minimizer of (10.8) was established in [80], in the sense of weak convergence and also of strong convergence, i.e. convergence in the norm of the Hilbert space $\mathcal{X}$. Notice that the convergence holds even if $p = 1$ and $\mathcal{N}(A) \neq \{0\}$, although the minimizer may not be unique in this case; then the limit can depend on the choice of the initial iterate f_0. When either $p > 1$ or $\mathcal{N}(A) = \{0\}$, however, the minimizer f_μ of Φ_μ is unique.

The complete convergence proof in [80] is too involved to be given here. However, a simpler proof holds for the special case of a finite-dimensional setting, when the operator/matrix A is full-rank (i.e. $\mathcal{N}(A) = \{0\}$). Indeed, in such case the mapping $T_{\mu,\mathbf{w}}^{(p)}$ governing the ISTA iteration is a contraction and the convergence of the iterates to the fixed point of $T_{\mu,\mathbf{w}}^{(p)}$ then simply results from the elementary contractive mapping theorem (see Section 5.3).

REMARK 10.1   *This elementary proof goes as follows.*
Since $T_{\mu,\mathbf{w}}^{(p)}$ is the product of the shrinkage operator $S_{\mu,\mathbf{w}}^{(p)}$ by the Landweber iteration $Lf = f + A^*(g - Af)$, it is sufficient to show that $S_{\mu,\mathbf{w}}^{(p)}$ is nonexpansive whereas L is a contraction (the product of a contraction by a nonexpansive mapping being contractive). The fact that the Landweber iteration for a full-rank matrix is a contraction is easily proved by means of the SVD expansion of A. On the other hand, as seen from its graph, the scalar shrinkage operator (10.7) for $p = 1$ satisfies $|S_\tau(x) - S_\tau(y)| \leq |x - y|$ for all x and y in $\mathbb{R}$. It is an easy exercise to show that the same holds true more generally for $1 < p \leq 2$. This property used component-wise implies that

$$\|S_{\mu,\mathbf{w}}^{(p)}(f) - S_{\mu,\mathbf{w}}^{(p)}(h)\| \leq \|f - h\| \tag{10.21}$$

i.e. that the shrinkage operator $S_{\mu,\mathbf{w}}^{(p)}$ is nonexpansive for $1 \leq p \leq 2$. Hence, the operator $T_{\mu,\mathbf{w}}^{(p)} = S_{\mu,\mathbf{w}}^{(p)}L$ is a contraction and the successive iterates $f_{k+1} = T_{\mu,\mathbf{w}}^{(p)}f_k$ converge to the unique fixed point $\bar{f}_\mu$ of $T_{\mu,\mathbf{w}}^{(p)}$, for any choice of the initial iterate f_0.
We still have to show that this fixed point is the minimizer of the functional (10.8)-(10.9). This can be done as follows. We will focus on the case $p = 1$, the other case being easier to treat. We first verify that, for all h,

$$\Phi_\mu^{SUR}(f + h; a) - \Phi_\mu^{SUR}(f; a) \geq \|h\|^2 \tag{10.22}$$

where f is the minimizer of the surrogate $\Phi_\mu^{SUR}(f; a)$, i.e. where $f = S_{\mu,\mathbf{w}}^{(1)}(a + A^*(g - Aa))$. Using the expression (10.19), this inequality is easily checked for each component γ, separating those with $f_\gamma = 0$ from those with $f_\gamma \neq 0$. Notice that the fixed point $\bar{f}_\mu = T_{\mu,\mathbf{w}}^{(p)}\bar{f}_\mu$ of $T_{\mu,\mathbf{w}}^{(p)} = S_{\mu,\mathbf{w}}^{(p)}L$ minimizes $\Phi_\mu^{SUR}(f; \bar{f}_\mu)$. Hence, by the inequality (10.22), we have, for all h,

$$\Phi_\mu^{SUR}(\bar{f}_\mu + h; \bar{f}_\mu) - \Phi_\mu^{SUR}(\bar{f}_\mu; \bar{f}_\mu) \geq \|h\|^2 \tag{10.23}$$

Now, since $\Phi_\mu^{SUR}(\bar{f}_\mu + h; \bar{f}_\mu) = \Phi_\mu(\bar{f}_\mu + h) + \|h\|^2 - \|Ah\|^2$ and $\Phi_\mu^{SUR}(\bar{f}_\mu; \bar{f}_\mu) = \Phi_\mu(\bar{f}_\mu)$, the inequality (10.23) reduces to

$$\Phi_\mu(\bar{f}_\mu + h) \geq \Phi_\mu(\bar{f}_\mu) + \|Ah\|^2 . \tag{10.24}$$

Since this holds for all h and since A is full-rank, we deduce that the fixed point $\bar{f}_\mu$ is indeed the minimizer f_μ of $\Phi_\mu(f)$ (which is unique when $\mathcal{N}(A) = \{0\}$). This concludes the proof.

The ISTA algorithm and sparsity-enforcing regularization, mostly for the case $p = 1$, have inspired a great abundance of subsequent work which would be impossible to review in an exhaustive fashion. For the interested reader, we only provide here a few pointers to the relevant literature and we refer to the book [22] and the references therein for a general overview of algorithms, especially first-order algorithms, and to [200] for a comparison of some of them.
First let us mention the two papers [263] and [118] which proposed finite-dimensional versions of ISTA independently and around the same time as [86, 80]. In [263], it is introduced

as an empirical algorithm for astronomical problems and its convergence is not proved. In [118], it is viewed as an EM algorithm. Later on, in [74], ISTA was reinterpreted in a convex analysis framework as a proximal algorithm.

ISTA is a gradient-descent method where all coordinates of the solution are updated simultaneously at each iteration. Iteration by blocks of variables is also possible, fixing some of them while updating the others either sequentially or in parallel. Monotonicity of the decrease of the cost function is lost in the latter case, and then search techniques are needed. In the limit of blocks of one variable, the algorithm is known as *Parallel Coordinate Descent* or as *Pathwise Coordinate Optimization*. Though very simple to implement, gradient methods are not in general particularly fast. Hence many efforts have been dedicated to derive accelerated schemes, the most well-known being *Fast ISTA* or FISTA [23], which relies on a general strategy due to Nesterov [220] and uses a simple but subtle recombination of two successive ISTA iterates allowing to transform the decay rate of the cost function $\Phi_\mu(f_k)$ from $1/k$ to $1/k^2$, where k is the number of iterations.

For really sparse solutions, i.e. when the number of non-zero components or active variables is very small, homotopy methods [223] and their implementation in statistics under the name *Least Angle Regression and Selection* (LARS) [110], which are recursive algorithms, can be very fast compared to iterative methods. In imaging problems, however, the number of active variables is usually too high for the LARS algorithm to be manageable.

Another way to accelerate the convergence of the algorithm is to reformulate it as an *Iteratively Reweighted Least Squares* (IRLS) problem where at each iteration one has to perform a pure ℓ^2-minimization, which can be made faster by means of conjugate gradient methods [300].

Primal-dual methods have also been widely advocated in the literature, in particular, as *split-Bregman* algorithms in the context of ℓ^1-constrained or ℓ^1-penalized inverse problems (see [224] for a review).

The ISTA algorithm has also been generalized in various ways, of which we only mention a few. The algorithm can be extended in a straightforward way to other separable penalties than weighted ℓ^p-norms (a separable penalty is a sum of terms involving only one component each). Then the surrogate used before for the least-squares term can still decouple the variational problem into one-dimensional minimizations. However, it is also desirable to deal with non-separable penalties such as the Total Variation (TV) penalty, or penalties involving a coupling operator within the ℓ^1-norm. Note that this is sometimes referred to as the *analysis approach* compared to the *synthesis approach* considered above, the two approaches being equivalent when using orthonormal bases. One possibility is to use a separable surrogate for the penalty as well, as done e.g. in [97]. In [201], a primal-dual algorithm has been proposed which reduces to ISTA in the absence of coupling in the penalty.

More sophisticated penalties have also been proposed, which enforce so-called *structured sparsity*. An early example is joint sparsity [119], also known as *group lasso* [311], where sparsity is promoted for groups of variables, with groups known in advance (such as the three channels in a color image), by means of a ℓ^2-penalty inside each group and of a weighted ℓ^1-penalty for combining those ℓ^2-norms. For an overview of more general penalties enforcing structured sparsity, we refer to e.g. [16].

While the original ISTA setting is made for orthonormal bases, it is interesting to generalize it to the case of *frames*, which are overcomplete (redundant) sets, in particular for wavelets [279]. The algorithm has also been generalized to cope with nonlinear inverse problems [240]. Finally, let us mention that nonconvex sparsity-enforcing penalties (e.g. involving ℓ^p-norms with $p < 1$) have also been used in [9] for denoising as well as for inverse problems. In this framework, ISTA has been generalized to *Iterative Hard Thresholding* [44].

10.4   Multiple components with mixed penalties

For some inverse problems, the object may be a superposition of several components, having different characteristics. They may be sparse in different bases or there might be some smooth components. For simplicity let us assume that the object is a superposition of only two components, $f = u + v$, the first one, u, being sparse in the orthonormal basis $(\varphi_\gamma)_{\gamma \in \Gamma}$ and the second one, v, being smooth and regularized by means of a quadratic penalty. In this setting, we can generalize the previous framework as follows. The functional (10.8) is replaced by

$$\Phi_{\lambda,\mu}(u,v) = \|A(u+v) - g\|^2 + \lambda\|Wv\|^2 + \mu\|u\|_{\mathbf{w},p}^p \tag{10.25}$$

where $\|u\|_{\mathbf{w},p}$ is the sparsity-enforcing norm of equation (10.9) and $\|Wv\|^2$ is a smoothness-enforcing quadratic penalty in the norm of $\mathcal{X}$, involving an operator $W : \mathcal{X} \to \mathcal{X}$ which is assumed to have a bounded inverse. We renormalize the operator A so that $\|A\| < 1/\sqrt{2}$ and we use the surrogate

$$\Phi_{\lambda,\mu}^{SUR}(u,v;a) = \Phi_{\lambda,\mu}(u,v) - \|A(u+v) - Aa\|^2 + \|u+v-a\|^2 . \tag{10.26}$$

This allows to derive the following iterative scheme, for $k = 0, 1, \ldots$ and arbitrary initial values u_0, v_0,

$$
\begin{aligned}
u_{k+1} &= S_{\mu,\mathbf{w}}^{(p)}\left(u_k + A^*(g - A(u_k + v_k))\right) \tag{10.27} \\
v_{k+1} &= (I + \lambda W^* W)^{-1}\left(v_k + A^*(g - A(u_k + v_k))\right) .
\end{aligned}
$$

This scheme can be shown to converge (strongly) to a minimizer of (10.25). We refer to [87] for the special finite-dimensional case where the mapping defining this iterative scheme is contractive and to [81] for the full proof in the infinite-dimensional case. Related papers are [82] with an application to texture/cartoon discrimination and [53] for a more general approach based on proximal algorithms. The extension to more than two components is straightforward.

A similar idea has been proposed in [264] (see also the book [267]) under the appellation *Morphological Component Analysis* (MCA). The penalties used, however, are all ℓ^1-penalties, but on the coefficients of the object on different sets of functions ('dictionaries'). Theoretical conditions on these dictionaries are then required of course to ensure that separation into the different components is possible.

Note that the approach outlined above can be revisited as follows [87]. Let us view the functionals (10.25) and (10.26) as depending on the unknowns u and f instead of u and v, and consider, taking $W = I$ here for simplicity,

$$\Phi_{\lambda,\mu}(u,f) = \|Af - g\|^2 + \lambda\|f - u\|^2 + \mu\|u\|_{\mathbf{w},p}^p \tag{10.28}$$

with the surrogate

$$\Phi_{\lambda,\mu}^{SUR}(u,f;a) = \Phi_{\lambda,\mu}(u,f) + \|f - a\|^2 - \|Af - Aa\|^2 . \tag{10.29}$$

The iterative minimization of (10.28) can then proceed as follows ($k = 0, 1, \ldots$; f_0 arbitrary):

$$
\begin{aligned}
u_k &= \arg\min_u \Phi_{\lambda,\mu}^{SUR}(u, f_k; f_k) \tag{10.30} \\
f_{k+1} &= \arg\min_f \Phi_{\lambda,\mu}^{SUR}(u_k, f; f_k)
\end{aligned}
$$

which yields the following explicit expressions for the minimizers

$$
\begin{aligned}
u_k &= S_{\mu/\lambda,\mathbf{w}}^{(p)}(f_k) \tag{10.31} \\
f_{k+1} &= u_k + (1+\lambda)^{-1}\left(f_k + A^*(g - Af_k) - u_k\right) .
\end{aligned}
$$

This algorithm can be rewritten just in terms of f as

$$f_{k+1} = (1 + \lambda)^{-1} \left(f_k + A^*(g - Af_k) + \lambda \, S^{(p)}_{\mu/\lambda, \mathbf{w}}(f_k) \right). \tag{10.32}$$

As for ISTA, a simple proof of convergence of the algorithm can be given when A is finite-dimensional and full-rank, in which case the mapping defining the iteration is a contraction. Interestingly, we can observe that we can minimize (10.28) for u, given f, by means of a simple denoising, obtaining the solution $u = S^{(p)}_{\mu/\lambda, \mathbf{w}}(f)$. In the special case $p = 1$ and with unit weights $w_\gamma = 1$, if we reinsert this solution into the functional, we see that it amounts to minimizing

$$\tilde{\Phi}_\mu(f) = \|Af - g\|^2 + \lambda \sum_\gamma \Pi(f_\gamma) \tag{10.33}$$

with the penalty

$$\Pi(f_\gamma) = \begin{cases} |f_\gamma|^2 & \text{if} \quad |f_\gamma| \le \mu/2\lambda \\ (\mu/\lambda)|f_\gamma| - (\mu/2\lambda)^2 & \text{if} \quad |f_\gamma| > \mu/2\lambda \,. \end{cases} \tag{10.34}$$

The penalty (10.34) is nothing else than the Huber function used in robust statistics to cope with data containing outliers. We can interpret it as follows: small components of f, likely due to noise, are penalized with a ℓ^2-norm, whereas the larger components, likely to reflect some structure in the object, are penalized with a ℓ^1-norm. Using the same surrogate as above for the least-squares misfit, and since the Huber penalty is separable, we can easily devise an iterative algorithm to minimize (10.33) along the lines explained above.

10.5 Examples of applications

In this section we consider a number of possible applications of sparsity regularization. Before implementing a sparse recovery algorithm like ISTA, one must of course specify in which basis the recovered object is expected to be sparse. The first instance which comes to mind is that of an object sparse in pixel space, the orthonormal basis being simply the set of the characteristic functions of each pixel.

10.5.1 Sparsity in pixel space

In this subsection we consider two applications, one coming from astronomy, discussed in Section 2.6.2, and the other from diffusion MRI discussed in Section 6.4.

(A) *Astronomical imaging*

To illustrate the benefits of sparsity we provide the results of a numerical simulation. In Figure 10.2, we display the model of an open star cluster, derived from a rescaled 512x512 image of the Pleiades. It consists of nine stars of different magnitudes concentrated on one pixel each and with values (intensities) ranging from 12393 to 958. We assume that the imaging system is an aberration-free ideal telescope with a circular mirror. Therefore, the imaging operator A is a convolution with a PSF which can be derived from equations (2.73) and (2.78) since we assume incoherent illumination. Therefore the PSF is proportional to the usual Airy pattern, given by

$$K(r) = \left(\frac{2J_1(\Omega r)}{\Omega r} \right)^2 \tag{10.35}$$

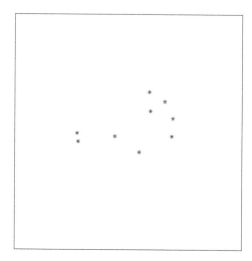

FIGURE 10.2   Star positions in the model of a star cluster.

with the bandwidth Ω of the telescope given by $\Omega = \pi D/\lambda$, where D is the diameter of the mirror, λ is the observation wavelength and r the angular distance from the axis of the telescope. Assuming reasonable values of these parameters, after discretization we find a discrete PSF such that the FWHM of the central peak is about 12 pixels while the Rayleigh distance is about 14 pixels. The image is corrupted by additive white Gaussian noise with a standard deviation of 100. In the left upper panel of Figure 10.3 we show the blurred and noisy image. We use a square-root grayscale for enhancing the visualization of the rings around the central peak of the PSF.

For object reconstruction we first used both the iterative Landweber method with early stopping and Tikhonov's regularization with the value of $\mu = 10^{-3}$, given by Morozov's discrepancy principle. The two reconstructions, shown in the right panel of the first row and in the left panel of the second one, are pretty similar and are affected by strong ringing effects, as discussed in Section 4.5. If one looks at the structure of the ringing, which is the result of the ringing coming from the different stars, one realizes that it is essentially the same in the two images, confirming the practical equivalence of the two methods. If one looks at the peaks corresponding to the different stars, one finds a shrinking with respect to the peaks appearing in the original image and, of course, the ringing artifacts degrade the quality of the image both from the visual and from the quantitative point of view.

Next, we reconstruct the object by means of ℓ^1-penalized regularization using the ISTA scheme (10.20) including soft thresholding with $p = 1$, unit weights w_γ and $\mu = 4$ (chosen according to Morozov's principle). We remark that the pure Landweber algorithm is also the scheme (10.20) but with no thresholding applied. In both cases, the number of iterations is 10000, a rather high number due to the difficulty of restoring point-like objects. This simulation involving a clearly sparse object demonstrates the advantage of sparsity-enforcing regularization for its restoration compared to classical quadratic Tikhonov regularization.

(B) *Diffusion* MRI

Another instance of sparsity in pixel space, but now on the unit sphere, is provided by the diffusion MRI problem considered in Section 6.4 and which, as seen, reduces to a spherical deconvolution problem. The inverse problem consists in the reconstruction of the fiber orientation vector (FOF) $\mathbf{f}$, whose components f_i, $i = 1, \ldots, M$, are the partial volumes of the fibers. It is a nonnegative vector and its ℓ^1-norm is set to one: $\sum_{i=1}^{M} f_i = 1$. This is expected

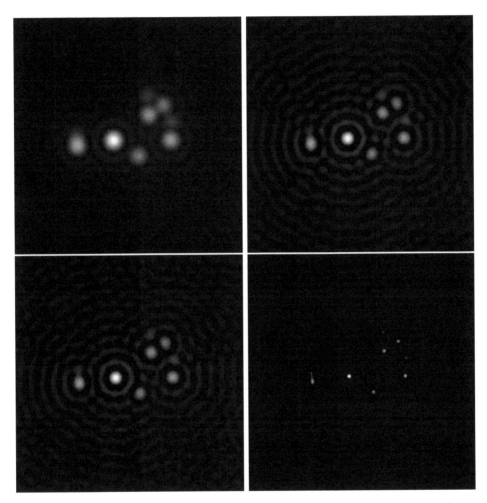

FIGURE 10.3 From left to right and from top to bottom, in a square-root grayscale: blurred and noisy image; reconstruction by pure Landweber iteration ($\mu = 0$); ℓ^2- regularized reconstruction ($\mu = 10^{-3}$) and ℓ^1- regularized reconstruction ($\mu = 4$ and unit weights).

to induce sparsity on the number of fibers through the given voxel. To demonstrate this phenomenon, we show the results of two numerical simulations. The first one corresponds to the crossing of two fibers within a given voxel and the second one to the crossing of three fibers.

The data are generated using the model of equation (6.17) with $b = 3000$ s/mm^2, a value typically used in HARDI acquisitions. Moreover, in both examples, it is assumed that the fibers have the same diffusivity characteristics, namely $\alpha = 1.5 \; 10^{-3}$mm^2/s and $\beta = 0.2 \; 10^{-3}$mm^2/s, values already given in Section 6.4; it is also assumed that the fibers have equal partial volumes, hence 0.5 in the case of two fibers and 0.33 in the case of three fibers.

We use the discretization of the spherical deconvolution introduced in [98]. Thanks to the symmetry of the kernel as a function of $\mathbf{s}$, the data are computed on the unit hemisphere where a sampling of 60 uniformly distributed points is used [172]. On the other hand a much finer sampling of the unit sphere is considered for the fiber directions and, in the simulations, the directions of the fibers do not coincide with sampling points. Noise-free data are computed using the previous discretization and, in view of obtaining data as realistic as possible, they are perturbed with Rician noise (defined in Section 11.1), given in terms of SNR values [154] (realistic values of 10, 20 and 30 dB are selected). Moreover, the dependence of the solution on the noise is estimated by repeating the computations for 25 different realizations of the noise. The results are represented by displaying the convex hull of the corresponding 25 solutions.

The results obtained in the case of two fibers are shown in Figure 10.4. It is evident that for a crossing angle of 15° the method is unable to resolve the two fibers which are reconstructed essentially as a single fiber for any SNR value. In the case of a crossing angle of 30°, the two fibers are resolved if the SNR is greater than 20 dB while for crossing angles greater than 45° the two fibers are resolved for any value of SNR $\geq$ 10 dB with an error in angle reconstruction decreasing with increasing SNR.

To analyze the performance of the method on configurations with more than two fibers, we consider a configuration of three fibers, of equal partial volumes, with crossing angles $\theta_{1,2} = 67°$, $\theta_{2,3} = 40°$ and $\theta_{1,3} = 60°$. Values of SNR of 20 and 30 dB are considered and again, for each case, 25 noise realizations are generated.

In Figure 10.5, we display the results represented, as in the previous case, by means of the convex hull of the 25 reconstructions. Since we have a 3D configuration of the fibers we show the projection onto the plane formed by the fibers $\{1, 2\}$. It is interesting to see that, even in the case of three fibers, the results are consistent with those obtained in the case of two fibers: indeed, the three angles are all reconstructed with a satisfactory accuracy, as it could be expected since they are all greater than the resolution limit of about 30° found in the case of two fibers. This shows that the presence of several fibers does not significantly worsen the performance of the method.

10.5.2 Denoising with shearlets

As argued at the beginning of this chapter, most applications will use sparsity in wavelets rather than in pixel space. In more than one dimension, orthonormal bases of wavelets can be easily obtained by means of tensor products of one-dimensional wavelets like compactly supported Daubechies wavelets [79]. However, such bases are known to be sub-optimal to represent directional features (such as oblique edges) arising typically, say, in natural images. Various schemes have been investigated as remedies and have led to sets of directional wavelets including *ridgelets*, *curvelets* or *shearlets*. We will provide examples using the latter ones which are also the most recently constructed and which can be implemented through a useful Matlab© software package called ShearLab. [184]. Such directional wavelets do not

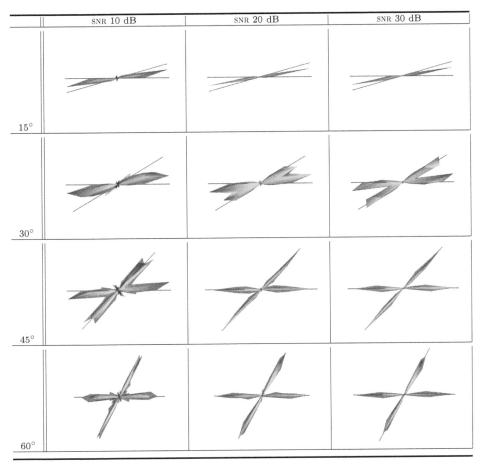

FIGURE 10.4   Reconstructions of two fibers (represented in the plane of the fibers) – In the rows, for a given crossing angle, we show the convex hull of 25 reconstructions, corresponding to 25 different realizations of the noise; in the columns we show the reconstructions for a given SNR value. The values of the angles and of the SNR's are as indicated in the figure.

FIGURE 10.5   Reconstructions obtained in the case of three fibers for two values of the SNR, 20 and 30 dB. We show the projection on the plane formed by the fibers $\{1, 2\}$.

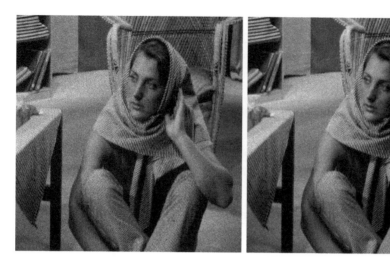

FIGURE 10.6   Left: noisy image (Gaussian noise $\sigma = 10$), PSNR $= 28.3$ dB; right: denoised image, PSNR $= 33.6$ dB.

form orthonormal bases but instead overcomplete sets called frames. A detailed description of shearlets and of the extension of ISTA to frames goes beyond the scope of this book and we refer the interested reader to [279].

A first illustration we provide in Figure 10.6 concerns the denoising (i.e. the case of an operator A equal to the identity) of the standard image 'Barbara' by means of a soft thresholding of the shearlets coefficients, using the ShearLab package (and in particular, Shearlab3Dv11 and SL2D1, using 4 scales with a redundancy factor of 25). White Gaussian noise with standard deviation $\sigma = 10$ was added (the image takes values between 0 and 255). The denoised image is given by equations (10.6) and (10.7). The weights w_γ were set to 1 for the first scale and to $2, 4$ and 8 for the next (finer) three scales and $\mu = 6$ was chosen according to the discrepancy principle.

10.5.3   Inpainting with shearlets

A second illustration is a case of so-called *inpainting*, which consists in restoring an image where some parts are missing or are corrupted by unwanted features. The operator A to be inverted is then simply a binary mask setting to zero all values in the missing or corrupted pixels and keeping the other values unchanged. Thanks to the wavelet multiscale structure, when enforcing sparsity in the wavelet/shearlet domain through a ℓ^1-penalty on the wavelet coefficients, the missing parts can be recovered often with spectacular success! In Figure 10.7, we provide an example of inpainting of the Barbara image where the missing parts are represented in black on the figure. No noise is added to the image. We use the ISTA algorithm adapted to a shearlet frame as implemented in the ShearLab package (using Shearlab3Dv11 and SL2D2). The weights were set as for the denoising example and the regularization parameter is given by $\mu = 0.005$. The number of iterations is 30.

10.5.4   Deblurring with shearlets

Our last simulations concern a true deblurring problem. Here we use the blurred and noisy images of Barbara already used in Section 9.3.2 with one difference: the original image is not rescaled to take values between 0 and 1 but it takes the original values between 0 and 255. Therefore, when we add the noise to the blurred versions we must adjust its variance in

FIGURE 10.7 Left: masked image; right: inpainted image after 30 iterations.

such a way that the PSNR – defined in equation (2.7) – of the two blurred and noisy images essentially coincides with the PSNR of the corresponding images considered in Section 9.3.2. First, from the original Barbara image we obtain two blurred images by convolving with the same Gaussian PSF's as in that section: the first one with a standard deviation of 2 pixels, the second one with a standard deviation of 5 pixels. Then both are corrupted with additive white Gaussian noise with variance 10. The object is reconstructed by means of the ISTA algorithm implemented in shearlet frames as in the previous examples. The thresholding is performed with weights set as above and the regularization parameter μ is estimated by means of the discrepancy principle. Its values are reported in the figure.

In Figure 10.8, we show our results to be compared with the results provided by TV regularization and shown in Figure 9.18. A visual comparison of the two figures leads to the following remarks. In the case of the narrower PSF, the two reconstructions are very similar. In the TV case the boundaries of the table are sharper but other details are more precise in the sparsity case. In both cases no detail of the scarf of Barbara is reconstructed. In the case of the broader PSF, the difference between the two images is considerable: no cartoon effect appears in the sparsity case but the TV image is much sharper. On the other hand in the sparsity case the face of Barbara is not distorted even if it is still affected by some blur, of course not as strong as before the deblurring. As a conclusion, in such a case no miracle can be obtained by the methods presented in this book. We ignore whether some miraculous method exists.

10.5.5 Compressed sensing

Sparse recovery algorithms also play an important role in the field of *compressed sensing* or else *compressive sampling*, the foundations of which were first set in the pioneering papers [59, 106, 58]. The idea is that when designing measurement matrices $\mathbf{A}$ to sense an unknown N-dimensional vector $\mathbf{f}$, if one knows that this vector is sparse, say with K non-zero components, the number of measurements could be reduced from N to a number of the order of K. Theoretical results guarantee indeed recovery from $K\log(N/K)$ measurements but at the price of using properly randomized measurements, i.e. a random matrix $\mathbf{A}$ which is then well conditioned, a property which is not valid in general for the case of inverse problems. The recovery of the sparse vector $\mathbf{f}$ can be done along the lines described in this chapter, by means of a ℓ^1-penalty or constraint. These questions are still motivating

FIGURE 10.8   First row: first blurred (standard deviation 2 pixels) and noisy image (PSNR = 23.4 dB) and reconstructed object with $\mu = 0.75$ (PSNR = 24.3 dB). Second row: second blurred (standard deviation 5 pixels) and noisy image (PSNR = 20.3 dB) and reconstructed object with $\mu = 4$ (PSNR = 22.1 dB).

extensive research (see e.g. the books [120] for recent overviews and [267] for applications in imaging).

A nice application of compressed sensing to Magnetic Resonance Imaging (MRI) has been proposed in [205] and makes use of sparsity in the Fourier domain (the so-called k-space) in order to reduce the acquisition time and has been successfully approved for clinical practice [107].

10.5.6   Sparsity in statistics and machine learning

Finally, let us briefly mention that sparsity-encouraging penalty terms are also important in statistics and in statistical learning. In the statistical literature, a least-squares linear regression problems involving an ℓ^p–penalty on the vector of regression coefficients, with $1 \leq p \leq 2$, is known as *bridge regression* [121, 126], with as special cases *ridge regression* for $p = 2$ [159] and *lasso regression* for $p = 1$ [281]. Several applications to learning problems are described in the books [150, 151]. More generally, in nonparametric statistics, one considers regression functions having a sparse expansion on a basis (e.g. wavelets), a situation close to that described above at least from a computational point of view. However, the setting is no longer deterministic: g is random and in some instances the (*design*) operator A is itself random. The framework of learning theory with random design operators is considered in [85], where consistency results are derived with the help of an additional ℓ^2–penalty, i.e. of a so-called *elastic net* strategy.

We can stress the fact that since the early days of computer vision [233], [38], the concept of regularization has played an important role in the development of learning theory, mainly in the classical Tikhonov sense of quadratic penalties, and interactions with the field of inverse problems have been emphasized and exploited. We can refer the interested reader to the papers [117], [116], [94] and [21].

III

Statistical Methods

11

Statistical approaches to linear inverse problems

Statistical approaches were proposed since the beginning of the inverse problems era. Let us mention at least three pioneering papers: firstly, the papers by Strand and Westwater [273] and by Turchin, Koslov and Malkevich [289], where only problems with discrete data and Gaussian random variables are considered; next, the paper by Franklin [122] where random processes over Hilbert spaces are introduced for describing the variables involved in the problems. In this third part of the book we consider only discrete approaches because we believe that, with the exception of few noise models, continuous approaches are not yet completely and satisfactorily developed.

In this chapter we first discuss different noise models of interest for different important applications. Next we show how *prior* information can be introduced in a statistical model for inverse problems and finally we introduce the concepts of *maximum likelihood* (ML) and *maximum a posteriori* (MAP) solutions.

11.1 Examples of noise and data statistics

In the first two parts of the book it is assumed that it is possible to take into account the effect of the noise on the data by means of the simple model of equation (2.3). We commented this equation by stating that the noise term $w(\mathbf{x})$ should be merely considered as the difference between the noisy and the noise-free data so that, in principle, it could be object dependent. However, in many important instances such as, for example, the definition of a *linear regularization method*, it is assumed that the noise term $w(\mathbf{x})$ is additive, signal-independent and bounded (in norm) by a constant ε. This assumption holds true in some cases but is clearly not valid in others discussed in the present section.

Even if in the next two chapters we only consider two of the noise models presented hereafter, we also introduce the others because, according to the general philosophy of inverse and ill-posed problems, any kind of information about the process of image formation can be useful and must be taken into account in the formulation of the problem. Indeed, since these models can contribute to reduce the extremely wide set of approximate solutions of the linear equation $Af = g$, they can be useful to the readers confronted with problems where the noise models discussed in detail in this book are not applicable. In particular, it must be taken into account that the noise model, suitable for a particular application, typically contributes to the definition of the discrepancy between noisy and noise-free data. Since we consider only discrete problems, we simplify the notation by assuming that images and objects are transformed into vectors by a lexicographic ordering of the pixels or voxels, so that the imaging matrix $\mathbf{A}$ has only two entries. Accordingly, we denote by $A_{m,n}$ its matrix elements and by g_m, w_m and f_n the components of the image, noise and object, respectively.

We assume that the notion of *random variable* (r.v.) is known to the reader [225]. Greek letters are used for denoting such variables which are functions with domain in a probability space and values in a set of possible outcomes of an experiment. These values, also called realizations of the r.v., are, in general, real numbers, at least in the applications we have in mind. In particular, we denote by η_m and ν_m the random variables whose realizations are g_m and w_m, respectively. Moreover, we denote by $\boldsymbol{\eta}$ and $\boldsymbol{\nu}$ the vector-valued random variables with components η_m and ν_m ($m = 1, 2, ..., M$).

(A) *Additive white Gaussian noise*

Our first example of noise is the paradigm underlying the model of equation (2.3) and was already repeatedly used in the simulations described above. It is a basic model frequently used in information theory (e.g. channel description) and imaging. The noise is said to be *additive* because it is added to the signal and to any other noise (we show an example in a moment) and *Gaussian* because its distribution is normal. Moreover, with the term *white*, we mean that the noise contributions, corrupting different pixels, are uncorrelated and have zero mean and the same variance σ^2 (incidentally, we remark that, in the following, we use the symbol σ^2 for indicating the variance of a r.v.). Therefore, from the relation $g_m = g_m^{(0)} + w_m$ implied by equation (2.3), we get the following relationship between the corresponding r.v.'s ν_m and η_m

$$p_{\nu_m}(w_m) = \frac{1}{\sqrt{2\pi}\,\sigma} \exp\left(-\frac{w_m^2}{2\sigma^2}\right) \quad \rightarrow \quad p_{\eta_m}(g_m) = \frac{1}{\sqrt{2\pi}\,\sigma} \exp\left(-\frac{(g_m - g_m^{(0)})^2}{2\sigma^2}\right) \quad (11.1)$$

where p_{ν_m} denotes the probability distribution of the noise. As it is well known, the expectation value (or mean value) of η_m is then $g_m^{(0)}$ while its *variance* is σ^2 and σ its *standard deviation* (again a symbol we use for any r. v.). We remark that, if σ tends to zero, then the corresponding sequence of Gaussian distributions $p_{\nu_m}(w_m)$ tends to the delta distribution, i.e. to the r.v. which takes the value 0 with probability 1; this is the precise meaning of the expression *'noise tending to zero'*. More general forms of additive Gaussian noise are considered in Section 12.1.

(B) *Rician noise*

This is a model of noise deriving, in a way, from the previous one, even if with completely different properties. It is important in *magnetic resonance imaging* (MRI) and it is not considered in this book but it constitutes an interesting example of noise which does not satisfy equation (2.3). In MRI the real and imaginary parts of the image are obtained by Fourier transforming the acquired data which are assumed to be affected by additive noise with a Gaussian distribution and zero mean. Since the Gaussian characteristic of the noise is preserved by Fourier transform, one can assume that the real and imaginary parts of the image are corrupted by additive and white Gaussian noise with zero mean and variance σ^2. However, in MRI, it is the modulus of the complex image which is usually displayed in order to discard phase artifacts. Since this is a nonlinear operation the Gaussian distribution is lost and the distribution of the image data is given by a Rician distribution which can be written as follows [142]

$$p_{\eta_m}(g_m) = \frac{g_m}{\sigma^2} \exp\left\{-\frac{1}{2\sigma^2}\left(g_m^2 + [g_m^{(0)}]^2\right)\right\} I_0\left(\frac{g_m\, g_m^{(0)}}{\sigma^2}\right) \tag{11.2}$$

where $I_0(x) = J_0(ix)$ is the modified Bessel function of the first kind. For small values of the ratio $g_m^{(0)}/\sigma$ this distribution is very far from a Gaussian distribution. In particular, for $g_m^{(0)} = 0$ (i.e. in the absence of signal so that g_m consists only of noise), it becomes a *Rayleigh distribution*, given by [225]

$$p_{\eta_m}(g_m) = \frac{g_m}{\sigma^2} \exp\left\{-\frac{1}{2\sigma^2}\, g_m^2\right\} \tag{11.3}$$

with mean and variance given by $\sigma\sqrt{\pi/2}$ and $(2 - \pi/2)\,\sigma^2$, respectively. Hence equation (11.2) is used for small values of $g_m^{(0)}/\sigma$. Moreover, the distribution introduces a bias because the expected value does not coincide with $g_m^{(0)}$. On the other hand, for values $g_m^{(0)}/\sigma \geq 3$ the Rician distribution starts to approximate a Gaussian distribution given by [142]

$$p_{\eta_m}(g_m) = \frac{1}{\sqrt{2\pi}\,\sigma} \exp\left\{-\frac{1}{2\sigma^2}\left(g_m - \sqrt{[g_m(0)]^2 + \sigma^2}\right)^2\right\} \tag{11.4}$$

with mean $\sqrt{[g_m(0)]^2 + \sigma^2}$ and variance σ^2. Hence a bias exists also for moderate values of $g_m^{(0)}/\sigma$. Only for very large values of this quantity we re-obtain the normal distribution of equation (11.1), an approximation very frequently used in the treatment of MRI data.

(C) *Poisson noise*

In electronics, this kind of noise originates from the discrete nature of the electric charge and is called *shot noise* whereas, in image processing, it originates from the discrete nature of the light, the photons. It describes the fluctuations in the counting of the photons that hit the detector and, for this reason, it is also called *photon noise* or *photon counting noise*. More precisely, the fluctuations in the counting are due to the fluctuations in the emission of photons by the source, if the detector efficiency is one. This noise is basic in important applications of inverse problems described in this book, namely fluorescence microscopy, astronomy and emission tomography.

The detected value of the image in pixel m, g_m, is a natural number (the number of detected photons) and, if the process of emission/counting is time independent, it is the realization of a Poisson r.v. η_m with a given expectation value λ_m; its probability distribution is given by

$$p_{\eta_m}(g_m) = e^{-\lambda_m} \frac{(\lambda_m)^{g_m}}{(g_m)!} \tag{11.5}$$

where g_m can take the values $0, 1, 2, \ldots$. A basic assumption in imaging is that the expectation value is the noise-free datum $g_m^{(0)}$, which has the physical meaning of average number of detected photons. We also recall that, as one can easily verify, λ_m is both the expectation value and the variance of η_m

$$E\{\eta_m\} = \lambda_m \;,\quad \sigma^2 = E\{(\eta_m - \lambda_m)^2\} = \lambda_m\;. \tag{11.6}$$

The average number of photons emitted by the source and recorded by the detector depends on the *observation time* τ: if we first observe the source during 1 second and then during 2

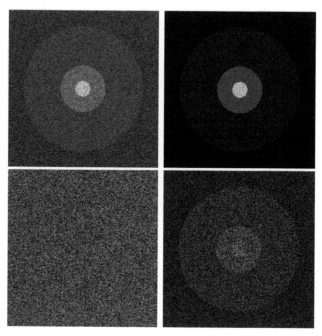

FIGURE 11.1 Upper panels: the image of the original object corrupted by additive white Gaussian noise (left) and by Poisson noise (right). Lower panels: the corresponding differences between noisy and noise-free images. It is evident that, in the Poisson case, the noise has a memory of the underlying image.

seconds and if the process is stationary, in the second observation the number of detected photons is approximately twice the number observed in the first one. In the second case the expected value of the number of photons is exactly twice that of the first one. In other words, the expected value is proportional to the observation time τ and we can write $\lambda_m = \tau g_m^{(0)}$, where $g_m^{(0)}$ is the average number of photons per unit time. Then, as follows from equation (11.6), also the variance σ^2 is given by $\tau g_m^{(0)}$ so that the standard deviation σ is $\sqrt{\tau g_m^{(0)}}$. It follows that the SNR, as defined in equation (2.5) also increases as $\sqrt{\tau}$. This remark suggests that, in principle, it is convenient to increase, as far as possible, the observation time. This is possible in astronomy but is not convenient in medical imaging because, in such a case, to get a similar effect, one would need to increase the number of photons and therefore also the dose administered to the patient.

From a theoretical point of view, the introduction of the observation time allows to give a precise meaning to the expression 'noise tending to zero' in the case of Poisson noise [161]. Indeed, let $g_m^{(\tau)}$, $(m = 0, 1, \dots)$, denote the realizations of a Poisson r.v. with expected value $\lambda_m = \tau g_m^{(0)}$. On the other hand, let us consider the quantities $g_m^{(\tau)}/\tau$ which are the realizations of a r.v. which can be called a *temporally normalized* r.v. and has $g_m^{(0)}$ as expectation value and $g_m^{(0)}/\tau$ as variance. Hence the relative standard deviation $\sigma/g_m^{(0)}$ is given by $1/\sqrt{\tau g_m^{(0)}}$. In other words, when the observation time tends to infinity, the sequence of the temporally-normalized data tends to concentrate around $g_m^{(0)}$, so that it provides an estimate of $g_m^{(0)}$ with an error tending to zero. This behavior can be interpreted as a 'noise tending to zero' and can be investigated by looking at the behavior of the temporally-normalized data in the case of an *observation time tending to infinity*.

In Figure 11.1, we highlight the difference between the perturbation of an image due to

Poisson noise and that due to additive white Gaussian noise. We show the two noisy images and the corresponding images of the noise, i.e. of the difference between the noisy and noise-free images given by $\mathbf{w} = \mathbf{g} - \mathbf{g}^{(0)}$. It is clear that in the image of Poisson noise the original object appears as a phantom immersed in a sort of fog, demonstrating that Poisson noise is object dependent.

REMARK 11.1 *If the value of λ_m is large, for values of g_m close to λ_m the Poisson distribution (11.5) can be approximated by a Gaussian distribution with expectation and variance given by equation (11.6)*

$$p_{\eta_m}(g_m) \simeq \frac{1}{\sqrt{2\pi\lambda_m}} \exp\left[-\frac{1}{2\lambda_m}(g_m - \lambda_m)^2\right] . \tag{11.7}$$

This approximation is frequently used and it is shortly discussed in the following. Remark that the signal dependence of the Gaussian probability distribution is expressed by the fact that the expected value and the variance coincide with the signal.

(D) *Mixed Poisson-Gaussian noise*

In astronomy as in other imaging applications, photons arriving in one pixel are detected by a *charged coupled device* (CCD) and the detector consists of an array of these elements. Since their outputs are statistically independent, we can consider a probabilistic model for a single element. Each CCD accumulates a number of photons during the observation time. At the end, an electronic system is used for reading the content of each element but this process perturbs the result by adding a white Gaussian noise, so that the r.v. describing the detected datum in the pixel m is the sum of two independent r.v.'s, one Poisson and one Gaussian, the second one corresponding to the so-called *read-out noise* (RON). As a consequence, the probability distribution is the convolution of a Poisson distribution with expected value λ_m and a Gaussian distribution with expected value zero and variance σ^2

$$p_{\eta_m}(g_m) = \frac{1}{\sqrt{2\pi}\,\sigma} \sum_{n=0}^{\infty} e^{-\lambda_m} \frac{(\lambda_m)^n}{n!} \exp\left(-\frac{(g_m - n)^2}{2\sigma^2}\right) . \tag{11.8}$$

Since in this case the expected value is the sum of the expected values of the two r.v.'s, it coincides with the expected value of the Poisson r.v., i.e. λ_m, so that we can assume that it is given by the noise-free data $g_m^{(0)}$.

(E) *Multiplicative noise*

Multiplicative noise occurs when the noise-free image $g_m^{(0)}$ is multiplied by a random variable, i.e. $g_m = g_m^{(0)} w_m$, where w_m is the realization of the random variable. Also in this case one assumes that r.v.'s associated to different pixels are statistically independent and identically distributed. An important example is provided by *synthetic aperture radar* (SAR) imaging. In this case the noise is also called *speckle noise*. A SAR image is obtained as follows. A radar placed on an aircraft sends a coherent wave towards the ground. If the surface is coarse, then the reflected image is degraded by a noise with large amplitude which gives a speckled aspect to the image. In general, the SAR image is obtained as the summation of N different images. Then the standard model is to assume that the multiplicative r.v. ν_m with realization w_m has the following Gamma distribution

$$p_{\nu_m}(w_m) = \frac{N^N}{\Gamma(N)} w_m^{N-1} \exp(-Nw_m) , \quad w_m \geq 0 \tag{11.9}$$

where $\Gamma(N) = (N-1)!$. For more details and a variational method for noise removal see, for instance, [13].

(F) *Salt-and-Pepper noise*

This kind of noise, also known as *impulse noise* can be generated by sharp perturbations introduced in the image, the result being a number of white and black pixels sparsely distributed within the image domain. It can be removed by a median filter. However, if the image is also blurred, the application of deconvolution methods after that kind of noise removal can produce inaccurate results. Indeed, the deconvolution process can amplify the distortions introduced in the image by the median filter.

Since this kind of noise is additive and sparse, it is natural to assume a noise distribution inducing sparsity, a point reconsidered in the next section; therefore, one can assume statistical independence of the different pixel values and a probability distribution in a single pixel given by

$$p_{\eta_m}(g_m) = \frac{1}{2\beta} \exp\left\{-\beta|g_m - g_m^{(0)}|\right\}. \tag{11.10}$$

If a non-negativity constraint is not introduced, the expectation value is the noise-free image. A variational approach related to this assumption is proposed in [222] for image denoising and in [19] for image deblurring.

11.2 Priors as additional information on the solution

Once again we recall that the main feature of inverse problems is their ill-posedness, due to a lack of information on the solution. Its consequence is the ill-conditioning of their discrete versions, which increases when refining the discretization. The information on data statistics which, as we show in the next section, is usually implemented in maximum likelihood approaches, is not sufficient to remove the numerical instability of the discrete solutions. Therefore, some *a priori* information on the unknown object is required for obtaining sensible approximate solutions of the problem of image reconstruction. In the methods considered in the previous chapters, the additional information takes the form of rigid constraints on the solution. They consists, for instance, in requiring that the approximate solutions belong to some fixed domain in the object space, such as the set of nonnegative objects or the set of objects with gradient bounded by a fixed quantity and so on. Some of these constraints, combined with the least-squares approach, lead to regularization methods.

In a statistical approach it is also possible to introduce *a priori* information on the unknown object by assuming that it is a realization of a random variable and that it is possible to guess a reasonable probability distribution of its values; therefore the unknown object $\mathbf{f}^{(0)}$ is the realization of a vector-valued r.v. ϕ whose density function is given. It can be obtained, for instance, by a complete use of previous experience about the possible objects to be restored. For example in medical imaging this can be provided by a large database of CT or PET images of a particular organ of a multitude of people. Even if such kind of information is not available, the probabilistic approach can be used to formulate assumptions about some features of the unknown objects, such as smoothness, sparsity or others. We illustrate this point in a moment.

A given probability distribution of the unknown objects is usually called a *prior*. According to a suggestion in [127], the most frequently used priors are the so-called Gibbs priors which are given in terms of Gaussian functions. More precisely they take the following form, indicating the dependence on a generic object $\mathbf{f}$

$$p_\phi(\mathbf{f}) = \frac{1}{Z} \exp\left\{-\mu \mathcal{G}(\mathbf{f})\right\}, \tag{11.11}$$

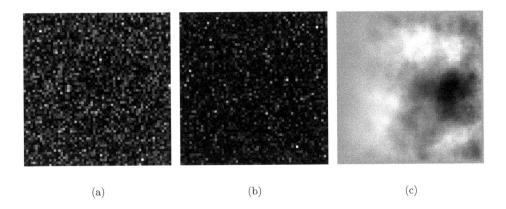

(a)   (b)   (c)

FIGURE 11.2 Examples of realizations of three priors: white noise (a); sparsity in pixel space (b); Laplacian regularization (c).

where ϕ is the r.v. with realizations $\mathbf{f}$ and Z is a normalization constant. The domain can be restricted to the nonnegative orthant in the object space if it is known that the objects are nonnegative. The function $\mathcal{G}(\mathbf{f})$ is nonnegative and is usually called the *energy function* or the *potential function* of the objects. We call it *regularization function* because of an analogy which will become clear in the following. Moreover, the parameter μ is a positive number, in general called *hyper-parameter*, and which we will also call *regularization parameter* because of the same analogy.

Indeed, most of the potential functions which are used are precisely regularization functions taken from the various application of the Tikhonov regularization theory discussed in the previous chapters. Let us consider for instance the Tikhonov regularizer; if we take as energy the square of the ℓ^2-norm of $\mathbf{f}$, we obtain the prior

$$\mathcal{G}(\mathbf{f}) = \|\mathbf{f}\|^2 \;\rightarrow\; p_\phi(\mathbf{f}) = \left(\sqrt{\frac{2}{\pi}}\mu\right)^N \exp(-\mu\|\mathbf{f}\|^2) \,. \tag{11.12}$$

This is precisely the white Gaussian noise distribution with $\sigma^2 = 1/2\mu$ and therefore one could say that it contains no information at all. This is not completely correct because, if we take into account the fact that, in the case of a Gaussian distribution, the probability of obtaining values outside the interval $(-5\sigma, +5\sigma)$ is practically zero, the information contained in this prior is that the objects are bounded and can not take values with an absolute value greater than $5/\sqrt{2\mu}$. As it is clear, the bound increases as μ decreases because the distribution then becomes very flat.

Similarly, if we know that our objects are smooth, we can look at the differential regularizers introduced in Section 4.4. We can take, for instance, as potential the squared norm of the modulus of the discrete gradient

$$\mathcal{G}(\mathbf{f}) = \sum_{n=1}^{N} |\nabla \mathbf{f}_n|^2 \,, \tag{11.13}$$

or the squared norm of the discrete Laplacian

$$\mathcal{G}(\mathbf{f}) = \sum_{n=1}^{N} |\Delta \mathbf{f}_n|^2 \,. \tag{11.14}$$

On the other hand, if we know that our objects have edges and discontinuities, then we can take the TV-smooth regularizer, defined in equation (9.30), which reduces for $\delta = 0$ to the discrete TV given in equation (9.29).

Finally, if we know that the object is sparse in some wavelet basis or tight frame, we can take the ℓ^1-norm of the object coefficients in this basis, obtained by applying a transform P to the object $\mathbf{f}$

$$\mathcal{G}(\mathbf{f}) = \sum_{n=1}^{N} |(P\mathbf{f})_{\mathbf{n}}| \, . \tag{11.15}$$

If the transform P is the identity, then the prior corresponds to sparsity in pixel space.

In order to visualize how quantitative information encoded in a prior represents a qualitative information on the unknown object, we show in Figure 11.2, realizations of the priors corresponding to equations (11.12), (11.15) (in the case $P = I$, i.e. of sparsity in pixel space) and (11.14), all with $\mu = 1$. These realizations are computed using the method indicated in [175], section 3.3. We remark that in the first case no structure of the function is assumed. Sparsity regularization implies the existence of localized sources while a prior in terms of the Laplacian implies realizations looking as views of a sky with clouds, i.e. corresponding to very smooth objects.

11.3   Maximum likelihood and Bayesian regularization

All the examples of data and noise statistics described in Section 11.1 are characterized by probability distributions, continuous or discrete, with an explicit expression as functions of g_m and $g_m^{(0)}$. When $g_m^{(0)}$, the m-th component of the noise-free image, is given, the probability of all noisy versions of this component can be easily computed.

In the case of a linear inverse problem, we know that the model of image formation is given by $\mathbf{g}^{(0)} = \mathbf{A}\mathbf{f}^{(0)}$. If we insert this model into the expression of the probability distribution of the noisy data, we obtain again an explicit expression as a function of $\mathbf{f}^{(0)}$ and $\mathbf{A}$. We remark that now, in a single pixel, the dependence is not on a single components of $\mathbf{f}$, as in the case of denoising, but on all its components since they are mixed by the action of the matrix $\mathbf{A}$. Again, if we insert $\mathbf{f}^{(0)}$ and $\mathbf{A}$ into this expression, we obtain the probability distribution of all possible noisy components corresponding to this particular choice of object and imaging matrix.

We summarize the main properties of the probability distributions of the data which hold true not only in the case of the particular examples discussed in the next chapters but also in the case of other examples described in Section 11.1 (and not further discussed in this book).

- The detected value in a given pixel m is the realization of a r.v. whose probability distribution has an explicit and known dependence on the detected value g_m and on the unknown object $\mathbf{f}^{(0)}$, as well as on any other admissible object $\mathbf{f}$: if we replace $\mathbf{f}^{(0)}$ with $\mathbf{f}$ we obtain the distribution of the corresponding noisy images. In other words we have a family of r.v.'s depending on the admissible objects $\mathbf{f}$; it can be denoted by $\eta_m(\mathbf{f})$ with a probability distribution $p_{\eta_m}(g_m|\mathbf{f})$. In the case of continuous random variables, it is a nonnegative function whose integral with respect to g_m is one; in the case of discrete random variables, as in the case of Poisson noise, it is a distribution, more precisely a linear combination with positive coefficients, of delta distributions concentrated on the values of the random variable. In such a case the sum of the coefficients is one.

  We observe that the notation $p_{\eta_m}(g_m|\mathbf{f})$ is similar to the notation used in probability theory for a conditional probability. The justification of this choice will be clear soon.

- The r.v.'s associated with different pixels are assumed to be statistically independent, so that the probability distribution of the complete image for a given $\mathbf{f}$ is given by

$$p_{\boldsymbol{\eta}}(\mathbf{g}|\mathbf{f}) = \prod_{m=1}^{M} p_{\eta_m}(g_m|\mathbf{f}) \ . \tag{11.16}$$

- The third assumption is that the expectation value of the random vector $\boldsymbol{\eta}(\mathbf{f})$ is $\mathbf{Af}$. Therefore, in the case of continuous random variables, $p_{\boldsymbol{\eta}}(\mathbf{g}|\mathbf{f})$ is a nonnegative function normalized to 1, as specified above, and such that

$$E[\boldsymbol{\eta}] = \int \mathbf{g} \, p_{\boldsymbol{\eta}}(\mathbf{g}|\mathbf{f}) \, \mathbf{dg} = \mathbf{Af} \ . \tag{11.17}$$

Remark that this condition is satisfied by Gaussian and Poisson noise, as well as by mixed Poisson-Gaussian noise and salt-and-pepper noise. With the exception of Poisson noise, the assumption fails to be valid if the domain of the probability distribution is restricted to the nonnegative orthant. Notice that also in the case of Rician noise the assumption is not valid.

The previous items characterize the probabilistic models considered in this book. The probability distributions of the data depend explicitly on a set of parameters, the components of the object $\mathbf{f}$. Once this is given, we can compute, in principle, all the quantities of interest about all possible realizations of the detected image from its probability distribution.
However, the object $\mathbf{f}$ is unknown and its estimation is precisely our problem which can be formulated as follows: given a noisy image $\mathbf{g}$, which is an observed value (realization) of an unknown r.v. $\boldsymbol{\eta}(\mathbf{f}^{(0)})$, which member of the family of r.v.'s $\boldsymbol{\eta}(\mathbf{f})$ is the most likely candidate to represent $\boldsymbol{\eta}(\mathbf{f}^{(0)})$? This is a typical estimation problem in statistics whose solution provides an estimate of $\mathbf{f}^{(0)}$.
The *method of maximum likelihood* (ML *method*) is useful for solving this kind of problems [239]. In our particular case it can be formulated as follows. Let $\mathbf{g}$ be the observed value, i.e. a noisy image of the object $\mathbf{f}^{(0)}$. Then the probability distribution of the r.v. $\boldsymbol{\eta}(\mathbf{f})$ takes in $\mathbf{g}$ the following form

$$L_{\mathbf{g}}(\mathbf{f}) = p_{\boldsymbol{\eta}}(\mathbf{g}|\mathbf{f}) \ , \tag{11.18}$$

where $\mathbf{g}$ is given and $L_{\mathbf{g}}(\mathbf{f})$ is a well-defined function of $\mathbf{f}$. This function is called the *likelihood* (or *likelihood function*) and is defined over the class of all admissible objects. In the case of $N \times N$ objects, it is a function of N^2 variables.
A maximum-likelihood estimate of $\mathbf{f}^{(0)}$ *is any object* $\tilde{\mathbf{f}}$ *which maximizes* $L_{\mathbf{g}}(\mathbf{f})$ *on its domain of definition.* In other words a maximum-likelihood estimate is an object which maximizes the probability of obtaining the observed image. This is the basic argument of the method. In order to show how the method works we consider two simple examples inspired by the two important cases considered in the next chapters.

Example 11.1

Assume that η is a Gaussian random variable with known variance σ^2 and unknown expectation value λ. If a given experiment provides the value g of η, then the likelihood function associated with this experiment is given by

$$L_g(\lambda) = (2\pi\sigma^2)^{-1/2}\exp\left[-\frac{1}{2\sigma^2}(g-\lambda)^2\right] \ . \tag{11.19}$$

It is obvious that the maximum of the likelihood function is reached at $\lambda = g$ and there-fore the maximum-likelihood estimate of the parameter coincides with the result of the experiment. Moreover, it is easy to verify that, if we have q values $g^{(1)}, \ldots, g^{(q)}$, then the maximum likelihood estimate is the arithmetic mean of these values.

Example 11.2

Assume that η is a Poisson random variable with unknown parameter λ. If the result of the experiment is the integer value g of η, then the likelihood function is

$$L_g(\lambda) = e^{-\lambda} \frac{\lambda^g}{(g)!} . \tag{11.20}$$

By deriving the logarithm of $L(\lambda)$ with respect to λ and setting the result to zero we obtain the condition $-1 + g/\lambda = 0$; therefore the maximum is reached at $\lambda = g$. Also in this case the maximum-likelihood estimate of the parameter coincides with the result of the experiment.

In these very simple examples the likelihood function has only one maximum and the max-imum can be easily computed. In general, this is not the case. In image reconstruction the likelihood is a function of N, N^2 and N^3 variables in the 1D, 2D and 3D cases, respectively, and it may have several global maxima as well as local maxima. Of course, one prefers a global one, i.e. that with the largest value of the likelihood function, but, anyhow, the computations can become very complex. In the case of the particular models which we will discuss, all the maxima are global. If several global maxima exist, then we can hope that they are very similar.

To determine the position of the maximum (maxima) of $L_{\mathbf{g}}(\mathbf{f})$ one can use the standard methods of differential calculus, i.e. compute the first derivatives of $L_{\mathbf{g}}(\mathbf{f})$ with respect to the components of $\mathbf{f}$ and set them to zero, look at the Hessian matrix, etc. However, when the components of $\boldsymbol{\eta}(\mathbf{f})$ are independent random variables, the density function of $\boldsymbol{\eta}(\mathbf{f})$ is the product of a large number of functions, as shown for instance in equation (11.16). Since the derivatives of a product with many factors are quite cumbersome to handle, it is useful to consider the logarithm of the likelihood function; more precisely, we consider the negative logarithm (neg-log) because, in such a way the maximization problem is transformed into the minimization of a function which is the sum of M terms. If we denote by $l(\mathbf{f}|\mathbf{g})$ the neg-log of the likelihood, from equation (11.16) we get

$$l(\mathbf{f}|\mathbf{g}) = -\ln L_{\mathbf{g}}(\mathbf{f}) = -\ln p_{\boldsymbol{\eta}}(\mathbf{g}|\mathbf{f}) = -\sum_{m=1}^{M} \ln p_{\eta_m}(g_m|\mathbf{f}) . \tag{11.21}$$

This function is called the *neg-log likelihood function* and its stationary points coincide with the stationary points of $L_{\mathbf{g}}(\mathbf{f})$.

It is important to point out that in all examples discussed in Section 11.1 and satisfying the conditions stated above, the neg-log likelihood is a convex function and therefore all its minima are global so that all the ML solutions correspond to global maxima.

For instance, using equation (11.17) and neglecting irrelevant terms independent of $\mathbf{f}$, in the case of white Gaussian noise we re-obtain the least-squares function given by the ℓ^2-norm

$$l(\mathbf{f}|\mathbf{g}) = \|\mathbf{g} - \mathbf{A}\mathbf{f}\|^2 \tag{11.22}$$

while in the case of Poisson noise we obtain the function

$$l(\mathbf{f}|\mathbf{g}) = \sum_{m=1}^{M} [(\mathbf{A}\mathbf{f})_m - g_m \ln(\mathbf{A}\mathbf{f})_m] \tag{11.23}$$

whose convexity is proved in Chapter 13. Moreover, in the case of mixed Poisson-Gaussian noise, we obtain the function

$$l(\mathbf{f}|\mathbf{g}) = - \sum_{m=1}^{M} \ln \sum_{n=0}^{\infty} e^{-(\mathbf{Af})_m} \frac{(\mathbf{Af})_m^n}{n!} \exp\left(-\frac{(g_m - n)^2}{2\sigma^2}\right) , \tag{11.24}$$

whose convexity is proved in [28, 29] and finally, in the case of salt-and-pepper noise, the neg-log likelihood is given by the ℓ^1-norm

$$l(\mathbf{f}|\mathbf{g}) = \|\mathbf{g} - \mathbf{Af}\|_1 , \tag{11.25}$$

whose convexity is obvious.

However, the minimization of the previous functions is in general an ill-conditioned problem (it is obvious in the first and fourth case); in other words the use of information about data statistics is not sufficient to cure the illness of the problem. As we already discussed in Section 11.2, the remedy consists in introducing *a priori* information on the solution by assuming that the possible objects $\mathbf{f}$ are realizations of a given r.v. ϕ. Two assumptions are required: the first is that the probability distribution of the data for a given object can be interpreted as a conditional probability; the second is that the probability distribution of the object is known with a well-defined dependence on $\mathbf{f}$.

The use of probabilistic *a priori* information is the basic feature of the so-called *Bayesian methods*, which can, in principle, remove the numerical instability of the ML methods. We recall that the Bayesian paradigm was proposed by Geman and Geman [127] in image processing and was subsequently applied to SPECT imaging by Geman and McClure [128]. Given the conditional probability of η for a given realization of ϕ and given the probability distribution of ϕ, then, according to the definition of conditional probability, the joint probability distribution of the two vector-valued random variables η and ϕ, denoted by $p_{\phi\eta}(\mathbf{f}, \mathbf{g})$, is given by the product of the likelihood function and of the prior of ϕ

$$p_{\phi\eta}(\mathbf{f}, \mathbf{g}) = p_\eta(\mathbf{g}|\mathbf{f}) \, p_\phi(\mathbf{f}) . \tag{11.26}$$

We remark that the density function $p_\phi(\mathbf{f})$ of ϕ is the *marginal density function* of $p_{\phi\eta}(\mathbf{f}, \mathbf{g})$

$$p_\phi(\mathbf{f}) = \int p_{\phi\eta}(\mathbf{f}, \mathbf{g}) \, d\mathbf{g} . \tag{11.27}$$

The result of an imaging experiment is a realization $\mathbf{g}$ of the random vector η. Therefore the obvious question is the following: what can we say about the r.v. ϕ when a value of η is known? The answer is provided by the conditional probability density of ϕ for given $\mathbf{g}$ which, using the so-called Bayes formula [239]

$$p_\phi(\mathbf{f}|\mathbf{g}) \, p_\eta(\mathbf{g}) = p_\eta(\mathbf{g}|\mathbf{f}) \, p_\phi(\mathbf{f}) \tag{11.28}$$

is given by

$$p_\phi(\mathbf{f}|\mathbf{g}) = \frac{p_{\phi\eta}(\mathbf{f}, \mathbf{g})}{p_\eta(\mathbf{g})} = \frac{p_\eta(\mathbf{g}|\mathbf{f}) \, p_\phi(\mathbf{f})}{\int p_\eta(\mathbf{g}|\mathbf{f}) \, p_\phi(\mathbf{f}) \, d\mathbf{f}} . \tag{11.29}$$

This conditional density function is also called the *a posteriori* (or *posterior*) density function of the random vector ϕ.

The density function $p_\phi(\mathbf{f}|\mathbf{g})$ contains all information we need about the unknown object. In this way, we do not have a unique estimate of the unknown object but rather a set of possible estimates with different probabilities. Since, in practical applications, the usual requirement is to provide a reconstruction of the unknown object, we can use $p_\phi(\mathbf{f}|\mathbf{g})$ for computing various significant estimates of $\mathbf{f}^{(0)}$.

The most frequently used in practice are the following estimates.

- The estimate provided by the *a posteriori* expectation value of ϕ or *conditional mean estimate* (CM):

$$\tilde{\mathbf{f}} = E\{\phi|\mathbf{g}\} = \int \mathbf{f} \, p_\phi(\mathbf{f}|\mathbf{g}) \, d\mathbf{f} \ . \tag{11.30}$$

- The estimate (or estimates) provided by the global maximum (or maxima) of the *a posteriori* density function $p_\phi(\mathbf{f}|\mathbf{g})$. These are called the *maximum a posteriori* (MAP) estimates.

If $p_\phi(\mathbf{f}|\mathbf{g})$ has a unique global maximum and is concentrated in a small neighborhood of this maximum, then the two estimates defined above can be very close. However, nowadays images contain millions of pixels, so that the computation of the integral in equation (11.30) can be a difficult task; on the other hand, the computation of MAP estimates can be reduced to a minimization problem and in such a case efficient methods of numerical optimization can be used for their computation. Indeed, by taking the neg-log of the posterior density function, equation (11.29), neglecting terms independent of $\mathbf{f}$ and recalling the definition (11.21), the problem is transformed into the minimization of the following function

$$l_B(\mathbf{f}|\mathbf{g}) = -\ln p_\eta(\mathbf{g}|\mathbf{f}) - \ln p_\phi(\mathbf{f}) = l(\mathbf{f}|\mathbf{g}) - \ln p_\phi(\mathbf{f}) \ . \tag{11.31}$$

From the assumption on data statistics and with the use of Gibbs priors, equation (11.11), it follows that the function takes the following form

$$l_B(\mathbf{f}|\mathbf{g}) = -\sum_{m=1}^{M} \ln p_{\eta_m}(g_m|\mathbf{f}) + \mu \, \mathcal{G}(\mathbf{f}) \ . \tag{11.32}$$

Therefore the role of the prior is to penalize the neg-log of the likelihood function. This remark justifies our use of the term regularization function for the potential function of a Gibbs prior.

12

Statistical methods in the case of additive Gaussian noise

In Section 11.1 we discussed the case of additive and white Gaussian noise and we observed that, in this case, the neg-log of the likelihood function coincides, except for irrelevant constants, with the least-squares function; therefore, the use of a Bayesian approach and the focus on MAP estimates lead precisely to minimization problems already considered in the first two parts of the book. We can say that this approach provides a statistical foundation to Tikhonov regularization theory. Nevertheless, for some applications, it may be useful to extend this framework to the case where spatial correlation of the noise is relevant. This is true, for instance, in some applications of medical imaging [216]. Therefore, in this chapter, we briefly discuss the case of additive and correlated Gaussian noise.

12.1 Maximum likelihood and weighted least-squares

For the case of additive and correlated Gaussian noise the usual assumption of a zero expectation value of the noise $\boldsymbol{\nu}$ is also made, namely

$$E\{\boldsymbol{\nu}\} = \mathbf{0} \tag{12.1}$$

and this assumption implies again equation (11.17). Then, the joint density function of the normally distributed r.v.'s ν_m has the following form

$$p_{\boldsymbol{\nu}}(\mathbf{w}) = [(2\pi)^M |\mathbf{S}_{\boldsymbol{\nu}}|]^{-1/2} \exp\left[-\frac{1}{2}\left(\mathbf{S}_{\boldsymbol{\nu}}^{-1}\mathbf{w} \cdot \mathbf{w}\right)\right] \tag{12.2}$$

where $\mathbf{S}_{\boldsymbol{\nu}}$ is the *covariance matrix* of the r.v.'s ν_m, $|\mathbf{S}_{\boldsymbol{\nu}}|$ is the determinant of $\mathbf{S}_{\boldsymbol{\nu}}$ and the scalar product is the canonical one.
We recall that, in the case of r.v.'s ν_m with zero expectation value, the covariance matrix is given by

$$(\mathbf{S}_{\boldsymbol{\nu}})_{m,m'} = E\{\nu_m \nu_{m'}^*\} , \tag{12.3}$$

the complex conjugation being used in the case of complex-valued random variables. This matrix is positive semi-definite because its definition implies that

$$(\mathbf{S}_{\boldsymbol{\nu}}\mathbf{w} \cdot \mathbf{w}) = E\left\{|(\boldsymbol{\nu} \cdot \mathbf{w})|^2\right\} \geq 0 \tag{12.4}$$

for any vector (or array) $\mathbf{w}$. Moreover, this matrix is positive definite and hence invertible, as required in equation (12.2), if and only if no component of the noise is zero with probability one. We make this assumption here and in the following. We also observe that any diagonal element of $\mathbf{S}_{\boldsymbol{\nu}}$, $(\mathbf{S}_{\boldsymbol{\nu}})_{m,m}$, is the variance σ_m^2 of the component ν_m of the noise. When the matrix is diagonal

$$(\mathbf{S}_{\boldsymbol{\nu}})_{m,m'} = \sigma_m^2 \, \delta_{m,m'} \, , \tag{12.5}$$

the components of the noise are uncorrelated. Since they are normally distributed, from equation (12.2) it follows that they are independent r.v.'s. In the particular case where $\mathbf{S}_{\boldsymbol{\nu}}$ is diagonal and all variances are equal, we re-obtain the case of white Gaussian noise. We remark that it is also a particular case of stationary noise because it is invariant with respect to translations, i.e. it is the same in all the pixels of the image.

REMARK 12.1   *The more general case of stationary and Gaussian noise, for 2D discrete $N \times N$ images (so that $M = N^2$), corresponds to a covariance matrix which is a circulant matrix with four entries, i.e. a block circulant matrix with circulant blocks*

$$(\mathbf{S}_{\boldsymbol{\nu}})_{m,n;m',n'} = (\bar{\mathbf{S}}_{\boldsymbol{\nu}})_{m-m',n-n'} \tag{12.6}$$

where the periodic array $(\bar{\mathbf{S}})_{m,n}$, with period N in both indices, is a discrete version of the correlation function of the noise. In particular, we observe that, if we put $\sigma^2 = (\bar{\mathbf{S}}_{\boldsymbol{\nu}})_{0,0}$, then from equation (12.6) we obtain $(\mathbf{S}_{\boldsymbol{\nu}})_{m,n;m,n} = \sigma^2$, i.e. all components of the noise have the same variance. The case of white noise corresponds to the following four-entries covariance matrix

$$(\mathbf{S}_{\boldsymbol{\nu}})_{m,n;m',n'} = \sigma^2 \delta_{m,m'} \, \delta_{n,n'} \, . \tag{12.7}$$

Finally, the DFT of the array $(\bar{\mathbf{S}}_{\boldsymbol{\nu}})_{m,n}$, i.e. $(\hat{\bar{\mathbf{S}}}_{\boldsymbol{\nu}})_{k,l}$, is an array with nonnegative elements, as follows from equation (12.4). It represents the power spectrum of the noise. In the case of white noise, the power spectrum is constant, i.e. $(\hat{\bar{\mathbf{S}}}_{\boldsymbol{\nu}})_{k,l} = \sigma^2$.

If the covariance matrix $\mathbf{S}_{\boldsymbol{\nu}}$ of the noise is known, then the density function (12.2) is completely determined and the density function of the r.v. $\boldsymbol{\eta}(\mathbf{f}) = \mathbf{Af} + \boldsymbol{\nu}$, can be obtained from equation (12.2) as

$$p_{\boldsymbol{\eta}}(\mathbf{g}|\mathbf{f}) = [(2\pi)^M |\mathbf{S}_{\boldsymbol{\nu}}|]^{-1/2} \exp\left[-\frac{1}{2} \left(\mathbf{S}_{\boldsymbol{\nu}}^{-1}(\mathbf{g} - \mathbf{Af}) \cdot (\mathbf{g} - \mathbf{Af}) \right) \right] \tag{12.8}$$

where, in general, $\mathbf{g}$ is a vector of length M, $\mathbf{f}$ a vector of length N and $\mathbf{A}$ a $M \times N$ matrix. It also follows that, if the noisy image $\mathbf{g}$ is given, the neg-log of the likelihood function (11.21) is given by

$$l(\mathbf{f}|\mathbf{g}) = \frac{1}{2}(\mathbf{S}_{\boldsymbol{\nu}}^{-1}(\mathbf{g} - \mathbf{Af}) \cdot (\mathbf{g} - \mathbf{Af})) + \frac{1}{2} \ln \left[(2\pi)^M |\mathbf{S}_{\boldsymbol{\nu}}| \right] \tag{12.9}$$

the last term being independent of $\mathbf{f}$, hence irrelevant from the point of view of the minimization. We can conclude that *the maximization of the likelihood function is equivalent to the minimization of the following generalized discrepancy functional*

$$\varepsilon_{\boldsymbol{\nu}}^2(\mathbf{f}; \mathbf{g}) = \left(\mathbf{S}_{\boldsymbol{\nu}}^{-1}(\mathbf{Af} - \mathbf{g}) \cdot (\mathbf{Af} - \mathbf{g}) \right) \, . \tag{12.10}$$

As already remarked, in the particular case of white noise we re-obtain the discrepancy functional which is basic for all methods discussed in the first two parts of the book. Therefore, *in the case of white Gaussian noise the maximum-likelihood estimates coincide with the least-squares solutions of the problem. As a consequence the* ML *method is affected by all the difficulties of the least-squares method such as non-uniqueness and numerical instability of the solution.*

If the noise is not white, i.e. if $\mathbf{S}_\nu$ is not a multiple of the identity matrix, then the solutions of the ML problem may not coincide with the solutions of the standard least-squares problem. For discussing this point we observe that the functional (12.10) is associated with the following weighted scalar product defined in the vector space of the images (for instance $N \times N$ arrays)

$$(\mathbf{g}^{(1)} \cdot \mathbf{g}^{(2)})_\nu = (\mathbf{S}_\nu^{-1}\mathbf{g}^{(1)} \cdot \mathbf{g}^{(2)}) \ , \tag{12.11}$$

the scalar product in the second member being the canonical one. Therefore, since the matrix $\mathbf{S}_\nu^{-1}$ is positive definite, the functional (12.10) is the norm, induced by this scalar product, of the vector $\mathbf{Af} - \mathbf{g}$. By elementary computations one can show that the gradient of this function is given by

$$\nabla_{\mathbf{f}} \, \varepsilon_\nu^2(\mathbf{f};\mathbf{g}) = 2\left(\mathbf{A}^*\mathbf{S}_\nu^{-1}\mathbf{Af} - \mathbf{A}^*\mathbf{S}_\nu^{-1}\mathbf{g}\right) \ , \tag{12.12}$$

so that the minimizers of the function are solutions of the Euler-Lagrange equation

$$\mathbf{A}^*\mathbf{S}_\nu^{-1}\mathbf{Af} = \mathbf{A}^*\mathbf{S}_\nu^{-1}\mathbf{g} \ . \tag{12.13}$$

This equation is a generalization of equation (3.40) and, as already remarked, the two equations coincide when the noise is white. Moreover, the equivalence holds true also in other cases. For instance, if the matrix $\mathbf{A}$ is circulant and the noise is stationary so that also the covariance matrix $\mathbf{S}_\nu$ is circulant, then the two matrices commute. In such a case, by multiplying both sides of equation (12.13) by $\mathbf{S}_\nu$, we re-obtain exactly equation (3.40). In the case of an invertible matrix $\mathbf{A}$, all the least-squares solutions coincide with the solutions of the equation $\mathbf{Af} = \mathbf{g}$.

Let us shortly comment about the ill-conditioning of the equation (12.13) when it is not equivalent to equation (3.40). To this purpose, let us look again at the functional (12.10). In the case of the standard least-squares problem, the condition number of the problem is given by the ratio between the maximum and the minimum singular value of the matrix $\mathbf{A}$ and therefore it depends only on the imaging system. When the minimization of the function is not reduced to the standard least-squares equation but to the solution of equation (12.13), then the condition number depends both on the imaging system and on the noise. The simplest case is that of a diagonal covariance matrix, i.e. the case where the components of the image are independent random variables with variance σ_m^2, so that the minimization of the functional (12.10) becomes

$$\sum_{m=1}^{M} \frac{1}{\sigma_m^2} \left|(\mathbf{Af})_m - g_m\right|^2 = \text{minimum} \ . \tag{12.14}$$

This problem can be reduced to the standard form of the least-squares problem if we introduce the following change of image and matrix

$$(g_\nu)_m = \frac{g_m}{\sigma_m} \ , \quad (\mathbf{A}_\nu)_{m,n} = \frac{1}{\sigma_m}A_{m,n} \ ; \tag{12.15}$$

then the matrix $\mathbf{A}_\nu$ depends on the imaging system through the matrix $\mathbf{A}$ and on the noise ν through the σ_m. The condition number of this matrix is the condition number of the minimization problem, so that it can be greater or smaller than the condition number of the matrix $\mathbf{A}$, according to the mutual behavior of the detector and the imaging device.

This argument can be extended to the general case. Indeed, since the matrix $\mathbf{S}_\nu^{-1}$ is positive definite, the matrix $\mathbf{S}_\nu^{-1/2}$ is well-defined as the matrix whose eigenvalues are the square roots of the eigenvalues of $\mathbf{S}_\nu^{-1}$ associated with the same eigenvectors; then the condition number of the problem is the condition number of the matrix $\mathbf{S}_\nu^{-1/2}\mathbf{A}$.

The ML method can also lead to constrained least-squares problems. Indeed, since one has to maximize the likelihood function over its domain of definition, one can restrict this domain by the introduction of constraints defining the set of permissible objects. For instance one can assume that the energy of the object cannot exceed a given quantity E^2 or that the objects must be nonnegative and so on. One easily realizes that in this way one re-obtains constrained problems already considered in the first two parts of the book or a generalization of them due to the presence of the covariance matrix in the generalized discrepancy (12.10). In other words one can inject 'deterministic' *a priori* information into the ML method by restricting in a suitable way the domain of definition of the likelihood function. This remark provides not only a statistical foundation but also a generalization of Tikhonov's regularization methods.

As concerns the computation of the solutions, the methods discussed in the first two parts of this book for the solution of the standard least-squares problem can be easily extended to the case of the minimization of the generalized discrepancy function (12.10). We leave the extension as an exercise to the reader. We only give the generalization of the Landweber method. If we remark that it is a gradient method with a fixed step in the descent direction, from equation (12.12) we deduce that the resulting iteration is given by

$$\mathbf{f}_{k+1} = \mathbf{f}_k + \tau \left(\mathbf{A}^*\mathbf{S}_\nu^{-1}\mathbf{g} - \mathbf{A}^*\mathbf{S}_\nu^{-1}\mathbf{A}\mathbf{f} \right). \qquad (12.16)$$

We also leave as an exercise the estimate of the interval of the values of τ assuring convergence of the iteration and the investigation of the semi-convergence of the algorithm. Obviously, it can also be modified into an algorithm for the estimation of constrained solutions, with a convex constraint, by projecting at each iteration on the convex set of the admissible solutions.

12.2 Bayesian methods with Gaussian priors and the Wiener filter

The analysis of the Bayesian method is rather simple in the case of additive Gaussian noise and of a Gaussian prior. For the reader's convenience we recall the basic assumptions:

- the noise $\boldsymbol{\nu}$ is a vector-valued Gaussian r.v. with zero expectation value and covariance matrix $\mathbf{S}_\nu$, so that its density function is given by equation (12.2);
- the object $\mathbf{f}$ is also the realization of a vector-valued Gaussian r.v. $\boldsymbol{\phi}$ with zero expectation value and covariance matrix $\mathbf{S}_\phi$, so that its density function is given again by equation (12.2) with $\mathbf{S}_\nu$ replaced by $\mathbf{S}_\phi$ and M by N;
- the noise $\boldsymbol{\nu}$ and the object $\boldsymbol{\phi}$ are independent random vectors so that their *cross-covariance matrix* $\mathbf{S}_{\phi\nu}$ is zero

$$(\mathbf{S}_{\phi\nu})_{n,m} = E\{\phi_n \nu_m^*\} = 0 . \qquad (12.17)$$

From the interpretation of equation (12.8) as the conditional density function of $\boldsymbol{\eta}$, for a given value $\mathbf{f}$ of $\boldsymbol{\phi}$, it follows that this conditional density function is Gaussian, with expectation value $E\{\boldsymbol{\eta}|\mathbf{f}\} = \mathbf{A}\mathbf{f}$ and covariance matrix (independent of $\mathbf{f}$) $\mathbf{S}_{\eta|\phi} = \mathbf{S}_\nu$. Then, from equation (11.26) it follows that

$$
\begin{aligned}
p_{\phi\eta}(\mathbf{f}, \mathbf{g}) &= p_\nu(\mathbf{g} - \mathbf{A}\mathbf{f})\, p_\phi(\mathbf{f}) = \qquad (12.18)\\
&= \left[(2\pi)^{MN} |\mathbf{S}_\nu|\, |\mathbf{S}_\phi| \right]^{-1/2} \exp\left[-\frac{1}{2}\, \Phi(\mathbf{f}; \mathbf{g}) \right]
\end{aligned}
$$

where

$$\Phi(\mathbf{f}; \mathbf{g}) = \left(\mathbf{S}_\nu^{-1}(\mathbf{Af} - \mathbf{g}) \cdot (\mathbf{Af} - \mathbf{g})\right) + \left(\mathbf{S}_\phi^{-1}\mathbf{f} \cdot \mathbf{f}\right) . \qquad (12.19)$$

These equations imply that the marginal density function of $\boldsymbol{\eta}$ is also Gaussian, so that it is completely characterized by the expectation value and the covariance matrix of $\boldsymbol{\eta}$, which can be obtained directly from the relation $\boldsymbol{\eta} = \mathbf{A}\boldsymbol{\phi} + \boldsymbol{\nu}$ and the assumption of statistical independence of $\boldsymbol{\nu}$ and $\boldsymbol{\phi}$. Indeed, from that relation, since both $\boldsymbol{\nu}$ and $\boldsymbol{\phi}$ have zero expectation value, we have that also the expectation value of $\boldsymbol{\eta}$ is zero

$$E\{\boldsymbol{\eta}\} = \mathbf{0} . \qquad (12.20)$$

Moreover, from the same relation and the statistical independence of $\boldsymbol{\nu}$ and $\boldsymbol{\phi}$ we can also obtain the covariance matrix of $\boldsymbol{\eta}$. The result is

$$\mathbf{S}_\eta = \mathbf{A}\,\mathbf{S}_\phi\,\mathbf{A}^* + \mathbf{S}_\nu . \qquad (12.21)$$

REMARK 12.2 *The derivation of equation (12.21) goes as follows. The independence of the r.v.'s ϕ_n and ν_m implies $E\{\phi_n \nu_m^*\} = E\{\nu_m \phi_n^*\} = 0$, so that from the relation between the three r.v.'s we obtain*

$$\begin{aligned}
(\mathbf{S}_\eta)_{m,m'} &= E\{\eta_m \eta_{m'}^*\} = \qquad\qquad\qquad (12.22)\\
&= \sum_{n,n'=1}^{N} E\{A_{m,n}\phi_n A_{m',n'}^* \phi_{n'}^*\} + E\{\nu_m \nu_{m'}^*\} = \\
&= \sum_{n,n'=1}^{N} A_{m,n}(\mathbf{S}_\phi)_{n,n'} A_{m',n'}^* + (\mathbf{S}_\nu)_{m,m'}
\end{aligned}$$

and this is precisely equation (12.21), where $\mathbf{A}^*$ is the adjoint matrix, defined by $(\mathbf{A}^*)_{n,m} = A_{m,n}^*$.

Equations (12.20) and (12.21) imply that the marginal density function of $\boldsymbol{\eta}$ is given by

$$p_\eta(\mathbf{g}) = \left[(2\pi)^M |\mathbf{S}_\eta|\right]^{-1/2} \exp\left[-\frac{1}{2}\left(\mathbf{S}_\eta^{-1}\mathbf{g} \cdot \mathbf{g}\right)\right] . \qquad (12.23)$$

By inserting this equation and (12.18) into (11.29), we obtain the conditional density function of $\boldsymbol{\phi}$ for a given value $\mathbf{g}$ of the image

$$p_\phi(\mathbf{f}|\mathbf{g}) = \left[(2\pi)^N \frac{|\mathbf{S}_\nu||\mathbf{S}_\phi|}{|\mathbf{S}_\eta|}\right]^{-1/2} \exp\left[-\frac{1}{2}\Phi(\mathbf{f}; \mathbf{g}) + \frac{1}{2}\left(\mathbf{S}_\eta^{-1}\mathbf{g} \cdot \mathbf{g}\right)\right] . \qquad (12.24)$$

It is obvious that this conditional density function is also Gaussian and this remark has a first important consequence: the CM estimate provided by the *a posteriori* expectation value of $\boldsymbol{\phi}$ and that provided by the MAP estimate of $\boldsymbol{\phi}$ coincide. If we denote by $\tilde{\mathbf{f}}$ this Bayesian estimate and by $\mathbf{S}_{\phi|\eta}$ the *a posteriori* covariance matrix, equation (12.24) takes the following form

$$p_\phi(\mathbf{f}|\mathbf{g}) = \left[(2\pi)^N |\mathbf{S}_{\phi|\eta}|\right]^{-1/2} \exp\left[-\frac{1}{2}\left(\mathbf{S}_{\phi|\eta}^{-1}(\mathbf{f} - \tilde{\mathbf{f}}) \cdot (\mathbf{f} - \tilde{\mathbf{f}})\right)\right] . \qquad (12.25)$$

We can obtain $\tilde{\mathbf{f}}$ and $\mathbf{S}_{\phi|\eta}$ by algebraic manipulations of the exponent of equation (12.24). A much simpler way is the following:

- compute $\mathbf{S}_{\phi|\eta}$ by equating the quadratic terms in $\mathbf{f}$ in the exponents of equations (12.24) and (12.25);
- compute $\tilde{\mathbf{f}}$ by minimizing the exponent in equation (12.24).

Indeed, by equating the quadratic terms we get

$$\left(\mathbf{S}_{\phi|\eta}^{-1}\mathbf{f} \cdot \mathbf{f}\right) = \left(\mathbf{S}_{\nu}^{-1}\mathbf{Af} \cdot \mathbf{Af}\right) + \left(\mathbf{S}_{\phi}^{-1}\mathbf{f} \cdot \mathbf{f}\right) \ . \tag{12.26}$$

Since this equality must hold for any $\mathbf{f}$, we obtain

$$\mathbf{S}_{\phi|\eta} = \left(\mathbf{A}^*\mathbf{S}_{\nu}^{-1}\mathbf{A} + \mathbf{S}_{\phi}^{-1}\right)^{-1} \ . \tag{12.27}$$

As concerns the determination of $\tilde{\mathbf{f}}$, we first observe that the minimization of the exponent of equation (12.24) is equivalent to the minimization of the functional (12.19). This functional coincides with the basic functional of regularization theory, equation (4.6), when both $\mathbf{S}_{\nu}$ and $\mathbf{S}_{\phi}$ are proportional to the identity matrix. When this condition is not satisfied, its minimizers are solutions of an equation which is a generalization of equation (4.23). It can be easily obtained as follows. By taking into account equation (12.12), one easily computes the gradient of the functional (12.19); by setting the result to zero one obtains that $\tilde{\mathbf{f}}$ is the solution of the following equation

$$\left(\mathbf{A}^*\mathbf{S}_{\nu}^{-1}\mathbf{A} + \mathbf{S}_{\phi}^{-1}\right)\tilde{\mathbf{f}} = \mathbf{A}^*\mathbf{S}_{\nu}^{-1}\mathbf{g} \tag{12.28}$$

so that it is given by

$$\tilde{\mathbf{f}} = \mathbf{S}_{\phi|\eta}\,\mathbf{A}^*\mathbf{S}_{\nu}^{-1}\mathbf{g} \ . \tag{12.29}$$

In conclusion we can summarize the main results of the previous analysis as follows:

- the *a posteriori* density function of ϕ is given in equation (12.25) with $\tilde{\mathbf{f}}$ and $\mathbf{S}_{\phi|\eta}$ given respectively by (12.29) and (12.27);
- the two Bayesian estimates (CM and MAP), introduced at the end of the Section 11.3, coincide and are given by $\tilde{\mathbf{f}}$;
- the conditional covariance matrix $\mathbf{S}_{\phi|\eta}$ is independent of the value $\mathbf{g}$ of η;
- the diagonal elements of $\mathbf{S}_{\phi|\eta}$ give a measure of the reliability of the estimates of the various components of the object as provided by $\tilde{\mathbf{f}}$;
- the estimate $\tilde{\mathbf{f}}$ is obtained by applying to the image $\mathbf{g}$ the matrix

$$\mathbf{R}_0 = \left(\mathbf{A}^*\mathbf{S}_{\nu}^{-1}\mathbf{A} + \mathbf{S}_{\phi}^{-1}\right)^{-1}\mathbf{A}^*\mathbf{S}_{\nu}^{-1} \ . \tag{12.30}$$

By using the following identity

$$\mathbf{A}^*\mathbf{S}_{\nu}^{-1}\left(\mathbf{AS}_{\phi}\mathbf{A}^* + \mathbf{S}_{\nu}\right) = \left(\mathbf{A}^*\mathbf{S}_{\nu}^{-1}\mathbf{A} + \mathbf{S}_{\phi}^{-1}\right)\mathbf{S}_{\phi}\mathbf{A}^* \tag{12.31}$$

we find that the matrix $\mathbf{R}_0$ can also be written in the following form

$$\mathbf{R}_0 = \mathbf{S}_{\phi}\mathbf{A}^*(\mathbf{A}S_{\phi}\mathbf{A}^* + \mathbf{S}_{\nu})^{-1} \ . \tag{12.32}$$

The matrix $\mathbf{R}_0$ is known under the name *Wiener filter*.

REMARK 12.3 *In the special case of white processes, i.e. if we assume that*

$$\mathbf{S}_{\nu} = \varepsilon^2\mathbf{I} \ , \quad \mathbf{S}_{\phi} = E^2\mathbf{I} \tag{12.33}$$

where $\mathbf{I}$ is the identity matrix, then the form (12.30) reduces to

$$\mathbf{R}_0 = \left(\mathbf{A}^* \mathbf{A} + \left(\frac{\varepsilon}{E} \right)^2 \mathbf{I} \right)^{-1} \mathbf{A}^* \tag{12.34}$$

which is the Tikhonov regularizer with the choice of the regularization parameter provided by the Miller method – see Section 4.6.

As it is known, the Wiener filter provides the optimal solution of a linear estimation problem. Assume that we wish to estimate the random vector ϕ by applying a matrix $\mathbf{R}$ to the random vector η. The reliability of the estimate can be measured by the quantity

$$\varrho(\mathbf{R}) = E\{\|\mathbf{R}\eta - \phi\|^2\} \tag{12.35}$$

and the optimal estimate is provided by the matrix $\mathbf{R}_0$ which minimizes $\varrho(\mathbf{R})$.
In order to solve this problem we do not need the joint density function of the random vectors ϕ, η; we only need their covariance matrices $\mathbf{S}_\phi$ and $\mathbf{S}_\eta$ as well as their cross-covariance matrix $\mathbf{S}_{\phi\eta} = \mathbf{S}_{\eta\phi}^*$. Indeed, by means of simple computations we find that

$$\begin{aligned} E\{\|\mathbf{R}\eta - \phi\|^2\} &= E\{(\mathbf{R}\eta \cdot \mathbf{R}\eta) - (\mathbf{R}\eta \cdot \phi) - (\phi \cdot \mathbf{R}\eta) + (\phi \cdot \phi)\} = \\ &= \mathrm{Trace}\{\mathbf{R}\mathbf{S}_\eta \mathbf{R}^* - \mathbf{R}\mathbf{S}_{\eta\phi} - \mathbf{S}_{\phi\eta}\mathbf{R}^* + \mathbf{S}_\phi\} \ . \end{aligned} \tag{12.36}$$

If we introduce the matrix

$$\mathbf{R}_0 = \mathbf{S}_{\phi\eta}\, \mathbf{S}_\eta^{-1} \tag{12.37}$$

the following identity can be easily proved

$$E\{\|\mathbf{R}\eta - \phi\|^2\} = \mathrm{Trace}\{(\mathbf{R} - \mathbf{R}_0)\mathbf{S}_\eta(\mathbf{R} - \mathbf{R}_0)^* + \mathbf{S}_\phi - \mathbf{S}_{\phi\eta}\, \mathbf{S}_\eta^{-1}\, \mathbf{S}_{\eta\phi}\} \tag{12.38}$$

and therefore the minimum is reached when $\mathbf{R} = \mathbf{R}_0$. The matrix $\mathbf{R}_0$ is called the *optimum linear filter* or also the *Wiener filter* .
Now, in the case of the random vectors ϕ, η related by $\eta = \mathbf{A}\phi + \nu$, assuming that $\mathbf{S}_{\phi\nu} = \mathbf{0}$, we find that $\mathbf{S}_\eta$ is given by equation (12.21) and that $\mathbf{S}_{\phi\eta}$ is given by

$$\mathbf{S}_{\phi\eta} = \mathbf{S}_\phi \mathbf{A}^* \ . \tag{12.39}$$

We conclude that $\mathbf{R}_0$, as defined by equation (12.37), coincides with $\mathbf{R}_0$ as defined by equation (12.32).

12.3   A simple method for nonnegative ML and MAP estimates

At the end of Section 12.1 we discussed the extension of the Landweber method to the minimization of the functional (12.10) and we recalled that it can be modified in a simple way to provide the solution of the constrained minimization of this functional in the presence of a convex constraint, in particular of non-negativity.
In this section we would like to discuss another simple method, specifically designed for the computation of nonnegative solutions, which can be easily implemented and used for testing the effect of non-negativity on the quality of the solution, when non-negativity is a natural property of the object to be reconstructed. It is an iterative method which does not require a projection on the nonnegative orthant because, if it is initialized with a nonnegative object, then at each iteration it automatically provides a nonnegative approximation of the solution. It is the so-called *image space restoration algorithm* (ISRA) proposed in [78],

whose convergence is proved in [90]. We provide two derivations of this algorithm, first a heuristic one and then as a MM algorithm. We also describe a generalization to the case of the minimization of a penalized likelihood.

In this section, we assume that the covariance matrix of the noise is the identity matrix, $\mathbf{S}_\nu = \mathbf{I}$, although some of the results below could be extended, in a quite straightforward way, to the case of uncorrelated noise with a diagonal covariance matrix. We also assume that all elements of the matrix $\mathbf{A}$ are real and nonnegative, as well as the components of the data vector $\mathbf{g}$. Let us then consider the problem of the minimization of the discrepancy function (12.10), which is equivalent to the minimization of the neg-log of the likelihood function (12.9), under the constraint of non-negativity on $\mathbf{f}$ (and with $\mathbf{S}_\nu = \mathbf{I}$)

$$\min_{\mathbf{f}} l(\mathbf{f}|\mathbf{g}) \ , \quad \text{subject to} \ \ \mathbf{f} \geq 0 \ . \tag{12.40}$$

Since the function is convex, we recall the well-known *Karush-Kuhn-Tucker conditions* (KKT conditions) [313] which, in this case, are not only necessary but also sufficient conditions for an object $\tilde{\mathbf{f}}$ to be a constrained minimizer of the discrepancy function. We write the conditions for a generic function $l(\mathbf{f}|\mathbf{g})$, hence in a form which will be useful also in the next chapter. If the function is restricted to the nonnegative orthant, then the KKT conditions for a minimizer $\tilde{\mathbf{f}}$ can be written in the following form:

$$\tilde{f}_n \frac{\partial l(\mathbf{f}|\mathbf{g})}{\partial f_n}\Big|_{\mathbf{f}=\tilde{\mathbf{f}}} = 0 \ , \quad n = 1, 2, \ldots, N \ , \tag{12.41}$$

$$\frac{\partial l(\mathbf{f}|\mathbf{g})}{\partial f_n}\Big|_{\mathbf{f}=\tilde{\mathbf{f}}} \geq 0 \ , \quad \text{if} \ \ \tilde{f}_n = 0 \ . \tag{12.42}$$

In particular, the first condition is necessary for any minimizer and also sufficient when the minimizer is an interior point. From the first KKT condition and the expression of the gradient (12.12), we get

$$0 = \tilde{\mathbf{f}} \circ \nabla_{\mathbf{f}} \varepsilon_\nu^2(\mathbf{f};\mathbf{g})|_{\mathbf{f}=\tilde{\mathbf{f}}} = 2\tilde{\mathbf{f}} \circ \left(\mathbf{A}^*\mathbf{A}\tilde{\mathbf{f}} - \mathbf{A}^*\mathbf{g}\right) \ , \tag{12.43}$$

where $\circ$ denotes the Hadamard product, i.e. the component by component (or pixel by pixel) product. By a simple manipulation one verifies that this condition implies the fixed-point equation

$$\tilde{\mathbf{f}} = \tilde{\mathbf{f}} \circ \frac{\mathbf{A}^*\mathbf{g}}{\mathbf{A}^*\mathbf{A}\tilde{\mathbf{f}}} \ , \tag{12.44}$$

where the quotient is also intended component by component (or pixel by pixel). Finally, by applying the method of successive approximations to the solution of this fixed-point equation, we obtain the ISRA iteration

$$\mathbf{f}_{k+1} = \mathbf{f}_k \circ \frac{\mathbf{A}^*\mathbf{g}}{\mathbf{A}^*\mathbf{A}\mathbf{f}_k} \ . \tag{12.45}$$

It is easy to prove by induction that all iterates are nonnegative if the initial guess is nonnegative. Indeed, if $\mathbf{f}_0 \geq 0$, then from equation (12.45) one deduces that $\mathbf{f}_1 \geq 0$; similarly one deduces that, if $\mathbf{f}_k \geq 0$ then $\mathbf{f}_{k+1} \geq 0$. One can also observe that, if $\mathbf{f}_0$ is zero in one pixel, then all the iterates will be zero in that same pixel. This is an easy way to introduce, for instance, a constraint on the support of the unknown object (see Section 8.5).

Another nice property of the algorithm is that the decrease of the cost function (discrepancy or neg-log likelihood) is monotonic throughout the iteration. This property results from the fact that the ISRA algorithm can also be viewed as a majorization-minimization or MM algorithm as defined in Section 10.3. Indeed, the iteration (12.45) can be rewritten as

$$\mathbf{f}_{k+1} = \arg\min_{\mathbf{f}} \Phi^{SUR}(\mathbf{f};\mathbf{f}_k) \tag{12.46}$$

with the surrogate functional, for $\mathbf{f}$ and $\mathbf{a}$ nonnegative,

$$\Phi^{SUR}(\mathbf{f};\mathbf{a}) = \sum_{m=1}^{M} \frac{1}{(\mathbf{Aa})_m} \sum_{n=1}^{N} A_{m,n} a_n \left[g_m - \frac{f_n}{a_n}(\mathbf{Aa})_m \right]^2 \tag{12.47}$$

REMARK 12.4 *The minimization of the separable surrogate (12.47) is easily done by hand. To establish the majorization property, let us define the weights $c_{m,n} = A_{m,n} a_n / (\mathbf{Aa})_m$ which are clearly nonnegative and sum to one, $\sum_n c_{m,n} = 1$, for any fixed m. Then the surrogate (12.47) can be written as $\sum_m [\sum_n c_{m,n}(d_{m,n})^2]$ where $d_{m,n} = g_m - (f_n/a_n)(\mathbf{Aa})_m$. Thanks to the convexity of the function $h(x) = x^2$, we have from the so-called Jensen inequality*

$$h\left(\sum_{i=1}^{n} c_i x_i\right) \le \sum_{i=1}^{n} c_i h(x_i) \tag{12.48}$$

that $\sum_m [\sum_n c_{m,n} d_{m,n}]^2 \le \sum_m [\sum_n c_{m,n}(d_{m,n})^2]$, the l.h.s. being simply the discrepancy $(\mathbf{Af} - \mathbf{g}) \cdot (\mathbf{Af} - \mathbf{g})$. Equality holds for $\mathbf{f} = \mathbf{a}$, so that the anchoring requirement is also fulfilled.

As already stated the limit points of the iterations are ML solutions, depending on the initial guess; if the ML solution is unique then the sequence converges precisely to this solution. Since we know that, in general, the ML solutions are not acceptable in practice, we conclude that it is not a good idea to push the iterations to convergence. However, as follows from numerical practice, the method appears to have the semi-convergence property (although, we do not know of a proof of this property) so that an early stopping of the iterations can provide sensible solutions. Again, as far as we know, no generally accepted stopping rule has been proposed.

It is easy to verify that one can rewrite equation (12.45) in the following interesting form

$$\mathbf{f}_{k+1} = \mathbf{f}_k - \frac{\mathbf{f}_k}{\mathbf{A}^*\mathbf{Af}_k} \circ (\mathbf{A}^*\mathbf{Af}_k - \mathbf{A}^*\mathbf{g}) \tag{12.49}$$

showing that ISRA is a gradient method, more precisely a scaled gradient method with a 'scaling' given by $\mathbf{f}_k/\mathbf{A}^*\mathbf{Af}_k$. This remark opens the way to fast numerical methods [30] based on a general approach known as *scaled gradient projection method* (SGP).

This scaling property of ISRA combined with an idea proposed in [194, 193] also allows the design of scaled gradient projection methods for the computation of nonnegative minimizers of functions derived from the Bayesian approach in the case of Gaussian priors with a differentiable potential functions with the following general structure

$$l_B(\mathbf{f}|\mathbf{g}) = l(\mathbf{f}|\mathbf{g}) + \mu \mathcal{G}(\mathbf{f}) . \tag{12.50}$$

From the first KKT condition we get that any minimizer $\tilde{\mathbf{f}}$ of this function must satisfy the condition

$$0 = \tilde{\mathbf{f}} \circ \nabla_{\mathbf{f}} \left(l(\mathbf{f}|\mathbf{g}) + \mu \mathcal{G}(\mathbf{f}) \right)|_{\mathbf{f}=\tilde{\mathbf{f}}} \tag{12.51}$$

which, by absorbing a factor 2 in the regularization parameter, leads to the equation

$$0 = \tilde{\mathbf{f}} \circ \left(\mathbf{A}^*\mathbf{A}\tilde{\mathbf{f}} - \mathbf{A}^*\mathbf{g} + \mu\nabla_{\mathbf{f}}\mathcal{G}(\mathbf{f})|_{\mathbf{f}=\tilde{\mathbf{f}}}\right) . \tag{12.52}$$

Now, the idea proposed in [194, 193] consists in writing the gradient of the regularizer as the difference of two nonnegative arrays

$$-\nabla_{\mathbf{f}} \mathcal{G}(\mathbf{f}) = U(\mathbf{f}) - V(\mathbf{f}) . \tag{12.53}$$

Such a decomposition always exists even if it is not unique; however, in practice, the interesting fact is that, when one write explicitly the gradient of the regularizer, for most of the differentiable regularizers used in the applications, a quite natural decomposition such as (12.53) is possible on the nonnegative orthant.

If we insert equation (12.53) into equation (12.52) we obtain the equation

$$\tilde{\mathbf{f}} \circ \left(\mathbf{A}^* \mathbf{A} \tilde{\mathbf{f}} + \mu V(\tilde{\mathbf{f}}) \right) = \tilde{\mathbf{f}} \circ \left(\mathbf{A}^* \mathbf{g} + \mu U(\tilde{\mathbf{f}}) \right) \tag{12.54}$$

which implies for $\tilde{\mathbf{f}}$ the following fixed-point equation

$$\tilde{\mathbf{f}} = \tilde{\mathbf{f}} \circ \frac{\mathbf{A}^* \mathbf{g} + \mu U(\tilde{\mathbf{f}})}{\mathbf{A}^* \mathbf{A} \tilde{\mathbf{f}} + \mu V(\tilde{\mathbf{f}})} . \tag{12.55}$$

Finally, by applying the method of successive approximations, we obtain the iteration

$$\mathbf{f}_{k+1} = \mathbf{f}_k \circ \frac{\mathbf{A}^* \mathbf{g} + \mu U(\mathbf{f}_k)}{\mathbf{A}^* \mathbf{A} \mathbf{f}_k + \mu V(\mathbf{f}_k)} , \tag{12.56}$$

which provides a simple generalization of the ISRA iteration (12.45) for the computation of MAP estimates. As far as we know the convergence of this algorithm is not proved except in a few simple cases. However, it is easy to verify that it can also be written in the following form

$$\mathbf{f}_{k+1} = \mathbf{f}_k - \frac{\mathbf{f}_k}{\mathbf{A}^* \mathbf{A} \mathbf{f}_k + \mu V(\mathbf{f}_k)} \circ \left\{ \mathbf{A}^* \mathbf{A} \mathbf{f}_k - \mathbf{A}^* \mathbf{g} + \mu \nabla_{\mathbf{f}} \, \mathcal{G}(\mathbf{f}_k) \right\} \tag{12.57}$$

which shows that the algorithm (12.56) is a scaled gradient method. This remark allows for the introduction of a relaxed version of the algorithm by introducing a step α_k – instead of the step 1 – along the descent direction; moreover a choice of step lengths ensuring the convergence of the relaxed version is possible thanks to a result proved by Lange [190] in the case of Poisson noise (see also [33]), a result which can be extended to the present case. However, the main advantage implied by this remark consists in the possibility of applying to the problem the SGP method [48], already mentioned, which can lead to very fast algorithms.

We provide, as an example, one possible decomposition (12.53) for the gradient of the TV-smooth regularizer which was widely discussed in Chapter 9. The computations are lengthy but elementary. At the end, one finds that the result can be written, in a quite natural way, as the difference of the following arrays

$$[U(\mathbf{f})]_{m,n} = \frac{f_{m+1,n} + f_{m,n+1}}{\sqrt{\delta^2 + |(D\mathbf{f})_{m,n}|^2}} + \frac{f_{m-1,n}}{\sqrt{\delta^2 + |(D\mathbf{f})_{m-1,n}|^2}} + \frac{f_{m,n-1}}{\sqrt{\delta^2 + |(D\mathbf{f})_{m,n-1}|^2}} \tag{12.58}$$

$$[V(\mathbf{f})]_{m,n} = \frac{2\,f_{m,n}}{\sqrt{\delta^2 + |(D\mathbf{f})_{m,n}|^2}} + \frac{f_{m,n}}{\sqrt{\delta^2 + |(D\mathbf{f})_{m-1,n}|^2}} + \frac{f_{m,n}}{\sqrt{\delta^2 + |(D\mathbf{f}_{m,n-1}|^2}} .$$

The two arrays are nonnegative on the nonnegative orthant. Therefore, if we initialize the iterations with a nonnegative $\mathbf{f}_0$, then also $\mathbf{f}_1$, as given by equation (12.56), is nonnegative so that also $U(\mathbf{f}_1)$ and $V(\mathbf{f}_1)$ are nonnegative. It follows that also $\mathbf{f}_2$ is nonnegative and so on. In conclusion, one proves by induction that the algorithm (12.56) is well defined for any k if the functions U, V are given by equation (12.58). As concerns boundary conditions one can use those which are the most appropriate for the particular application under consideration even if periodic boundary conditions can always be used.

13

Statistical methods in the case of Poisson data

In many imaging modalities images are formed by counting the numbers of photons emitted by the source and arriving on the detector. Important examples are fluorescence microscopy, optical and infrared astronomy and emission tomography, a fundamental modality of nuclear medicine, including PET and SPECT imaging. In these applications the components g_m of the image are essentially integer numbers. If the emission process is stationary in time, then the numbers of collected photons satisfy Poisson statistics. This feature requires completely different methods for the computation of ML and MAP estimates, both for denoising and for deblurring.

In this chapter we first show that the statistical model implies that the ML approach leads to a minimization problem which goes beyond the standard least-squares method; similarly the application of the Bayesian approach leads to minimization problems which go beyond the standard Tikhonov theory. Reconstruction methods are discussed and examples of reconstructions are given. Since, in our opinion, a completely satisfactory formulation in functional spaces is not yet available, we only consider fully discrete problems. All quantities in this chapter are real and, accordingly, the adjoint matrix $\mathbf{A}^*$ is replaced by the transpose matrix $\mathbf{A}^T$. Moreover, we assume that images may be non-squared.

13.1 The ML approach and the generalized Kullback-Leibler divergence

In the problems for which the methods discussed in this chapter apply, the basic imaging model has to be slightly modified. The simple linear model $\mathbf{A}\mathbf{f}^{(0)} = \mathbf{g}^{(0)}$ is replaced by a linear affine model which is more appropriate as shown in [258]

$$\mathbf{A}\mathbf{f}^{(0)} + \mathbf{b} = \mathbf{g}^{(0)} , \qquad (13.1)$$

where the vector $\mathbf{b}$, in general well approximated by a constant and positive vector, is the expected value of the so-called background emission, with different physical origin in

DOI: 10.1201/9781003032755-13

different cases. For instance, in astronomy it is due to the sky emission, in microscopy to self-fluorescence, in SPECT and PET to the scattering of photons by the traversed material, etc. As concerns the other quantities appearing in the equation, the component $f_n^{(0)}$ of the object is the mean value of the number of photons emitted by the source at the pixel (or voxel) n of the object domain. Analogously the component $(\mathbf{A}\mathbf{f}^{(0)})_m$ of the noise-free image is the mean value of the number of photons arriving from the source and detected at the pixel (or voxel) m of the image domain. Moreover, the matrix element $A_{m,n}$ is the fraction of photons emitted by the source at the pixel n and detected at the pixel m of the image domain. Therefore, all matrix elements are nonnegative and smaller than 1: they may be interpreted as probabilities. In particular, if all the emitted photons are detected with probability 1, then a normalization condition holds true for all the columns of the matrix $\mathbf{A}$, namely

$$\sum_{m=1}^{M} A_{m,n} = 1 \; ; \tag{13.2}$$

otherwise, this sum is smaller than 1. In addition, we assume that each row and column of the matrix contains at least one non-zero element: a column of zeros would imply that one pixel/voxel of the object does not contribute to the image while a row of zeros would imply that one pixel/voxel of the image domain does not receive photons with probability one. Therefore this row or column can be removed in the formulation of the problem. In conclusion, all quantities $f_n^{(0)}$, $(\mathbf{A}\mathbf{f}^{(0)})_m$ and $A_{m,n}$ can be assumed to be nonnegative. If g_m is the number of photons detected at pixel m, then g_m is a nonnegative integer number.

Also in the case of X-ray tomography the raw data are numbers of photons; however, the use of their logarithms as data for the image reconstruction problem modifies the statistics. If one intends to reconstruct the unknown object from the raw data, then one must solve a nonlinear problem, as discussed by Lange and Carson [191]. Therefore, this discussion is beyond the scope of this book and only emission tomography is considered in this chapter. We must also recall that some features of the detection system can slightly perturb the data values; for instance, the model of mixed Poisson-Gaussian noise, discussed in Section 11.1, is more accurate if the detector is a CCD camera because of the additional read-out noise. However, these perturbations are, in general, small so that they can be neglected in the statistical model of data acquisition or approximated by a suitable Poisson noise, as shown in [260]. Therefore, if the emission process is stationary, we can assume that the fluctuations in the photon counting and in the read-out noise can be described by Poisson statistics: g_m is the realization of a Poisson r.v. η_m with expectation value $\lambda_m = (\mathbf{A}\mathbf{f}^{(0)})_m + \mathbf{b}_m$ and probability distribution given by

$$p_{\eta_m}(g_m) = e^{-\lambda_m} \frac{(\lambda_m)^{g_m}}{(g_m)!} \; . \tag{13.3}$$

We recall that λ_m is both the expectation value and the variance of η_m

$$E[\eta_m] = \lambda_m \; , \quad \sigma_m^2 = E[(\eta_m - \lambda_m)^2] = \lambda_m \; . \tag{13.4}$$

It is obvious that, if the data are generated by another object $\mathbf{f}$, then their distribution is obtained by replacing $\mathbf{f}^{(0)}$ with $\mathbf{f}$ in the previous equation.

Assuming statistical independence of the photon distributions in different pixels, in the case of a generic admissible object $\mathbf{f}$, the joint probability distribution of the image components is given by the product of the probability distributions given in equation (13.3)

$$p_\eta(\mathbf{g}|\mathbf{f}) = \prod_{m=1}^{M} e^{-(\mathbf{A}\mathbf{f}+\mathbf{b})_m} \frac{(\mathbf{A}\mathbf{f}+\mathbf{b})_m^{g_m}}{(g_m)!} \; . \tag{13.5}$$

Given $\mathbf{A}, \mathbf{b}$ and $\mathbf{f}$, the probability distribution of the detected data is well defined and, in principle, all quantities of interest concerning the data can be computed from this distribution. On the other hand, in the case of an inverse problem the typical situation is that $\mathbf{A}$ and $\mathbf{b}$ are available from the model of the imaging instrument while only one realization $\mathbf{g}$ of the data is available. Then, according to the ML approach discussed in Section 11.3, one considers the likelihood function which is given now by equation (13.5) where $\mathbf{A}, \mathbf{b}$ and $\mathbf{g}$ are given while $\mathbf{f}$ is unknown. This function of $\mathbf{f}$ is restricted to the set of the admissible objects (in the present case the nonnegative ones), i.e. restricted to the nonnegative orthant of the object space. Since the natural domain of the likelihood function is broader than the nonnegative orthant, its maximization is a constrained problem.

Finally, by taking the neg-log of the likelihood as derived from equation (13.5), the ML approach is reduced to the constrained minimization of the function given in equation (11.23). However, by modifying the terms depending only on $\mathbf{g}$, we prefer to write it in the following form because, as shown below, we obtain what can be called a *divergence* of the noisy data $\mathbf{g}$ from the noise-free data $\mathbf{Af} + \mathbf{b}$

$$l(\mathbf{f}|\mathbf{g}) = \sum_{m=1}^{M} \left\{ g_m \ln \frac{g_m}{(\mathbf{Af} + \mathbf{b})_m} + (\mathbf{Af} + \mathbf{b})_m - g_m \right\} . \tag{13.6}$$

Since the natural domain of this function is broader than the nonnegative orthant, the constrained minimizer can lie on the boundary of the nonnegative orthant, i.e. some of the components of the ML estimates can be zero. In practice, since the problems we are considering are ill-posed, several components are zero and in fact all minimizers lie on the boundary as we will show.

REMARK 13.1   *In Remark 11.1 it is observed that, when the expected values λ_m are large, the Poisson distributions can be approximated by Gaussian distributions. If we use this approximation for defining the likelihood we obtain the following minimization problem for the ML estimates*

$$\sum_{m=1}^{M} \frac{1}{\lambda_m} |(\mathbf{Af})_m + b_m - g_m|^2 = \text{minimum} , \tag{13.7}$$

where the λ_m depend on the unknown object $\mathbf{f}$ as indicated above. This problem is investigated in [262] where it is shown that the use of this approximation does not provide an improvement in efficiency or accuracy of the solution. However, the standard application of the Gaussian approximation consists in replacing the λ_m by their ML estimates. As shown in Example 11.2, if g_m is the realization of a Poisson r.v. with unknown expected value λ_m, then the ML estimate of λ_m is g_m. Therefore, if we replace λ_m with g_m in (13.7), we obtain a function with the same structure as that given in equation (12.14). It is obvious that, in such a case, we obtain a least-squares problem corresponding to a noise which is not stationary in space; it follows that, in the case of deconvolution problems, it is not possible to use FFT-based methods for the solution of the corresponding minimization problem.

The function (13.6) is called the *generalized Kullback-Leibler* (KL) *divergence* of $\mathbf{g}$ from $\mathbf{Af} + \mathbf{b}$, i.e. between noisy and noise-free data. The KL divergence was introduced by Kullback and Leibler in probability theory [182] as a divergence between two probability distributions. In that case the difference between the distributions is zero because of their normalization to 1 so that only the logarithmic term appears in the definition. This remark is at the origin of the qualification of *generalized* attributed to the function defined in equation (13.6).

A divergence is a function of two vectors with the following properties:

- it is nonnegative;
- it is zero if and only if the two vectors coincide.

Therefore, a divergence has the first property of a metric distance, which it is not because it is not symmetric with respect to the exchange of the two vectors and it does not satisfy the triangular inequality. Since a divergence is a function most similar to a metric distance, we modified the terms depending only on $\mathbf{g}$ in such a way as to obtain a divergence from the neg-log of the Poisson likelihood. For this reason we call the function defined in equation (13.6) the *Poisson discrepancy function*, since it can be taken as a measure for the discrepancy between the noisy and the noise-free data (in this order!). Anyway we still have to prove that $l(\mathbf{f}|\mathbf{g})$ is a divergence of $\mathbf{g}$ from $\mathbf{Af}+\mathbf{b}$. Since it is a sum of terms, all depending on only one component of $\mathbf{g}$, it is sufficient to prove the two defining properties for each of these terms. First, if $g_m = 0$ the non-negativity is obvious because we assume $u \ln u = 0$ if $u = 0$; next, if $g_m > 0$, then non-negativity follows from the elementary inequality $\ln u \leq u - 1$. Moreover, each term is zero when the two variables take the same value. Since, in our case, this implies $\mathbf{g} = \mathbf{Af} + \mathbf{b}$, from the ill-conditioning of the basic equation it follows that in general no nonnegative solution of this equation exists. As a consequence the Poisson discrepancy is strictly positive and has a positive lower bound.

It is important to investigate other properties of the Poisson discrepancy function. To this purpose, one can easily compute the gradient and the Hessian of this function. The results are the following

$$[\boldsymbol{\nabla}_{\mathbf{f}} \, l(\mathbf{f}|\mathbf{g})]_n = \frac{\partial l(\mathbf{f}|\mathbf{g})}{\partial f_n} = \alpha_n - \left(\mathbf{A}^T \frac{\mathbf{g}}{\mathbf{Af}+\mathbf{b}} \right)_n \, , \tag{13.8}$$

$$[\boldsymbol{\nabla}_{\mathbf{f}}^2 \, l(\mathbf{f}|\mathbf{g})]_{n,n'} = \frac{\partial^2 l(\mathbf{f}|\mathbf{g})}{\partial f_n \partial f_{n'}} = \sum_{m=1}^{M} \frac{A_{m,n} A_{m,n'} g_m}{(\mathbf{Af}+\mathbf{b})_m^2} \, . \tag{13.9}$$

In equation (13.8) the following notations are used: α_n is the sum of the elements of the n-th column of the matrix $\mathbf{A}$

$$\alpha_n = \sum_{m=1}^{M} A_{m,n} \tag{13.10}$$

with $\alpha_n \leq 1$ if the interpretation of the matrix elements of $\mathbf{A}$ as probabilities is used; the quotient of the two vectors $\mathbf{g}$ and $\mathbf{Af}$ is component-wise and therefore it is still a vector

$$\left(\frac{\mathbf{g}}{\mathbf{Af}+\mathbf{b}} \right)_m = \frac{g_m}{\sum_{n=1}^{N} A_{m,n} f_n + b_m} \, ; \tag{13.11}$$

finally $\mathbf{A}^T$ denotes the transposed matrix of $\mathbf{A}$, i.e. $(\mathbf{A}^T)_{n,m} = A_{m,n}$. These equations can be written in the more compact form

$$\boldsymbol{\nabla}_{\mathbf{f}} \, l(\mathbf{f}|\mathbf{g}) = \mathbf{A}^T \mathbf{1} - \mathbf{A}^T \frac{\mathbf{g}}{\mathbf{Af}+\mathbf{b}} \tag{13.12}$$

$$\boldsymbol{\nabla}_{\mathbf{f}}^2 \, l(\mathbf{f}|\mathbf{g}) = \mathbf{A}^T \mathrm{diag} \left\{ \frac{\mathbf{g}}{(\mathbf{Af}+\mathbf{b})^2} \right\} \mathbf{A}$$

where $\mathbf{1}$ is a vector of ones and 'diag' means the diagonal matrix formed by the elements of the vector. From equation (13.9) it follows that the Hessian matrix is positive semi-definite, because for any vector ξ we have

$$(\boldsymbol{\nabla}_{\mathbf{f}}^2 \, l(\mathbf{f}|\mathbf{g}) \xi \cdot \xi) = \sum_{m=1}^{M} \frac{g_m}{(\mathbf{Af}+\mathbf{b})_m^2} (\mathbf{A}\xi)_m^2 \geq 0 \, . \tag{13.13}$$

Therefore the discrepancy $l(\mathbf{f}|\mathbf{g})$, equation (13.6), is a *convex function*. From the previous expression it follows that it is *strictly convex* if and only if all the components g_m are strictly positive (possible in the case of background) and the null space of the matrix $\mathbf{A}$ is trivial. The function is also coercive. Indeed, let us consider the terms $(\mathbf{Af})_m$. By an exchange of the summation order we obtain

$$\sum_{m=1}^{M} (\mathbf{Af})_m = \sum_{n=1}^{N} \alpha_n f_n \ , \tag{13.14}$$

where the quantities α_n, defined in equation (13.10), are strictly positive thanks to the assumptions made on the imaging matrix $\mathbf{A}$. It follows that

$$\|\mathbf{f}\| \to \infty \ \text{ implies } \ \sum_{m=1}^{M} (\mathbf{Af})_m \to \infty \ \text{ implies } \ l(\mathbf{f}|\mathbf{g}) \to \infty \ , \tag{13.15}$$

which is precisely the property stated above. Coercivity combined with convexity implies the existence of at least one minimizer; moreover all minimizers are global.

Let us now summarize the properties of the Poisson discrepancy function and their consequences.

- If the matrix $\mathbf{A}$ is ill-conditioned, the function is strictly positive; moreover, it is differentiable and convex on the nonnegative orthant and it is strictly convex if and only if all data are positive and the null space of the imaging matrix is trivial.
- The function is coercive on the nonnegative orthant.
- There exists at least one nonnegative minimum point; all minimum points are global and form a convex set.
- The minimum point is unique if and only if the conditions for strict convexity of the function are satisfied.

We conclude this section by proving two general properties of the minimum points of the Poisson discrepancy function. Let $\tilde{\mathbf{f}}$ be a generic minimum point of this function; then the following properties hold true.

- Let $\mathbf{b} > 0$ and let $\mathbf{g} > \mathbf{b}$, then $\mathbf{f} = 0$ is not a minimum point of the Poisson discrepancy function.

Indeed, if $\mathbf{f} = 0$, then, from the expression of the gradient, equation (13.8), we obtain

$$\nabla_{\mathbf{f}} l(0|\mathbf{g}) = \mathbf{A}^T \left(1 - \frac{\mathbf{g}}{\mathbf{b}} \right) \ . \tag{13.16}$$

But, thanks to the assumption, the gradient has negative components and therefore the KKT condition (12.42) implies that $\mathbf{f} = 0$ cannot be a minimum point. Let us remark that the condition $\mathbf{g} > \mathbf{b}$ implies that the image does not consists uniquely of background and noise but that there exists a signal which lies above the background. Therefore the non-zero minimum point should capture this signal.

- If the minimum point $\tilde{\mathbf{f}}$ is interior to the nonnegative orthant, then it is a solution of the basic equation $\mathbf{A}\tilde{\mathbf{f}} + \mathbf{b} = \mathbf{g}$.

The proof of this statement is also based on the KKT conditions (12.41)-(12.42) which, thanks to the expression of the gradient (13.8), can be written as follows

$$\tilde{\mathbf{f}} \circ \left(\mathbf{A}^T \mathbf{1} - \mathbf{A}^T \frac{\mathbf{g}}{\mathbf{A}\tilde{\mathbf{f}} + \mathbf{b}} \right) = 0 \ , \tag{13.17}$$

$$\tilde{\mathbf{f}} \geq 0 \ , \quad \mathbf{A}^T \mathbf{1} - \mathbf{A}^T \frac{\mathbf{g}}{\mathbf{A}\tilde{\mathbf{f}} + \mathbf{b}} \geq 0 \ . \tag{13.18}$$

The second and third condition imply that the first one is equivalent to

$$\tilde{\mathbf{f}} \cdot \left(\mathbf{A}^T \mathbf{1} - \mathbf{A}^T \frac{\mathbf{g}}{\mathbf{A}\tilde{\mathbf{f}} + \mathbf{b}} \right) = 0 \tag{13.19}$$

or else

$$\mathbf{A}\tilde{\mathbf{f}} \cdot \left(1 - \frac{\mathbf{g}}{\mathbf{A}\tilde{\mathbf{f}} + \mathbf{b}} \right) = 0 \ . \tag{13.20}$$

If $\tilde{\mathbf{f}} > 0$ then also $\mathbf{A}\tilde{\mathbf{f}} > 0$ and the previous equality implies

$$1 - \frac{\mathbf{g}}{\mathbf{A}\tilde{\mathbf{f}} + \mathbf{b}} = 0 \tag{13.21}$$

which proves our statement.

The last result has important consequences. Indeed, since the imaging matrix is ill conditioned and possibly overdetermined, the minimization problem does not have in general a positive solution in the case of noisy data. Therefore any minimum point has at least one zero component and all minimum points lie on the boundary of the nonnegative orthant. As follows from numerical practice they have a lot of zero components and, even if their sparsity is not proved as far as we know, in many instances they look like sparse objects. This effect is sometimes called *checkerboard effect* [218] or *night-sky effect* [20]. The consequence of this property is that the ML approach does not provide in general reliable estimates of the true object (except perhaps for a sparse object).

13.2   The ML-EM or Richardson-Lucy algorithm

The problem is now to find methods for computing or approximating the minimum points of $l(\mathbf{f}|\mathbf{g})$. The benchmark of the methods designed to this purpose is the so-called ML-EM (Maximum Likelihood - Expectation Maximization) method introduced in the seminal paper of Shepp and Vardi [253] where the maximum likelihood approach is proposed. ML-EM, a terminology used in medical imaging, is an iterative method derived from the *Expectation Maximization method* introduced in statistics by Dempster, Laird and Rudin [99] for the computation of ML solutions from incomplete data. The same method is used in astronomy (and also in microscopy) under the name of Richardson-Lucy (RL) algorithm, after the papers by Richardson [241] and Lucy [203] who independently proposed the method even if using different arguments. In all these cited papers the method is introduced in the case $\mathbf{b} = 0$.

For the reader's convenience we first derive the method with $\mathbf{b} \geq 0$ in a heuristic way, similar to that used for the derivation of ISRA and afterwards we also establish it as a MM-algorithm. The starting point is again given by the KKT conditions which are necessary and sufficient conditions to be satisfied by a minimum point of the convex function $l(\mathbf{f}|\mathbf{g})$. In particular, by combining the first condition (12.41) with (13.8), we deduce that the minimum points of $l(\mathbf{f}|\mathbf{g})$ are solutions of the following nonlinear equation

$$\alpha_n \tilde{f}_n = \tilde{f}_n \left(\mathbf{A}^T \frac{\mathbf{g}}{\mathbf{A}\tilde{\mathbf{f}} + \mathbf{b}} \right)_n \ ; \quad n = 1, 2, \ldots, N \ , \tag{13.22}$$

which implies the fixed-point equation

$$\tilde{\mathbf{f}} = \frac{\tilde{\mathbf{f}}}{\boldsymbol{\alpha}} \circ \left(\mathbf{A}^T \frac{\mathbf{g}}{\mathbf{A}\tilde{\mathbf{f}} + \mathbf{b}} \right) , \tag{13.23}$$

where $\boldsymbol{\alpha}$ is the vector, with components $1 \geq \alpha_n > 0$, defined in equation (13.10). The converse, of course, is not true: not all the solutions of this equation are minimum points of the Poisson discrepancy function.

REMARK 13.2 *In the case* $\mathbf{b} = 0$, *it is easy to verify that equation (13.23) has at least* N *solutions* $\mathbf{f}^{(n)}$ *$(n = 1, 2, \ldots, N)$ given by*

$$(\mathbf{f}^{(n)})_n = \frac{1}{\alpha_n} \sum_{m=1}^{M} g_m \; ; \; (\mathbf{f}^{(n)})_{n'} = 0 \; ; \quad n' \neq n . \tag{13.24}$$

These fixed points, with only one non-zero component, are not, in general, ML *solutions.*

REMARK 13.3 *By summing both members of equation (13.22) with respect to n in the case* $\mathbf{b} = 0$ *we find*

$$\sum_{n=1}^{N} \left(\sum_{m=1}^{M} A_{m,n} \right) \tilde{f}_n = \sum_{n=1}^{N} \tilde{f}_n \sum_{m=1}^{M} A_{m,n} \frac{g_m}{\sum_{n'=1}^{N} A_{m,n'} \tilde{f}_{n'}} \tag{13.25}$$

so that, by exchanging the summation order, we get

$$\sum_{m=1}^{M} (\mathbf{A}\tilde{\mathbf{f}})_m = \sum_{m=1}^{M} g_m . \tag{13.26}$$

The physical meaning of this equation is the following: the total mean intensity of the noise-free image generated by $\tilde{\mathbf{f}}$ *coincides with the total number of counts of the noisy image of* $\mathbf{f}$, *the true image. The mathematical meaning is that all minimum points of* $l(\mathbf{f}|\mathbf{g})$ *lie in the intersection of the first quadrant with the hyperplane defined by the equation (13.26). This is a bounded domain in the object space.*

If we apply the method of successive approximations to the fixed-point equation (13.23), we obtain the following iterative algorithm

- select $\mathbf{f}_0 > 0$;
- for $k = 0, 1, \ldots,$ set

$$\mathbf{f}_{k+1} = \frac{\mathbf{f}_k}{\boldsymbol{\alpha}} \circ \left(\mathbf{A}^T \frac{\mathbf{g}}{\mathbf{A}\mathbf{f}_k + \mathbf{b}} \right) . \tag{13.27}$$

In the case $\mathbf{b} = 0$, the convergence of the iteration $\mathbf{f}_k$ to a minimizer of $l(\mathbf{f}|\mathbf{g})$ is proved by several authors [191, 295, 215, 167, 168]. The limit depends on the initial guess $\mathbf{f}_0$ if $l(\mathbf{f}|\mathbf{g})$ has several minimum points. A proof based on analytical techniques, derived from [215], is given in [219]. In all these proofs the flux condition (13.26) is a basic ingredient. Only recently convergence has been proved also in the case $\mathbf{b} \neq 0$ [245].
Let us summarize the main properties of the algorithm.

- Since $\mathbf{f}_0$ is positive each iterate $\mathbf{f}_k$ is also positive, as one can easily verify by induction; moreover, if one component of $\mathbf{f}_0$ is zero, then the corresponding component of all iterates is zero, a way for introducing a constraint on the support of the solution (see Section 8.5).

- In the case $\mathbf{b} = 0$, each iterate $\mathbf{f}_k$ satisfies condition (13.26) i.e. the total mean intensity of $\mathbf{A}\mathbf{f}_k$ coincides with the total number of photons of the image $\mathbf{g}$.
- The Poisson discrepancy function is non-increasing throughout the iteration: $l(\mathbf{f}_{k+1}|\mathbf{g}) \leq l(\mathbf{f}_k|\mathbf{g})$ (the proof is given in [253] and here below).
- The sequence $\mathbf{f}_k$ converges to a minimum point of $l(\mathbf{f}|\mathbf{g})$ (the proof is given e.g. in [295] as well as in the other papers mentioned above).

The non-increasing behavior of the Poisson discrepancy function results from the general properties of MM algorithms as stated in Section 10.3 if we notice that the iteration (13.27) can be rewritten as

$$\mathbf{f}_{k+1} = \arg \min_{\mathbf{f}} \Phi^{SUR}(\mathbf{f}; \mathbf{f}_k) \tag{13.28}$$

with the following separable surrogate functional, valid for $\mathbf{a} \geq 0$, $\mathbf{b} \geq 0$ (and under all other non-negativity assumptions made above),

$$
\begin{aligned}
\Phi^{SUR}(\mathbf{f}; \mathbf{a}) \;=\; \sum_{m=1}^{M} \Bigg[& g_m \ln g_m - g_m + (\mathbf{A}\mathbf{f} + \mathbf{b})_m \\
& - \frac{g_m}{(\mathbf{A}\mathbf{a} + \mathbf{b})_m} \sum_{n=1}^{N} A_{m,n} a_n \ln\left(\frac{f_n}{a_n} (\mathbf{A}\mathbf{a} + \mathbf{b})_m \right) \\
& - \frac{g_m b_m}{(\mathbf{A}\mathbf{a} + \mathbf{b})_m} \ln (\mathbf{A}\mathbf{a} + \mathbf{b})_m \Bigg] .
\end{aligned}
\tag{13.29}
$$

The anchoring property is easily checked: for $\mathbf{a} = \mathbf{f}$ the surrogate coincides with the Poisson discrepancy function (13.6). The majorization property results from arguments similar to those of Remark 12.4 and from the strict concavity of the logarithmic function. In the case $\mathbf{b} = 0$, the surrogate reduces to the one given in [91] by De Pierro who first noticed that the ML-EM algorithm as well as ISRA can be viewed as MM methods.

In the case of image deconvolution, the implementation of the algorithm is simple. First we observe that, for circulant matrices, all coefficients α_n, equation (13.10), are equal because the columns are obtained by means of permutations of the components of a vector (in 1D case) or of an array (in 2D or 3D case). In particular, in the 2D case we have

$$\alpha_{m,n} = \sum_{k,l=1}^{N} K_{m-k,n-l} = \sum_{k,l=1}^{N} K_{k,l} = \hat{K}_{0,0} \tag{13.30}$$

i.e. the coefficients coincide with the zero-frequency component of the TF. If we normalize the PSF in such a way that

$$\hat{K}_{0,0} = \sum_{k,l=1}^{N} K_{k,l} = 1 , \tag{13.31}$$

which is the standard normalization in microscopy and astronomy, then the EM algorithm for image deconvolution can be written as follows:

$$(\mathbf{f}_{k+1})_{m,n} = (\mathbf{f}_k)_{m,n} \left(\mathbf{K}^T * \frac{\mathbf{g}}{\mathbf{K} * \mathbf{f}_k + \mathbf{b}} \right)_{m,n} . \tag{13.32}$$

The implementation of the algorithm requires the computation of four FFT's at each iteration step: two for the computation of $\mathbf{K} * \mathbf{f}_k$ and two for the computation of the convolution of $\mathbf{K}^T$ with $\mathbf{g}/(\mathbf{K} * \mathbf{f}_k + \mathbf{b}_k)$. Therefore the computational cost of one step of this algorithm is higher than that of one step of the projected Landweber method or of ISRA, which requires

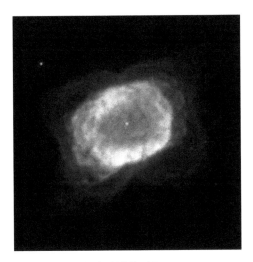

FIGURE 13.1   Image of the planetary nebula NGC 7027.

the computation of only two FFT's. Moreover, from numerical experience, it follows that also the convergence of the ML-EM algorithm is quite slow.

As already remarked, the limit of $\mathbf{f}_k$ may not be a sensible estimate of the true object $\mathbf{f}^{(0)}$. Numerical simulations indicate that the ML-EM-algorithm has the semi-convergence property typical of the iterative regularization methods discussed in the previous chapters. However a rigorous proof of this property is not available, as far as we know. In practice, sensible approximate solutions could be obtained by early stopping of the iterations. Some stopping rules have been proposed; see, for instance [199, 198, 34]. In conclusion, what we expect is to find a minimum in the reconstruction error, whatever its definition; moreover, with increasing noise the minimum error increases while the number of iterations required for reaching this value decreases, properties verified in numerical experiments.

We illustrate the previous conjectures with one numerical example. The object is a high-resolution image of the planetary nebula NGC 7027 shown in Figure 13.1. Note that this is a misnomer, derived from an old denomination, because this astronomical object has nothing to do with planets but it is the result of the ending of a star (also our Sun will form a planetary nebula at the end of its life cycle). The object is convolved with a diffraction-limited PSF with a FWHM of 12 pixels, simulating the observation of the object with a smaller telescope. Moreover, we consider two versions of the object with different total fluxes, respectively 1.77×10^{11} with a background of 6.75×10^4 and 1.77×10^9 with a background of 6.75×10^2. Since the PSF is normalized to 1 as usual, the noise-free images have exactly the same fluxes as the corresponding objects. In this way we simulate two observations, the first one obtained with a long observation time and the second one with a shorter observation time. Finally the results are perturbed with Poisson noise. The first image can be called a low-noise image and the second a high-noise image. Both images are reconstructed with ML-EM.

In Figure 13.2, we plot as a function of k the relative reconstruction error, defined as follows: $\|\mathbf{f}_k - \mathbf{f}^{(0)}\|/\|\mathbf{f}^{(0)}\|$. We observe that the curve corresponding to the high-noise case is always above that corresponding to the low-noise case. The first curve, after a steep descent up to about 100 iterations, becomes more flat and has a minimum after 231 iterations, with a reconstruction error of about 15 %. The second curve has also a steep descent and has a minimum (not very visible) after 637 iterations. However, around this value the curve is so flat that for about two hundred iterations the reconstructions are practically indistinguishable from each other. This is, in a sense, good news because it implies that the

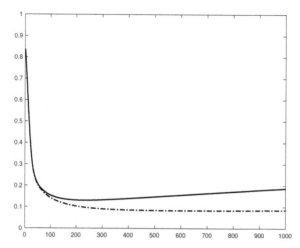

FIGURE 13.2   Behavior of the reconstruction error as a function of k. Full line: high-noise case; dash-dotted line: low-noise case.

choice of the number of iterations is not so critical at least in the case of low-noise images (even if, in our case, the noise is not so low since it is clearly visible in the observed image).

Finally, in Figure 13.3, we show the blurred and noisy images with the corresponding reconstructions. We remark that the stars visible in the object are completely lost in the reconstruction. They appear at a higher number of iterations when the reconstruction of the nebula is completely corrupted by the night-sky effect discussed above.

As a final remark we point out that also in the case of ML-EM, as for the other iterative methods discussed in this book, the basic operation needed for the implementation of the method is a matrix-vector multiplication, so that it can also be very useful in cases where the matrix **A** is large and sparse.

13.3   Acceleration of the ML-EM method

In this section we briefly discuss two different approaches for accelerating the ML-EM method; the goal is to reduce the number of iterations without increasing excessively the numerical cost per iteration. The first approach applies mainly to emission tomography and the second one to deconvolution or similar problems.

13.3.1   Ordered Subset Expectation Maximization (OSEM)

As already remarked the ML-EM or RL method is slow: in general, it requires a large number of iterations before reaching the best achievable approximation of the unknown object, as expected from its semi-convergence property. The method is widely used in astronomy and microscopy because its implementation is easy and is based on the FFT so that the cost of one iteration is low. It is not used in medical imaging, though it was originally proposed for emission tomography, because the cost of one iteration is high and the number of iterations can be large. However, a remarkable acceleration was proposed by Hudson and Larkin in 1994 [164].

The starting point is the particular structure of the data in tomography: they consists of a set of projections as defined in Section 6.2. Then the idea is to partition the data into disjoint ordered subsets of projections and to apply one ML-EM iteration to each subset; an iteration

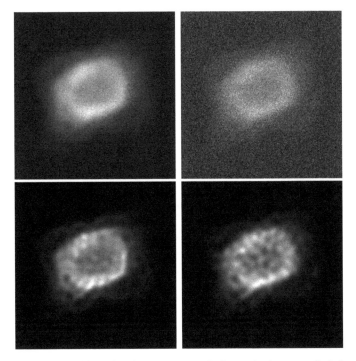

FIGURE 13.3   First line: the blurred and noisy images, the low-noise image to the left and the high-noise image to the right. Second line: the corresponding reconstructions; the error is about 9 % in the low-noise case and 15 % in the high-noise case.

of the method, called by the authors *Ordered Subset Expectation Maximization* (OSEM), consists in a cycle over the selected subsets. The efficiency gained with this approach is such that OSEM can be a practical method in CT [25] and PET [96]: very few iterations are sufficient for reaching a minimum restoration error.

Let us denote by $\mathcal{M} = \{1, 2, \ldots, M\}$ the set of indices characterizing the components of the image $\mathbf{g}$ and by $\mathcal{N} = \{1, 2, \ldots, N\}$ that of the object $\mathbf{f}$. Then the set $\mathcal{M}$ is partitioned into disjoint subsets $\mathcal{M}^{(l)}$, $l = 1, \ldots, L$, such that their union coincides with the total set. We denote by $\mathbf{g}^{(l)}$ the data vector with components g_m, $m \in \mathcal{M}^{(l)}$ and by $\mathbf{A}^{(l)}$ the block of the matrix $\mathbf{A}$ consisting of the rows with $m \in \mathcal{M}^{(l)}$. Moreover, for each l, we introduce the vector $\alpha^{(l)}$ defined by

$$\alpha_n^{(l)} = \sum_{m \in \mathcal{M}^{(l)}} A_{m,n}, \quad n \in \mathcal{N} . \tag{13.33}$$

Then the method consists of a cycle over the subsets $\mathcal{M}^{(l)}$ as follows

- choose $\mathbf{f}_0 > 0$ and a partition of $\mathcal{M}$;
- for $k = 0, 1, \ldots$ set $\mathbf{f}_k^{(0)} = \mathbf{f}_k$;
- for $l = 1, \ldots, L$

$$\mathbf{f}_k^{(l)} = \frac{\mathbf{f}_k^{(l-1)}}{\alpha^{(l)}} \circ (\mathbf{A}^{(l)})^T \left(\frac{\mathbf{g}^{(l)}}{\mathbf{A}^{(l)} \mathbf{f}_k^{(l-1)} + \mathbf{b}} \right) ; \tag{13.34}$$

- set $\mathbf{f}_{k+1} = \mathbf{f}_k^{(L)}$.

A first remark is that, when $\mathbf{b} = 0$, from the definition (13.33) and the iteration (13.34) for $\mathbf{f}_k^{(l)}$ one obtains, by an exchange of the summation order (see also equation (13.26)), the

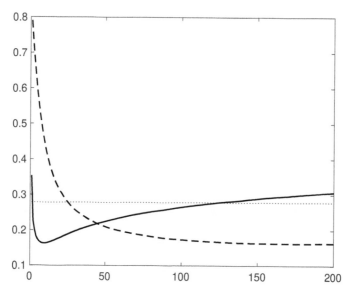

FIGURE 13.4 Plot of the reconstruction error as a function of the number of iterations in the case of ML-EM (16% error at iteration 200 – dashed line) and OSEM with 18 subsets (16% at iteration 10 – full line). The reconstruction error obtained by FBP (28%) is plotted as a dotted line.

following relationship

$$\sum_{m \in \mathcal{M}^{(l)}} (\mathbf{A} \mathbf{f}_k^{(l)})_m = \sum_{n \in \mathcal{N}} \alpha_n^{(l)} (\mathbf{f}_k^{(l)})_n = \sum_{m \in \mathcal{M}^{(l)}} g_m \; . \tag{13.35}$$

Therefore, in order to avoid excessive oscillations of the ℓ^1-norm of the computed data within one OSEM cycle, it is a good policy to select the subsets in such a way that the last term in equation (13.35) is approximately independent of l. We also remark that, the l-th step of the internal cycle is a ML-EM iteration for the minimization of the partial Poisson discrepancy function

$$l(\mathbf{f}|\mathbf{g}^{(l)}) = \sum_{m \in \mathcal{M}^{(l)}} \left\{ g_m \ln \frac{g_m}{(\mathbf{A}^{(l)}\mathbf{f} + \mathbf{b})_m} + (\mathbf{A}^{(l)}\mathbf{f} + \mathbf{b})_m - g_m \right\} \; . \tag{13.36}$$

A consequence of this remark is that a convergence proof of the algorithm should require that all these partial functions have the same minimum point $\tilde{\mathbf{f}}$. In fact, a convergence proof is given in [164] in the case $\mathbf{b} = 0$, by assuming that there exists a nonnegative solution $\tilde{\mathbf{f}}$ of the linear equation $\mathbf{A}\mathbf{f} = \mathbf{g}$ and that the matrices $\mathbf{A}^{(l)}$ are *balanced*, in the sense that the vectors $\alpha^{(l)}$ are independent of l. In other words the result applies to noise-free data with an additional condition on the imaging matrix. This is called the *consistent case*. We believe that convergence in the general case does not hold true.

The lack of a convergence guarantee does not prevent the practical use of this algorithm in tomography where each subset consists of a suitable number of projections. The important point is that OSEM, as ML-EM, has a semi-convergent behavior so that its acceleration effect means that we need a much smaller number of iterations for reaching a sensible solution similar to that provided by ML-EM. The reduction depends in a crucial way on the choice of the subsets and on their order in the internal cycle [164].

Let us also remark that the cost per iteration is approximately the same for the two algorithms, ML-EM and OSEM, at least in medical imaging which is the main application of OSEM. In particular, this is true in the specific case of SPECT (the application considered by

Hudson and Larkin), where data can be structured into a sequence of groups of projections. Hence a subset of OSEM can consist of a number of projections, possibly the same number for each subset. The number of subsets is called by Hudson and Larkin the OS-*level*. Then the main result, supported by theoretical arguments and numerical experiments, is that, in the sense of semi-convergence, the number of iterations required by OSEM for reaching approximately the same result as ML-EM is roughly the number of ML-EM iterations divided by the OS-level. Thanks to the previous remark this also represents the gain in computational time.

We give a simple example for illustrating the properties of OSEM. We consider the well-known Shepp-Logan phantom, compute the corresponding sinogram and perturb it with Poisson noise. Next we reconstruct the phantom by FBP, ML-EM and OSEM with 18 subsets obtained from 180 projections. The order is dictated by the idea that two subsequent subsets should contain very different information on the object. Therefore, if we order the projections from 1 to 180 degrees, starting from angle zero, the first subset contains the projections from 1 to 10, the second the projections from 91 to 100, the third from 11 to 20, the fourth from 111 to 120 and so on. For FBP and for each iterate of the iterative methods we compute the RMSE. The results are shown in Figure 13.4, where the dotted horizontal line is the error of FBP, the full line the error of OSEM as a function of the number of iterations and the dashed line the error of ML-EM also as a function of the number of iterations. It is evident that the two curves have a minimum and that the minimum value is the same in the two cases. To be honest, we did a number of attempts to optimize the number of subsets even if with other choices the reconstruction error is only slightly higher than that of ML-EM. We remark that the ratio between the numbers of iterations corresponding to the two minima is 20 and therefore slightly higher than the number of subsets used in OSEM, confirming the gain in number of iterations and computing time provided by this algorithm.

In several applications of tomography in industry and medical imaging, it is not possible to cover completely the 180 degrees required in CT or the 360 degrees required in SPECT; other important examples of incomplete tomographic information are breast tomosynthesis [251] and dental tomosynthesis [103]. The inverse problem arising in these applications is the so-called *limited-angle tomography*, a highly ill-posed problem with ill-posedness increasing with the size of the missing angle. The reconstructions based on FBP are affected by the so-called *streak artifacts* and one of the main problems is their reduction [125]. It is interesting to remark that, in the case of Poisson noise, ML-EM and hence also OSEM are able to reduce streak artifacts at least when the missing angle is not too large. To demonstrate this property we repeated the previous experiment with a missing angle of 60 degrees. In Figure 13.5, we show the Shepp-Logan phantom together with the FBP reconstruction and the OSEM reconstruction. The FBP reconstruction is clearly affected by streak artifacts while in the OSEM reconstruction they disappear even if the reconstruction is distorted along the directions of the missing angle.

13.3.2 Scaled gradient methods

Another way of accelerating the ML-EM method results from the following observation. As proved in [253] and also in the previous section, ML-EM is a descent method since the objective function $l(\mathbf{f}_k|\mathbf{g})$ is decreasing along the iterations. Then, by re-arranging the r.h.s. of equation (13.27) and remarking that $\mathbf{A}^T \mathbf{1} = \boldsymbol{\alpha}$, it is easy to show that it is a *scaled gradient method* since one ML-EM iteration can also be written in the following form

$$\mathbf{f}_{k+1} = \mathbf{f}_k - \frac{\mathbf{f}_k}{\boldsymbol{\alpha}} \circ \left(\mathbf{A}^T \mathbf{1} - \mathbf{A}^T \frac{\mathbf{g}}{\mathbf{A}\mathbf{f}_k + \mathbf{b}} \right) \ . \tag{13.37}$$

Therefore, ML-EM is a descent method with a descent direction given by the negative gra-

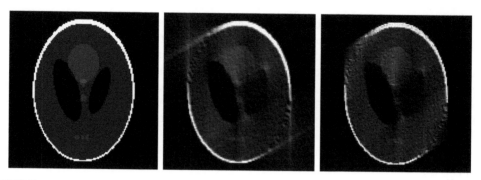

FIGURE 13.5   Reconstruction with a missing angle of 60 degrees. From left to right: the Shepp-Logan phantom; the FBP reconstruction and the OSEM reconstruction.

dient of the discrepancy function (see equation (13.8)) multiplied component-wise with the vector $\mathbf{f}_k/\alpha$. This vector is positive since α is positive and also $\mathbf{f}_k$ is positive if the initial guess is positive, as assumed. Moreover, the step-size in the descent direction is 1.

This remark suggests a way for reducing the number of iterations, namely to introduce a suitable step-size $\beta_k > 1$ depending on the iteration, which now takes the following form

$$\mathbf{f}_{k+1} = \mathbf{f}_k - \beta_k \,\frac{\mathbf{f}_k}{\alpha} \circ \left(\mathbf{A}^T \mathbf{1} - \mathbf{A}^T \frac{\mathbf{g}}{\mathbf{A}\mathbf{f}_k + \mathbf{b}} \right) . \tag{13.38}$$

This iteration is sometimes called the *relaxed* form of the ML-EM algorithm and was proposed independently in different applications: see, for instance, [162] in the case of microscopy, [5] in the case of astronomy and [194] in a more general context (in all cases for deconvolution problems). Accordingly, the parameter β_k is called the *relaxation parameter*. The approach can reduce the number of iterations but the computational cost per iteration increases because of the search for the value of the relaxation parameter, so that the final gain may not be very relevant.

It is easy to find an upper bound on β_k ensuring that, if $\mathbf{f}_k$ is nonnegative, then also $\mathbf{f}_{k+1}$ is nonnegative. Indeed, by looking at equation (13.38), one can first observe that, if $(\mathbf{f}_k)_n = 0$ then also $(\mathbf{f}_{k+1})_n = 0$; moreover, if

$$\left(\mathbf{A}^T \mathbf{1} - \mathbf{A}^T \frac{\mathbf{g}}{\mathbf{A}\mathbf{f}_k + \mathbf{b}} \right)_n \le 0 \,, \tag{13.39}$$

then $(\mathbf{f}_{k+1})_n \ge 0$, for any value of β_k. Therefore, we must only consider the components n for which the following conditions are simultaneously satisfied

$$(\mathbf{f}_k)_n > 0 \,, \quad \text{and} \quad \left(\alpha - \mathbf{A}^T \frac{\mathbf{g}}{\mathbf{A}\mathbf{f}_k + \mathbf{b}} \right)_n \ge 0 \,. \tag{13.40}$$

This case implies a condition on β_k: from the requirement $(\mathbf{f}_{k+1})_n \ge 0$ one easily derives from equation (13.38) that

$$\beta_k \le \min_{n \in \mathcal{C}_+} \frac{\alpha_n}{\left(\alpha - \mathbf{A}^T \frac{\mathbf{g}}{\mathbf{A}\mathbf{f}_k + \mathbf{b}} \right)_n} \,, \tag{13.41}$$

where $\alpha = \mathbf{A}^T \mathbf{1}$ and $\mathcal{C}_+$ is the set of the indices satisfying conditions (13.40). It is obvious that the upper bound is certainly greater than 1. Next, an optimal step-size can be obtained by a line search in the allowed interval. However, it is worthwhile to remark that standard rules such as the Armijo rule cannot be adapted to the present case because the scaling factor $\mathbf{f}_k$ does not have a positive lower bound; indeed, as we know, the limit is a maximum-likelihood solution with several null components. A criterion for the selection of the relaxation parameter, as well as a convergence proof is given in [33]. The proof, derived from a proof given in [190], applies to ML-EM as well as to other iterative methods with similar characteristics, namely descent methods with a descent direction given by a diagonal scaling of the negative gradient (other examples are given in the next section). However, it is important to remark that a crucial point in the proof is the strict convexity of the objective function, in the present case the Poisson discrepancy function. Since this condition is not always satisfied, the result does not apply to all cases.

Finally, a convergent algorithm, able to accelerate ML-EM with the scaling of the gradient indicated above, is the *Scaled Gradient Projection* (SGP) method proposed in [312]. The structure of the algorithm is as follows: given the parameters β_k, λ_k and the scaling vector $\mathbf{S}_k$ then the iteration is given by

$$\mathbf{d}_k = P_+ \left\{ \mathbf{f}_k - \beta_k \, \mathbf{S}_k \circ \left(\mathbf{A}^T \mathbf{1} - \mathbf{A}^T \frac{\mathbf{g}}{\mathbf{A}\mathbf{f}_k + \mathbf{b}} \right) \right\} \qquad (13.42)$$
$$\mathbf{f}_{k+1} = \mathbf{f}_k + \lambda_k (\mathbf{d}_k - \mathbf{f}_k)$$

where P_+ is the projection on the nonnegative orthant. In the case of the acceleration of ML-EM, the scaling is related to the scaling appearing in equation (13.37) and is given by

$$(\mathbf{S}_k)_n = \max \left\{ \frac{1}{L}, \min \left\{ L, \frac{(\mathbf{f}_k)_n}{(\mathbf{A}^T \mathbf{1})_n} \right\} \right\} . \qquad (13.43)$$

A complete description of the algorithm, as concerns the choice of the upper and lower bound of the scaling, the choice of β_k assuring that $\mathbf{d}_k$ is a descent direction and finally the choice of λ_k assuring convergence of the algorithm, is rather complex and technical. In the cited paper the authors show on numerical experiments the considerable gain in efficiency provided by this algorithm with respect to ML-EM.

13.4   Bayesian methods

The difficulties related to the ML solutions were soon recognized and an application of the method of *sieves*, proposed by Grenander [139], was used for obtaining reliable solutions in emission tomography [261]. On the other hand, the approach based on Bayesian methods can be considered as a statistical regularization of the ML method. The analogy is enforced by our choice of writing the neg-log of the Poisson likelihood as a divergence, thus introducing a function which can be interpreted as a discrepancy function between noisy and noise-free data. The introduction of prior information on the object in the form of its probability distribution and the computation of MAP estimates leads, as shown in Section 11.3, to the constrained minimization of a function which is given by the sum of the Poisson discrepancy function and of a penalty deriving from the prior of the object, in agreement with the general philosophy of inverse and ill-posed problems.

In this section, we focus on MAP estimates because the main request of a user of image reconstruction methods is to actually be able to produce an image derived from raw data, and MAP estimates satisfy such request, usually in a cheap way because their computation can be reduced to the solution of a minimization problem and can be obtained, in general, by means of efficient optimization methods.

The paper of Geman and Geman [127] is the starting point of the applications of Bayesian methods to imaging problems with Poisson data. Therefore, as suggested in that paper, the initial focus is on Gibbs distributions characterized by suitable potential functions: using the notation introduced in Chapter 11, the suggested potentials (or regularization functions) have the following form

$$\mathcal{G}(\mathbf{f}) = \sum_{n \in \mathcal{N}} \sum_{n' \in \mathcal{N}_n} \varphi \left(\frac{\mathbf{f}_n - \mathbf{f}_n'}{\varepsilon_{n,n'}} \right) , \tag{13.44}$$

where φ is a nonnegative, symmetric and differentiable function, $\mathcal{N}_n$ is a suitable neighborhood of the pixel/voxel n and $\varepsilon_{n,n'}$ is 1 for horizontal or vertical nearest neighbors n, n' and $\sqrt{2}$ for nearest diagonal neighbors n, n'.

As far as we know, a number of papers based on this idea were soon published [128, 129, 152] with application in particular to SPECT imaging. A few years later, particular interest was raised by a paper by Green [138], where a simple algorithm is proposed for the minimization of the function defined in equation (11.32). Indeed, the algorithm consists in a simple modification of ML-EM and is easily implementable; it is called by Green *One-Step Late* (OSL) and it is the first algorithm discussed in the next subsection.

13.4.1 MAP estimates: differentiable regularization

In this subsection we consider simple methods for the computation of the minimizers of the basic function

$$l_B(\mathbf{f}|\mathbf{g}) = l(\mathbf{f}|\mathbf{g}) + \mu \mathcal{G}(\mathbf{f}) , \quad \mathbf{f} \geq 0 , \tag{13.45}$$

where $l(\mathbf{f}|\mathbf{g})$ is the Poisson discrepancy function and $\mathcal{G}(\mathbf{f})$ is the regularization function which we assume to be differentiable. Let us give a simple and heuristic derivation of the OSL method, similar to the one given above for ISRA and ML-EM. Hence the starting point is the first KKT condition for the minimization of $l_B(\mathbf{f}|\mathbf{g})$

$$\tilde{\mathbf{f}} \circ \nabla_{\mathbf{f}} l_B(\tilde{\mathbf{f}}|\mathbf{g}) = 0 , \tag{13.46}$$

$\tilde{\mathbf{f}}$ being a constrained (nonnegative) minimizer of the objective function. Then, taking into account the expression of the gradient of the discrepancy function, equation (13.8), we obtain from the previous equation

$$\tilde{\mathbf{f}} \circ \left(\boldsymbol{\alpha} - \mathbf{A}^T \frac{\mathbf{g}}{\mathbf{A}\tilde{\mathbf{f}} + \mathbf{b}} + \mu \, \nabla_{\mathbf{f}} \mathcal{G}(\tilde{\mathbf{f}}) \right) = 0 \tag{13.47}$$

where $\boldsymbol{\alpha} = \mathbf{A}^T \mathbf{1}$, and after simple algebraic manipulations, we get the fixed-point equation

$$\tilde{\mathbf{f}} = \frac{\tilde{\mathbf{f}}}{\boldsymbol{\alpha} + \mu \, \nabla_{\mathbf{f}} \, \mathcal{G}(\tilde{\mathbf{f}})} \circ \mathbf{A}^T \frac{\mathbf{g}}{\mathbf{A}\tilde{\mathbf{f}} + \mathbf{b}} . \tag{13.48}$$

Finally, the OSL method is obtained by applying the method of successive approximations to this fixed-point equation

$$\mathbf{f}_{k+1} = \frac{\mathbf{f}_k}{\boldsymbol{\alpha} + \mu \, \nabla_{\mathbf{f}} \, \mathcal{G}(\mathbf{f}_k)} \circ \mathbf{A}^T \frac{\mathbf{g}}{\mathbf{A}\mathbf{f}_k + \mathbf{b}} . \tag{13.49}$$

If $\boldsymbol{\alpha} + \mu \, \nabla_{\mathbf{f}} \mathcal{G}(\mathbf{f}) > 0$ on the nonnegative orthant and if the initial guess $\mathbf{f}_0$ is positive, then, as in the case of ISRA and ML-EM, one can prove by induction that all the iterates are positive. The main difficulties with the OSL method are the following:

- For several regularization functions used in practice the gradient does not satisfy conditions ensuring the positivity of the denominator in equation (13.49).
- No convergence result is given in the original paper.

The first problem is already addressed in Green's paper; since the author has in mind a function with the structure given in equation (13.44), he remarks that, if the modulus of the derivative of the function φ is bounded, then, for a sufficiently small value of the regularization parameter μ, the denominator in equation (13.49) is positive. As a particular example, the function $\varphi(t) = \ln[\cosh(t)]$ is considered; it has all the good properties and the modulus of its derivative is bounded by 1. Moreover, it is easy to remark that the function behaves as t^2 for small t and as $|t|$ for large t so that it has a behavior similar to that of the TV-smooth regularization introduced in Chapter 9.

We also remark that in the case of both Tikhonov regularization $\mathcal{G}(\mathbf{f}) = \|\mathbf{f}\|^2$ and ℓ^1-regularization in pixel space $\mathcal{G}(\mathbf{f}) = \|\mathbf{f}\|_1$ – both restricted to the nonnegative orthant – the condition $\nabla_{\mathbf{f}} \mathcal{G}(\mathbf{f}) \geq 0$ is satisfied. Indeed, in the first case $\nabla_{\mathbf{f}} \mathcal{G}(\mathbf{f}) = 2\mathbf{f}$ and in the second case $\nabla_{\mathbf{f}} \mathcal{G}(\mathbf{f}) = \mathbf{1}$. Therefore in these cases the OSL algorithm is well defined. However, the gradient is not positive and not bounded in the case of the Tikhonov regularizers $\mathcal{G}(\mathbf{f}) = \|D\mathbf{f}\|^2$, were $D\mathbf{f}$ is the modulus of the discrete gradient of $\mathbf{f}$, equation (9.28), and $\mathcal{G}(\mathbf{f}) = \|\Delta\mathbf{f}\|^2$, where $\Delta\mathbf{f}$ is the discrete Laplacian of $\mathbf{f}$. Hence the OSL method is not applicable to these cases.

The two problems mentioned above are investigated in a paper by Lange [190] where a class of regularization functions with a bounded gradient is characterized and, moreover, a proof of convergence is given, starting from the remark that the OSL iteration can also be written in the following form

$$\mathbf{f}_{k+1} = \mathbf{f}_k - \frac{\mathbf{f}_k}{\alpha + \mu \nabla_{\mathbf{f}} \mathcal{G}(\mathbf{f}_k)} \circ \nabla_{\mathbf{f}} l_B(\mathbf{f}_k; \mathbf{g}) \; , \tag{13.50}$$

so that, OSL is another example of a scaled gradient algorithm. If the conditions assuring the positivity of the scaling are satisfied, then one can look for a relaxed version of OSL by introducing a relaxation parameter β_k and performing a line search along the descent direction in order to compute a suitable value of it

$$\mathbf{f}_{k+1} = \mathbf{f}_k - \beta_k \frac{\mathbf{f}_k}{\alpha + \mu \nabla_{\mathbf{f}} \mathcal{G}(\mathbf{f}_k)} \circ \nabla_{\mathbf{f}} l_B(\mathbf{f}_k; \mathbf{g}) \; . \tag{13.51}$$

In the paper by Lange an approximate line search is proposed such that the convergence of the iterates to a solution can be proved for a strictly convex $l_B(\mathbf{f}; \mathbf{g})$. This proof can be extended to the other relaxed algorithms considered in the book [33].

In the case of emission tomography and similar problems, the OSEM method can be applied to the OSL method: in the OSEM cycle, one ML-EM iteration is replaced by one OSL iteration. The same remark applies to the iterative methods introduced in the following. In all these cases, unfortunately, convergence proofs are not available.

The main drawback of OSL for an application to arbitrary differentiable regularization is the restriction to functions $\mathcal{G}(\mathbf{f})$ with a bounded gradient. This drawback does not affect a similar method proposed by Lantéri et al. [193]. The method is already mentioned in Section 12.3 and is based on the decomposition of the gradient indicated in equation (12.53) and discussed in that section, where U, V are suitable nonnegative arrays. An example of decomposition is also given in that section. In the case of Poisson noise, the method was successfully applied both to 3D fluorescence microscopy [297] and to astronomy [235], in the latter case with an acceleration provided by the SGP approach. The method can be derived heuristically in the usual way. If we insert equation (12.53) into the first KKT condition (13.47) we obtain

$$\tilde{\mathbf{f}} \circ \left(\alpha - \mathbf{A}^T \frac{\mathbf{g}}{\mathbf{A}\tilde{\mathbf{f}} + \mathbf{b}} - \mu \, U(\tilde{\mathbf{f}}) + \mu V(\tilde{\mathbf{f}}) \right) = 0 \; , \tag{13.52}$$

and from this condition, by simple algebraic manipulations, we derive the fixed-point equation

$$\tilde{\mathbf{f}} = \frac{\tilde{\mathbf{f}}}{\boldsymbol{\alpha} + \mu\, V(\tilde{\mathbf{f}})} \circ \left\{ \mathbf{A}^T \frac{\mathbf{g}}{\mathbf{A}\tilde{\mathbf{f}} + \mathbf{b}} + \mu\, U(\tilde{\mathbf{f}}) \right\} . \tag{13.53}$$

With a procedure which is now familiar, we obtain the iteration

$$\mathbf{f}_{k+1} = \frac{\mathbf{f}_k}{\boldsymbol{\alpha} + \mu\, V(\mathbf{f}_k)} \circ \left\{ \mathbf{A}^T \frac{\mathbf{g}}{\mathbf{A}\mathbf{f}_k + \mathbf{b}} + \mu\, U(\mathbf{f}_k) \right\} . \tag{13.54}$$

This iteration is slightly more complex than that of OSL but it is obvious that it can be used for any differentiable regularization function and that, if the iteration is initialized with a positive $\mathbf{f}_0$, then all the iterates are strictly positive. No convergence proof is available for this iterative method. However it can be transformed into a scaled gradient method, as in the case of the previous algorithms, if we remark that it can be written in the following form

$$\mathbf{f}_{k+1} = \mathbf{f}_k - \frac{\mathbf{f}_k}{\boldsymbol{\alpha} + \mu\, V(\mathbf{f}_k)} \circ \nabla_{\mathbf{f}}\, l_B(\mathbf{f}_k; \mathbf{g}) , \tag{13.55}$$

which leads to a relaxed form with the addition of a line search along the descent direction

$$\mathbf{f}_{k+1} = \mathbf{f}_k - \beta_k \frac{\mathbf{f}_k}{\boldsymbol{\alpha} + \mu\, V(\mathbf{f}_k)} \circ \nabla_{\mathbf{f}}\, l_B(\mathbf{f}_k; \mathbf{g}) . \tag{13.56}$$

If $l_B(\mathbf{f}; \mathbf{g})$ is strictly convex, the convergence proof given in [33] and borrowed from [190] applies also to this case. Moreover, a fast and convergent algorithm can be obtained using the scaling and the decomposition of the gradient for designing an appropriate SGP algorithm [48].

13.4.2   MAP estimates: TV regularization

Also in the case of Poisson data, edge-preserving methods for denoising and deblurring are important in all domains of application ranging from astronomy to medical imaging and microscopy. Since TV is the benchmark for these methods, most of this section is dedicated to this particular case and, as in Chapter 9, it is compared with TV-smooth regularization. In the case of TV, the problem is the minimization of the function (13.45) where now $\mathcal{G}(\mathbf{f})$ is the discrete TV regularizer given in equation (9.29). The main differences with respect to the discrete problem considered in Chapter 9 are the replacement of the least-squares function with the Poisson discrepancy and the noise corrupting the data. As concerns the choice of the regularization parameter μ, in Chapter 9 it is based on the so-called Morozov discrepancy principle which provides the solution of the ROF problem. In the case of Poisson data, we use the criterion proposed in [34] which, even if it does not have a sound theoretical basis – essentially due to the lack of a complete regularization theory for this case –, it presents an interesting analogy with Morozov's principle. Indeed, as the Morozov principle is based on the discrepancy function in the case of additive Gaussian noise, i.e. the least-squares function, this Poisson discrepancy principle is based on the Poisson discrepancy function; it is very simple because it does not require additional information on the noise; the information is already contained in the assumption of Poisson data. For a justification of the principle we refer the reader to the original paper. If we denote again as $\mathbf{f}_\mu$ the minimizer of the function (13.45) then the value of μ provided by the *Poisson discrepancy principle* is the solution of the following equation

$$\mathcal{D}(\mu) = \frac{1}{M}\, l(\mathbf{f}_\mu | \mathbf{g}) = 1 \tag{13.57}$$

FIGURE 13.6   Left panel: the image obtained with an observation time of 10 seconds. Right panel: the image obtained with an observation time of 1 second. The noise is clearly visible in this second image.

where M is the number of data. We remark that this equation is derived in the case of a large number of detected photons; however, it may already work when in each pixel the number of photons is greater than 20. In [34], it is also proved that $\mathcal{D}(\mu)$ is a non-decreasing function of μ so that, if a solution of equation (13.57) exists, it is unique. The criterion can be used both in the case of denoising and in the case of deblurring.

(A) *Image denoising*

As in Chapter 9 we use the alternating extragradient method (AEM), adapted to the case of Poisson data, for data denoising with the TV regularizer and SGP, also adapted to Poisson data, for the TV-smooth regularizer, as given in equation (9.30).

The first object we consider is similar to the first object with radial symmetry considered in Section 9.2.1. However, in the present case of Poisson noise, this particular object can have a 'physical interpretation'. Indeed, we can assume that it is a source of radiation (photons) with a background emission of intensity 10 (number of photons per second and per unit area), while the subsequent annular sources have respectively intensities 27 and 71; finally the central source, the brightest one, has intensity 161. A graphical representation of this object is quite similar to that of the object considered in Section 9.2.1. The previous values are also the mean values of the numbers of photons emitted and observed in several observations of the source with an observation time of 1 second (see Section 11.1, example (C)). The result of one of these observations is shown in the right panel of Figure 13.6 and this can be an example of *high noise*. In a second example we can consider the result achievable with an observation time of ten seconds. Then the mean values of the numbers of photons are the previous ones multiplied by 10. The result of one observation is shown in the left panel of Figure 13.6 and this can be called the case of *low noise*. The denominations *high noise* and *low noise* are justified by the representations of the horizontal radial cuts of the two images shown in Figure 13.7. In the caption of this figure we also give, for the two images, the mean values of the radiation emitted by the different regions of the source during the corresponding observation times. Moreover, in brackets, next to each value, the corresponding values of SNR$_\mathrm{dB}$, as defined in equations (2.6) and (2.5), are given. We remark that, since here the noise is Poisson, the mean values of the different regions coincide with the variance σ^2 so that the standard deviation σ is the square root of the mean value. For estimating the value of μ in the two cases, we use the Poisson discrepancy principle defined

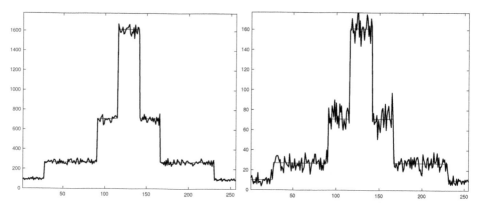

FIGURE 13.7   Horizontal radial cuts of the two noisy images. We give the mean values of the different sources (from top to down) and, in brackets, the corresponding values of the SNR in dB. Left panel: *low noise* – 1600 (16 dB), 710 (14.2 dB), 270 (12.2 dB) 100 (10 dB). Right panel: *high noise* – 160 (12 dB), 71 (9.3 dB), 27 (7.1 dB), 10 (5 db).

in equation (13.57). In Figure 13.8, we show the behavior of the Poisson discrepancy $\mathcal{D}(\mu)$ as a function of μ in the case of *low noise*; it is an increasing function of μ, as proved in [34], and it crosses the value 1 as expected. In the case of *high noise*, the corresponding curve lies below that of *low noise* so that the crossing of the value 1 occurs at a larger value of μ. We point out this result because it seems to indicate that the value of μ provided by this principle tends to zero when the noise tends to zero or, as discussed in Section 11.1, when the observation time tends to infinity.

In Figure 13.9, we show the reconstructions obtained with the values of μ provided by the Poisson discrepancy principle and indicated in the caption, where we also give the values of the SNR$_{dB}$, showing the improvement with respect to the noisy images. The results looks quite nice, even better than those obtained in the case of additive Gaussian noise, as shown in Figure 9.3. Remark that in the latter case the noise level is intermediate between the levels of low and high Poisson noise. In spite of that, in both cases the reconstruction of the flat regions is more accurate in the Poisson case than in the Gaussian case (except for the artifact in one edge in the case of *high noise*). However, since we consider only one example for comparison between Gaussian and Poisson denoising, we can not state that TV works better in the Poisson case than in the Gaussian case; we can only state that TV works well also in the Poisson case with similar artifacts. Indeed, in Figure 13.9, we can observe a loss of contrast which is more important in the *high noise* case than in the *low noise* case. In the first case we also observe a smoothing of one edge, as already remarked. In Figure 13.10, we display the corresponding images. If we apply TV-smooth we obtain reconstructions which coincide with those provided by TV if we use $\delta = 10^{-3}$, a value already used in the Gaussian case. For this reason we do not show the results obtained in this case.

As a further example we consider again the image of 'Barbara' shown in Figure 9.7. We do not attempt at a physical interpretation of what we are doing; it is only an exercise. What we do is to multiply the image of 'Barbara' by a factor such that, when perturbed by Poisson noise, its PSNR is approximately the same as for the noisy image of 'Barbara' shown in the left panel of the first line of Figure 9.8 (30 dB in that case and 30.2 dB in the present case). The noisy image is shown in the left panel of Figure 13.11 while the result of denoising is shown in the right panel. The value of μ provided by the Poisson discrepancy principle is indicated in the caption where we give also the value of the PSNR after denoising. We remark that it is the same in Poisson and in Gaussian noise, namely 32.2. We did not perform a search for a value of μ providing an improvement of the reconstruction in the Poisson

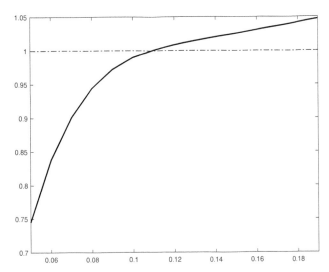

FIGURE 13.8 Behavior of the Poisson discrepancy as a function of μ in the case of *low noise*. The curve crosses the value 1 at $\mu = 0.11$.

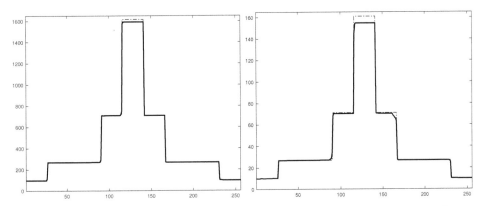

FIGURE 13.9 Horizontal radial cuts of the two denoised images obtained with $\mu = 0.11$ (*low noise*) and $\mu = 0.3$ (*high noise*). The dotted line represents the original object, highlighting a low loss of contrast. We also give the values of the SNR in dB for each region. Left panel: *low noise* – 18.41 dB, 18.52 dB, 19.62 dB, 18.57 dB. Right panel: *high noise* – 15.34 dB, 14.72 dB, 15.52 dB, 13.97 dB.

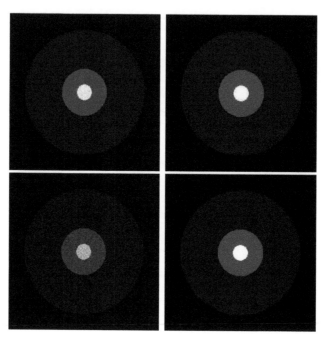

FIGURE 13.10   Comparison between noisy and denoised images; the noisy images are shown in the left column and the corresponding denoised ones in the right column. First row: *low noise*. Second row: *high noise*.

case. We observe a possible cartoon effect in the illuminated part of the tablecloth. Several small details are already visible in the noisy version and are reproduced in the denoised one.

(B) *Image deblurring*

As another illustration of the use of TV regularization in the case of Poisson data, we consider the problem of deblurring. The first example we consider is the one of the two piecewise constant objects introduced in the case of denoising and we convolve both with a Gaussian with a standard deviation of 2 pixels, the same PSF used for convolving a similar object in the case of Gaussian noise; the blurred images are perturbed with Poisson noise.

The interpretation of this numerical experiment is that we are observing the source described above for the denoising case with an instrument introducing a distortion on the path of the photons traveling from the source to the detector, distortion which is described by the Gaussian blur. As in the denoising case we consider two observations with two different observation times: 1 second in the *high noise* case and 10 seconds in the *low noise* case. Because of the additional blur we expect the reconstructions to be poorer than those obtained in the denoising case. The results are quite surprising because the reconstructions look quite good. The cuts of the blurred and noisy images along the radial horizontal line are shown in the left column of Figure 13.12, the *low noise* case in the upper panel and the *high noise* case in the lower one. The corresponding deconvolved images are shown in the right column. The values of μ are again obtained with the Poisson discrepancy principle and are indicated in the caption. Finally, in Figure 13.13, we show the corresponding and complete images. In both cases the locations of the discontinuities can be hardly estimated from the blurred and noisy images while they are correctly reproduced in the reconstructed images and in both cases. Only in the case of the region with the highest intensity the vertical lines are replaced by very steep ramps; moreover a small curvature of some edges can

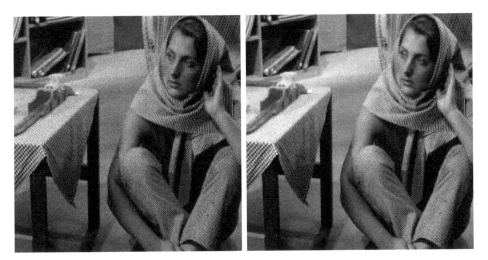

FIGURE 13.11 Comparison between the noisy image of 'Barbara' (left) and the denoised one (right), obtained with $\mu = 0.04$, the value provided by the Poisson discrepancy principle. The value of the PSNR, defined in equation (2.7), for the noisy image is 30.2 and 32.2 for the denoised one.

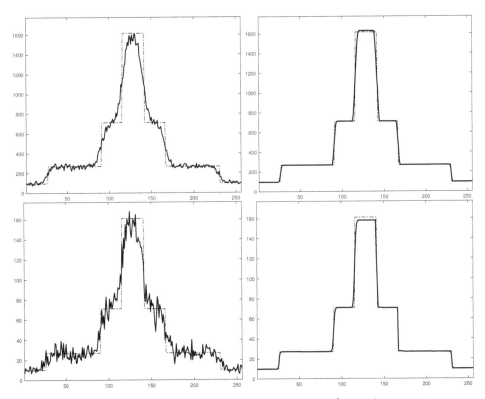

FIGURE 13.12 Left column: the blurred and noisy images, with the *low noise* case in the upper panel and the *high noise* case in the lower panel. Right column: the corresponding reconstructed images. The values of μ are respectively 0.04 and 0.05, again increasing with increasing noise.

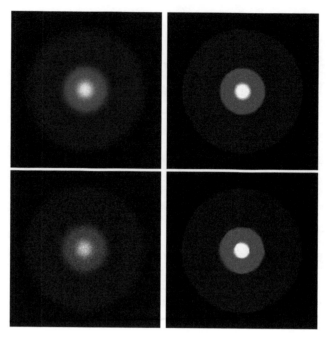

FIGURE 13.13   The images corresponding to the cuts of the previous figure, in the same order.

be observed as well as a weak loss of contrast, especially in the *high noise* case. However, in general, the quality of the reconstructions is comparable with that achieved in the denoising case, shown in Figure 13.9, and is higher than that achieved in the case of Gaussian noise, as can be derived from Figure 9.13.

It is not easy to understand this effect, which is certainly related to the different properties of the noise. Indeed, the additive Gaussian noise is independent of the signal so that the same amount of noise perturbs both bright and darker details while the Poisson noise depends on the signal and its perturbation of the bright details is lower than that of the darker ones, allowing a higher enhancement of the first ones in image reconstruction. We believe that this point could deserve further investigation.

We conclude this chapter with an application to real data in astronomy. Our example consists of infrared images of a volcanic eruption on Io, the inner one of the four Jupiter moons discovered in 1619 by Galileo Galilei and called by him *Astri Medicei*, in honor of Cosimo II de' Medici. The blurred image of a moon is a good test for edge-preserving deconvolution methods because of the presence of the limb, i.e. the curved edge of the apparent disk of the celestial body. In general, it is not very sharp due to the blurring introduced by the PSF of the telescope. Therefore it should be sharpened in order to improve the quality of the image.

Raw images were kindly provided to us by Imke de Pater, Professor of Astronomy at the University of California in Berkeley (USA). The images capture the eruption of one of Io volcanoes, Pillan, occurred on 14 August 2014 and observed with the near-infrared camera NIRC2, coupled to the adaptive optics system on the 10-meter *W. M. Keck II* telescope on Mauna Kea, Hawaii. The images are acquired using different filters corresponding to different infrared wavelengths; here we consider only those corresponding to the bands (in brackets the central wavelength) J (1.22 micron), H (1.63 micron) and K (2.19 micron). For each band, images of the star HD 161903 are also acquired to estimate the PSF. Details on acquisition and astronomical considerations can be found in [88, 89]. As it is known, Io is a highly volcanic moon of Jupiter and astronomers are interested in studying its temporal evolution.

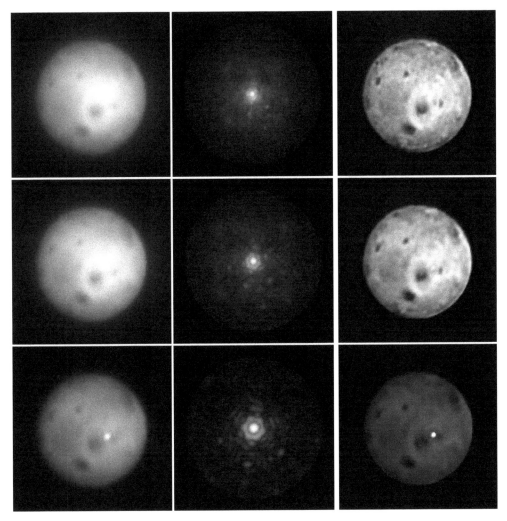

FIGURE 13.14   Left column: the original images (J, H, K band from top to down). Middle column: the corresponding PSF's (the central part of the Keck PSF's). Right column: the corresponding reconstructions.

The J, H and K band images with the corresponding PSF's (central parts of the Keck PSF) and their reconstructions with TV-smooth regularization are shown in Figure 13.14. The values of the parameters are obtained with a search of their values based on inspection of the corresponding reconstructions. The results look quite satisfactory. A very sharp reconstruction of the limb is obtained even if, as it is obvious, the PSF is affected by errors due to the processing of the raw data obtained from the observation of a star. But also the image of the interior of the moon is greatly improved and rich of details. In general, the darker spots correspond to volcanic regions not active in the instant of the observation. The eruption of Pillan, already visible in the original image in K band and clearly detected, in the reconstructed image, by the white spot close to the darker and larger region of the volcano Pele; however, we believe that it already appears as a small gray spot observable in the same location of the J band image, a spot which becomes weakly bright in the H band and finally very bright in the K band. This behavior is due to the fact that, roughly speaking, different wavelengths image different temperature distributions.

14

Conclusions

In this book we have presented a selection of the most frequently used methods for the solution of linear inverse problems. Our presentation should have made clear that we have at our disposal several methods to deal with the same inverse problem. This richness generates a quite natural question: how can we decide which method is the most appropriate for the specific problem we have to solve? In this concluding chapter we discuss possible answers to this question and possible strategies in the quest for an answer.

14.1 Is there an overall best method?

The first question we address is the one formulated in the title of this section, which will be very short. Indeed, the answer is very simple and it is negative: *there is no overall best method*. If we apply regularization methods with quadratic penalties to the reconstruction of natural images with edges, the result is unsatisfactory because the reconstruction is affected by Gibbs artifacts, bad reconstruction of the edges, etc. Similarly, if we apply methods with edge-preserving regularization to images without edges, such as images of nebulae in astronomy, the results are also unsatisfactory because of cartoon effects.

Any given regularization method can be applied, in principle, to any image of any object but the quality of the result depends on specific features of the object. This point explains the relevance, outlined several times in this book, of a priori information on the object to be reconstructed: this is the guide in the selection of the most appropriate regularization penalty.

14.2 A recap of regularization methods

In Section 4.3, a definition is given of a *regularization method in the sense of Tikhonov*, definition which can be also extended to other methods considered in the subsequent chapters. In practice, a regularization method is a method which provides an approximate solution of the basic equation $Af = g$, more precisely a family of approximate solutions depending on one parameter, the regularization parameter, for any given noisy data g; moreover, there exists a rule for selecting this parameter in such a way that, when the noise tends to zero (in an appropriate sense), then the selected value of the parameter tends to zero and the corresponding approximate solution tends to the exact solution of the problem. In the

DOI: 10.1201/9781003032755-14

literature, also the case of more than one parameter is investigated but it is not considered in this book.

Even if we repeat some already expressed concepts, it may be useful to summarize here the methods discussed in the book and satisfying these conditions. They belong to two classes:

- *Variational Methods*

  They consist in the minimization of a functional of the unknown object f with a specific structure. It is the sum of two terms: the first is a measure of the discrepancy between the '*computed data*' Af and the '*detected and noisy data*' g; the second is a term, depending only on f, which penalizes the first one and expresses properties of the unknown object.

  The first term, called in this book the '*discrepancy function*' (other names are used in the literature as 'data misfit', 'data fidelity function', etc.), can be a *metric distance* as in the case of additive Gaussian noise or a *divergence* as in the case of Poisson noise. Another example, not discussed in this book, is provided by the salt-and-pepper noise, in which case the discrepancy function is given by the ℓ^1-norm of the difference between computed and real data (see Section 11.1), hence also a metric distance.

  The second term, called in this book the '*regularization functional*' (or sometimes simply the 'penalty') has the role of restricting the domain of the functional by expressing in a quantitative way *a priori* information about the unknown object (smoothness, existence of edges, sparsity, etc.). It is multiplied by a parameter μ called the '*regularization parameter*' which quantifies the relative weight of the two terms. The minimizers of the sum of the two terms are called '*regularized solutions*' and depend both on g and on μ.

  Finally, we need a rule for the selection of μ for given data g; this rule should be such that, when the noise affecting the data tends to zero, the corresponding values of μ tend to zero and the corresponding regularized solutions converge to the true solution (or to the true generalized solution) of the problem. In the case of additive Gaussian noise such a rule can be the Morozov discrepancy principle.

- *Truncation Methods*

  This class contains the iterative methods with the '*semi-convergence*' property, such as the Landweber, Steepest Descent and Conjugate Gradient iterations, converging to least-squares solutions in the case of noise-free data; another method in this class is the ML-EM (or RL) iteration converging to minimizers of the Kullback-Leibler divergence. As we discussed, in all these methods, sensible solutions can be obtained by means of an early stopping of the iterations, so that one needs a selection rule for the number of iterations. Such rule provides a number which tends to infinity when the noise tends to zero, and the corresponding iteration converges to the true solution of the problem. In other words the 'regularization parameter' is the *inverse of the number of iterations*. The regularization properties of the three methods converging to least-squares solutions are proved in [114] (for a discussion of Landweber iteration see also Sections 5.1 and 8.4). We do not know however of any such proof in the case of ML-EM, even if numerical practice shows that, when the noise decreases (in such a case this means that the number of detected photons increases – see the discussion in Section 11.1 and the introduction of the *observation time* τ), an improvement in the reconstructed images can be obtained by increasing the number of iterations.

In this class we can also include truncated singular function expansions. Indeed, while the full expansions are not convergent in the case of noisy data, they can provide sensible solutions when taking only a suitable and finite number of terms. This number tends to infinity when the noise tends to zero and therefore the inverse of this number is the regularization parameter for this kind of methods.

In the case of variational methods, an additional difficulty arises from the requirement of computing the minimizer(s). When the methods discussed in Chapters 9 and 10 were proposed, they stimulated the use of methods in the field of *numerical optimization* for the computation of the corresponding regularized solutions. The purpose of these methods is to increase the efficiency in the computation of the minimizers and, in fact, a tremendous number of methods have been proposed. They are very technical because they require subtle concepts of numerical optimization and therefore they are beyond the scope of this book. Presumably only a few of them will survive after their testing in practical applications.

All methods of numerical optimization are iterative and must be pushed to convergence as they are methods for computing the minimizer(s) of a stable functional. Therefore they require a stopping rule which differs from the stopping rules of the truncation methods; indeed, it must guarantee that the result of the last iteration is sufficiently close to the minimizer of the functional. An additional advantage of the use of optimization methods is the possibility of introducing additional constraints in a particular variational approach: one can search for minimizers of the regularized functional while restricting its domain to a convex set $\mathcal{C}$. Since, in general, in the implementation of the method it is required to compute the projection on this set, it is clear that convex sets such that this projection is easily computable are privileged. Examples are the so-called box constraints, i.e. of an upper and/or lower bound on the solution (the most frequent case being the non-negativity of the solution), or linear constraints, for instance the requirement for nonnegative solutions to have a given ℓ^1-norm. A survey of optimization methods for image reconstruction with Poisson data can be found in [33].

Nevertheless, it is important to remark that also some of the semi-convergent iterative methods include constraints in a quite natural way or can be easily modified to implement constraints on the solution. For instance, in the case of additive Gaussian noise, the iterative method ISRA, introduced in Section 12.3, produces nonnegative iterates if the initial guess is nonnegative; moreover the method can be extended to compute nonnegative solutions of regularization methods with a differentiable penalty. Similarly, in the case of Poisson noise, the ML-EM method produces nonnegative iterates and can also be extended to the case of regularization methods with differentiable penalties. Moreover, in the case of a zero background i.e. $\mathbf{b} = 0$, each ML-EM iterate $\mathbf{f}_k$ satisfies the condition (derived from equation (13.26))

$$\sum_{m=1}^{M} (\mathbf{A}\mathbf{f}_k)_m = \sum_{m=1}^{M} g_m \tag{14.1}$$

which is a linear constraint on the solution. Moreover, if the matrix $\mathbf{A}$ satisfies the conditions

$$\sum_{m=1}^{M} A_{m,n} = 1 \, , \tag{14.2}$$

then equation (14.1) implies that the ℓ^1-norm of $\mathbf{f}_k$ coincides with the ℓ^1-norm of $\mathbf{g}$ for any k. Therefore ML-EM embeds both a non-negativity and a sparsity constraint. Finally, the Landweber method, as discussed in Sections 5.2 and 8.4, can be easily modified into the projected Landweber method by applying, at each iteration, the projection onto a convex set $\mathcal{C}$. Again the implementation is easy if the computation of this projection is easy. In such

	Tikhonov
Reconst. error	13.3%
μ_{opt}	$7.0 \ 10^{-3}$

	Landweber	Conjugate gradient	Landweber with positivity
Reconst. error	13.2%	13.6%	11.9%
k_{opt}	45	11	42

TABLE 14.1   The values of the minimum relative reconstruction error and of the corresponding optimal parameter (regularization parameter, μ_{opt}, or number of iterations, k_{opt}) for the methods listed above, when applied to the example of Figure 4.13.

a way we obtain a simple method for obtaining, by early stopping, a regularized solution which satisfies the additional constraints. Indeed, numerical practice suggests that also in this case the semi-convergence of the iterations holds true.

14.2.1   A few comparisons

A first comparison of the methods for image deconvolution introduced in Part I was performed by applying these methods to the example of Section 4.6, i.e. the reconstruction of an image blurred by uniform motion. In the case of the projected Landweber method, the constraint of non-negativity was used. The results obtained by means of the various methods have already been reported separately but, since they are scattered throughout the book, we collect them in Table 14.1 for the reader's convenience. In this table we give both the minimum reconstruction error and the value of the regularization parameter (or of the number of iterations) which is needed to obtain this minimum error. Tikhonov's method is given separately from the others because it is not an iterative method: the regularized solution is computed by means of the Fourier Transform.

A glance at this table is sufficient for realizing that all the methods provide essentially the same results. Indeed the various reconstructions look quite similar since a variation of about 1 or 2% in the reconstruction error has no visible effect. Some more specific comments are in order. First, all methods without the non-negativity constraint provide the same reconstruction error of around 13.3%. Since these methods can be applied to any linear inverse problem, they may appear as equivalent general-purpose methods. Hence the decision to choose one method or another cannot be based on the quality of the solution (they provide the same solution) but must rather rely on an analysis of the questions raised by the implementation of the various methods. For instance, if efficiency is an important issue, then the conjugate gradient method appears to be the most efficient one because, in general, it requires a rather small number of iterations. On the other hand, in such a case, the choice of the number of iterations is rather critical so that accurate criteria for the estimation of this number (stopping rules) are needed. For this reason, in some cases, a less efficient method may be convenient, for which the choice of the correct number of iterations would not be so crucial. The second comment concerns the constraint of non-negativity. In the example of Table 14.1, the object to be restored is certainly nonnegative: it represents a rather complex scene and its values are, in general, considerably larger than zero. In such a situation, the unconstrained methods already produce reconstructions which are essentially nonnegative and therefore the use of the constraint does not improve the reconstruction in a significant way (even if a small improvement is obtained). We conclude that the reconstruction of this object, obtainable with the methods discussed in the first part of the book, is not very good. Since we attempt to reconstruct a natural image with several edges, edge-preserving regularization methods could provide better results.

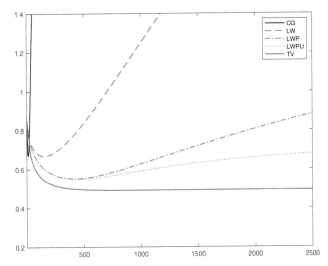

FIGURE 14.1   Behavior of the relative reconstruction error for the iterative methods used in the reconstruction of the object of Figure 14.2. The symbols have the following meaning: CG = conjugate gradient, LW = Landweber, LWP = Landweber with positivity, LWPU = Landweber with positivity and upper bound, TV = total variation.

In any case, it is important to remark that this discussion is based on the definition of *relative reconstruction error*, given in equation (4.42) for the variational methods and, similarly, in equation (5.36) for the semi-convergent iterative methods. This quantity can be computed in the case of simulations and is an attempt at quantifying the diversity or else the similarity of two solutions. As we will see in the following, we can get the same relative error for two solutions which appear visually as very different. In fact, the relative reconstruction error is but one possible *figure of merit* (FOM) among the many ones which are proposed to quantify the similarity of two functions [302]. A brief account is given in [33], Section 4.4.

Let us now consider a second example. The object, with size 512×512, takes values in the interval $[0, 255]$. It is a white inscription over a black background; it takes only two values: 0 for the background and 20 for the inscription, so that it is a binary object. It is shown in the upper left panel (a) of Figure 14.2. It is convolved with a PSF corresponding to out-of-focus blur, with COC diameter $D = 26$ pixels (see Section 2.5). Moreover, the blurred image is perturbed by white Gaussian noise with $\sigma = 1.5$; the corresponding PSNR is 21.5 dB, which is very low, meaning that the blurred image is perturbed by a high noise level. As shown in Figure 14.2(b), the inscription is not readable in its blurred and noisy image. Therefore we cannot expect a very good reconstruction. The main question is: can we read the inscription after deconvolution?

We apply all the methods used in the previous example; in addition, we also consider the projected Landweber method with lower and upper bound on the solution (i.e. 0 and 20) and the Total Variation method. We point out again that a specificity of the Landweber method is that it allows to easily implement additional *a priori* information, in this case an upper bound on the solution. For all different versions of Landweber's iteration the value of the relaxation parameter is $\tau = 1$. For the various iterative methods, the behavior of the reconstruction error, as a function of the number of iterations, is plotted in Figure 14.1. The figure confirms that all the iterative methods are semi-convergent and that the most efficient one is CG. In all cases the iterations can be stopped when the minimum reconstruction error is reached. In the figure we also plot the behavior of the reconstruction error for the iterative method used for TV regularization (see Chapter 9); in such a case the value

	Tikhonov	Total Variation
Reconst. error	68%	49%
μ_{opt}	$3.5 10^{-4}$	0.04
	Conjugate gradient	Landweber
Reconst. error	67%	67%
k_{opt}	13	156
	Landweber with pos.	Landweber with pos. and upper bound
Reconst. error	55%	55%
k_{opt}	423	452

TABLE 14.2   Relative reconstruction errors in the example of Figure 14.2.

of μ is derived from Morozov's discrepancy principle and corresponds to the solution of the ROF problem. The plot exhibits a convergent behavior; indeed, for a given value of μ, the iterations must be pushed to convergence because one needs the regularized solution corresponding to that value of μ. In the case of Tikhonov's method, no iteration is required because the solution can be computed directly by means of the FFT. Even if we estimated the value of μ by means of the discrepancy principle, we would like to recall that the reconstruction error as a function of μ exhibits a semi-convergent behavior both in the case of Tikhonov's and of TV regularization: the reconstruction error starts from a very large value for $\mu = 0$, first decreases when increasing μ, reaches a minimum value and after that increases again. We also remark that it could be possible to consider constrained minimization both of the Tikhonov and of the TV functional. In the Tikhonov case it could be possible to use a modified version of the projected Landweber iteration while in the TV case refined optimization methods would be required.

In Table 14.2, we report the values of the reconstruction errors obtained for this second example and, as expected, they are very large. As in the previous example, all methods without the constraint of non-negativity are equivalent, while improvement is obtained with the two constrained Landweber methods and with TV. The reconstruction errors of the two constrained Landweber cases coincide and, as we will see, also the reconstructions are practically identical; clearly, in this example the constraint on the upper bound is not very active: it slightly increases the number of iterations without obtaining a significant improvement of the solution. On the other hand, the very interesting feature of TV is that the non-negativity constraint is not required and therefore, in this example, it is definitely superior to the other unconstrained methods and provides an improvement also with respect to the constrained ones.

Some of the reconstructions we have obtained are shown in Figure 14.2; in this figure the images are represented using a linear grayscale with *black* corresponding to the minimum value of the image and *white* to its maximum value. Since all the unconstrained iterative methods used in the first example provide the same reconstruction error also in this second example, in Figure 14.2(c), we only show the result given by the Landweber method. We observe that, even if the reconstruction error is very large and the image is affected by strong artifacts, the inscription is readable and therefore, from the point of view of the interpretation of the image, the unconstrained methods are satisfactory. If this is the unique goal of the reconstruction, then Tikhonov regularization is the most efficient one since it does not require the use of iterations but only of FFT. Therefore the other methods and in particular TV only improve the reconstructed image from the visual point of view. Obviously this property could be very important in particular applications. We can add that in cases of lower noise, the reconstructions by projected Landweber methods and TV are very nice. The properties of the different reconstructions are further highlighted and explained by Figure 14.3 where we show their cuts along a piece of a line crossing the left leg of the letter **N** in the word **IN** of the original inscription. The effect of the unconstrained methods

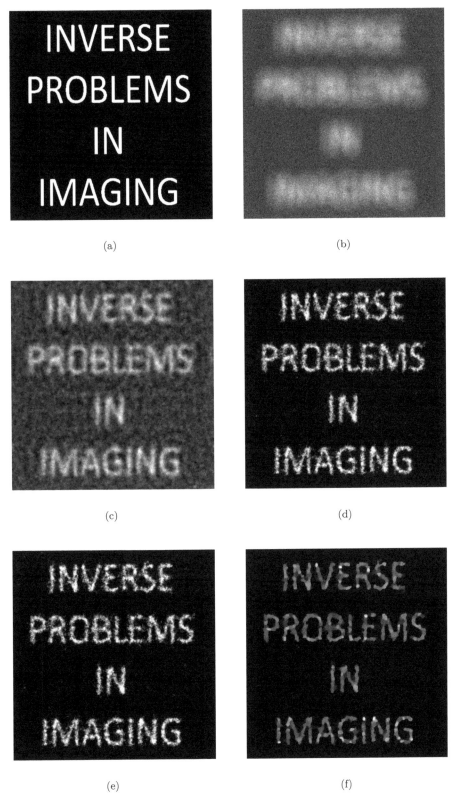

FIGURE 14.2 Comparison of (a) the object; (b) the blurred and noisy image (out-of focus blur); (c) the reconstruction provided by the Landweber method; (d) by the Landweber method with non-negativity; (e) by the Landweber method with non-negativity and upper bound; (f) by the TV method.

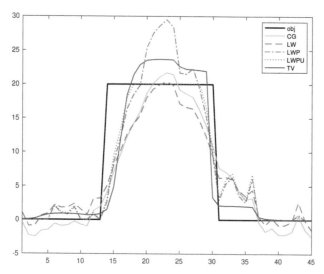

FIGURE 14.3 Plot of the cut along a piece of horizontal line of the original object and of its reconstructions shown in Figure 14.2. The black full line corresponds to the profile of the object. The symbols have the following meaning: CG = conjugate gradient; LW = Landweber; LWP = Landweber with positivity, LWPU = Landweber with positivity and upper bound; TV = total variation.

(Landweber and CG) is a sort of shrinking of the letter with negative side lobes, especially relevant in the case of CG (a price to be payed to the fast semi-convergence of the method); the constrained LWP reconstruction takes values greater than 20 in the region of the letter while the effect of LWPU is to lower these values to 20 without changing significantly the rest of the reconstruction. For this reason a significant improvement is not obtained. Finally, the effect of TV is a slight broadening of the reconstruction of the support of the letter and a clear detection of the discontinuity to the right (even if its position is not completely correct), which is flanked by small positive artifacts. This behavior can explain the improvement in the reconstruction error

We now consider a third example similar to the previous one but perturbed by Poisson noise. In such a case we introduce a background, otherwise the noise on a zero background is zero. Moreover, in microscopy and astronomy a background exists due to physical phenomena, as explained in Chapter 2. Also in medical imaging a background may exist due to scattering effects; sometimes a small 'artificial' background is even introduced in order to exclude possible zeros in the denominator of the ML-EM or OSEM method. Hence, we consider a background corresponding to an average value of 5 photons with the addition of the same inscription as in the previous example, with an average emission of 20 photons. The result is again a binary object which takes the two values 5 and 25. It is shown in Figure 14.4(a). The object is convolved with the same PSF as in the previous example and the result is perturbed with Poisson noise. The PSNR is now 21.6 dB since we adjusted the variance of the Gaussian noise of the previous example in order to get approximately the same PSNR as for the Poisson case. The result is shown in Figure 14.4(b) and clearly also in this case the inscription is not readable.

For deconvolution we used two methods presented in Chapter 13: ML-EM and TV for the regularization of the Kullback-Leibler divergence (we call it in short the TV-Poisson method). The first is a semi-convergent iterative method and also in this case it exhibits this behavior. The other method is a variational one and the value of μ is selected by means of the Poisson discrepancy principle given in equation (13.57); we did not search for a possible alternative value of μ, possibly providing a better reconstruction. Moreover, in order to see the action

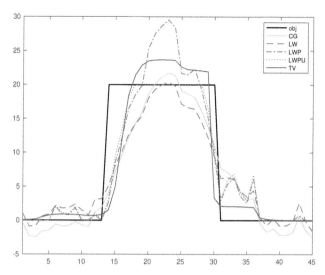

FIGURE 14.3   Plot of the cut along a piece of horizontal line of the original object and of its reconstructions shown in Figure 14.2. The black full line corresponds to the profile of the object. The symbols have the following meaning: CG = conjugate gradient; LW = Landweber; LWP = Landweber with positivity, LWPU = Landweber with positivity and upper bound; TV = total variation.

(Landweber and CG) is a sort of shrinking of the letter with negative side lobes, especially relevant in the case of CG (a price to be payed to the fast semi-convergence of the method); the constrained LWP reconstruction takes values greater than 20 in the region of the letter while the effect of LWPU is to lower these values to 20 without changing significantly the rest of the reconstruction. For this reason a significant improvement is not obtained. Finally, the effect of TV is a slight broadening of the reconstruction of the support of the letter and a clear detection of the discontinuity to the right (even if its position is not completely correct), which is flanked by small positive artifacts. This behavior can explain the improvement in the reconstruction error

We now consider a third example similar to the previous one but perturbed by Poisson noise. In such a case we introduce a background, otherwise the noise on a zero background is zero. Moreover, in microscopy and astronomy a background exists due to physical phenomena, as explained in Chapter 2. Also in medical imaging a background may exist due to scattering effects; sometimes a small 'artificial' background is even introduced in order to exclude possible zeros in the denominator of the ML-EM or OSEM method. Hence, we consider a background corresponding to an average value of 5 photons with the addition of the same inscription as in the previous example, with an average emission of 20 photons. The result is again a binary object which takes the two values 5 and 25. It is shown in Figure 14.4(a). The object is convolved with the same PSF as in the previous example and the result is perturbed with Poisson noise. The PSNR is now 21.6 dB since we adjusted the variance of the Gaussian noise of the previous example in order to get approximately the same PSNR as for the Poisson case. The result is shown in Figure 14.4(b) and clearly also in this case the inscription is not readable.

For deconvolution we used two methods presented in Chapter 13: ML-EM and TV for the regularization of the Kullback-Leibler divergence (we call it in short the TV-Poisson method). The first is a semi-convergent iterative method and also in this case it exhibits this behavior. The other method is a variational one and the value of μ is selected by means of the Poisson discrepancy principle given in equation (13.57); we did not search for a possible alternative value of μ, possibly providing a better reconstruction. Moreover, in order to see the action

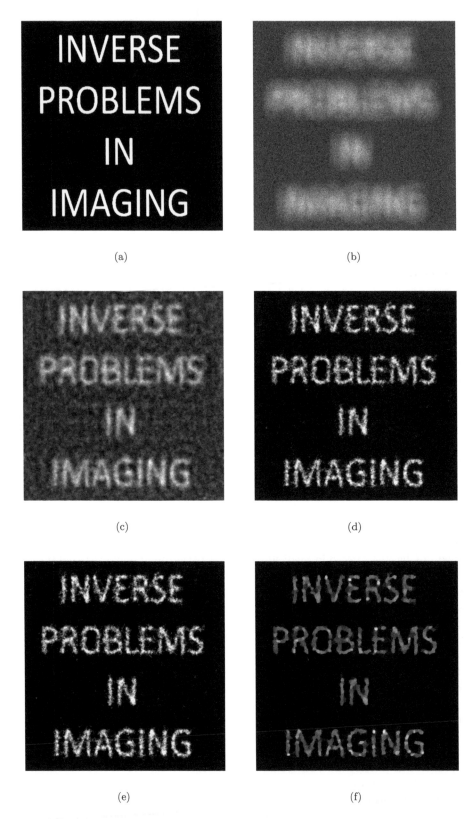

FIGURE 14.2 Comparison of (a) the object; (b) the blurred and noisy image (out-of focus blur); (c) the reconstruction provided by the Landweber method; (d) by the Landweber method with non-negativity; (e) by the Landweber method with non-negativity and upper bound; (f) by the TV method.

e	TV-Poisson	TV-Gauss
Reconst. error	65%	67%
μ_{opt}	0.01	0.1
	ML-EM	Landweber
Reconst. error	69%	72%
k_{opt}	163	80
	Landweber with pos.	Landweber with pos. and upper bound
Reconst. error	69%	69%
k_{opt}	174	174

TABLE 14.3   Relative reconstruction errors in the case of Poisson data.

on Poisson data of methods designed for the case of additive Gaussian noise, we used the previous three Landweber versions and also TV-Gauss, the method investigated in Chapter 9. For these four methods the data are obtained by subtraction of the average background of 5 so that the resulting noisy and blurred image contains also negative values. In all cases the methods provide reconstructions of the original inscription which takes the values 0 and 20.

In Table 14.3, we report the values of the reconstruction errors obtained with the six methods we considered. In Figure 14.4, we show some of the reconstructions; we use the same gray level scale as in Figure 14.2.

Also in this case the reconstruction errors are very large, even larger than those obtained in the previous case. The minimum error is provided by TV-Poisson with 65%; on the other hand, ML-EM provides an error which is smaller than that of the Landweber method but comparable with that of the two constrained Landweber methods. All methods provide a readable reconstruction of the inscription but it is difficult to decide which method is the best from the visual point of view.

For a further insight into the different reconstructions of Figure 14.4, in Figure 14.5 we plot their cuts along the same piece of the same line considered in Figure 14.3. We observe that the ML-EM method provides a shrinking of the letter accompanied by a peak much higher than 20 and this effect is presumably responsible for the white spots visible on the reconstructed letters. Landweber with positivity produces a similar shrinking with a lower peak but surrounded by small artifacts with an overall effect of producing an image which still looks a bit noisy. As concerns the two TV methods, they have different effects: while TV-Poisson produces a broadening of the reconstruction with at least one edge reconstructed correctly (but with a small error in the position), the TV-Gauss produces again a shrinking of the reconstruction.

A weak point in the previous analysis could be that, for stopping the iteration, we used the minimum RMSE. It is possible that this figure of merit is not the most appropriate one in the case of Poisson noise but we do not know of a generally accepted substitute. As already remarked, by looking at the reconstructions, it is difficult to decide which one is the best. We can only say that the TV reconstructions are 'cleaner' than the reconstructions provided by the iterative methods; in particular that provided by Landweber's method with non-negativity is still affected by noise artifacts while that provided by ML-EM looks affected by a loss of contrast.

From this unique example it is difficult to decide whether the methods designed for additive Gaussian noise can be successfully used also in the case of Poisson noise. To this purpose we recall that, as observed in previous chapters, the Gaussian approximation to the Poisson distribution is used by several authors with the purpose of replacing the Kullback-Leibler divergence with a weighted least-squares functional. The main reason for using this approximation is that in such a way one has to deal with a more familiar problem and one can use, with simple modifications, all the methods developed for the regularization of least-squares

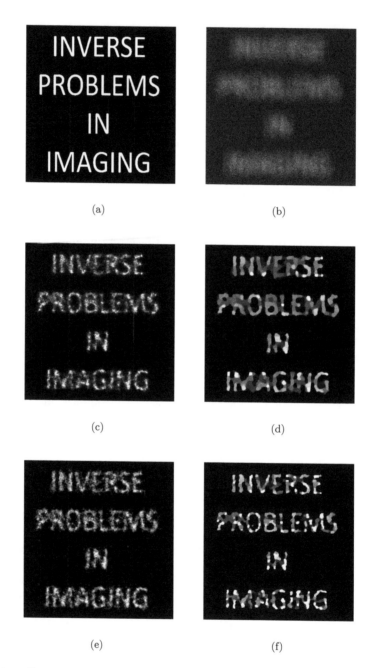

FIGURE 14.4 Comparison of: (a) the object; (b) the image; (c) ML-EM reconstruction (163 iterations and 69% relative RMSE); (d) reconstruction by TV-Poisson ($\mu = 0.01$ and 65% error); (e) reconstruction by the Landweber method with positivity and upper bound (174 iterations and 69% error); (f) reconstruction by TV-Gauss ($\mu = 0.1$ and 67% error).

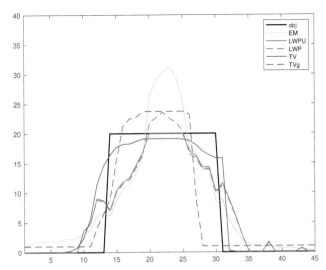

FIGURE 14.5 Plot of the cuts, along a piece of the same horizontal line as in Figure 14.3, of the original and reconstructed objects shown in Figure 14.4. The symbols have the following meaning: EM = ML-EM; LWP = Landweber with positivity; LWPU = Landweber with positivity and upper bound; TV = TV-Poisson; TVg = TV-Gauss.

problems. Another reason for considering least-squares problems in the case of Poisson data is based on the so-called *Anscombe transform* [8] which provides a homogenization of the variance over the image domain and is frequently used in astronomy [266]. On the other hand methods such as OSEM are already used in medical imaging [25]. As a conclusion of this short discussion, we should say that the Gaussian approximation leading to weighted least-squares problems can be certainly used in the case of images with large numbers of photons per pixel but, in the case of small numbers, hence very noisy images, the methods derived from the regularization of the maximum likelihood approach, hence of the Kulback-Leibler divergence, should provide superior results. Observations with small numbers of photons are frequent in microscopy, when applied to the imaging of living cells, and in medical imaging (for instance SPECT and PET).

14.3 In praise of simulation

In the usual problems of mathematical physics and applied mathematics such as problems of diffusion, wave propagation and scattering, one has, in general, to solve well-posed boundary value or initial value problems for partial differential equations. Since the solution exists, is unique and depends continuously on the data, the main problem is to find accurate discretization schemes and efficient algorithms for computing accurate approximations of the solution.

In the case of inverse problems, however, the situation is quite different. When formulated in terms of functions depending on continuous variables, an inverse problem is, in general, ill-posed so that its solution does not exist in the case of noisy data, and noise is an unavoidable perturbation of the data in practical problems. Of course, one can consider a regularized version of the problem as provided e.g. by quadratic Tikhonov's regularization, by Total Variation regularization or by sparsity-enforcing regularization; in such a case, one can look again for accurate discretization schemes and efficient numerical methods for computing the solution. Such an effort, however, may not be very useful in those cases where the regularized solutions are intrinsically poor approximations of the true one, because of large noise and/or small information content of the data. The last situation occurs, for instance, when the effective band of the imaging system is small so that no information is available on the high-frequency content of the object (implying a loss of information on small details), or when the number of significant singular values of the imaging operator is also small (here, significant means not too small with respect to the largest singular value). In these cases, one does not need very accurate numerical methods and a rough discretization of the problem can be sufficient.

Another typical feature of inverse problems, already discussed in the previous section, is that we have at our disposal many different approximations of the solution provided by the various methods proposed in the literature. As a consequence, to reconstruct different types of objects provided by the same imaging system, it may be appropriate to use different methods. An example is provided by astronomical images: different methods must be used to reconstruct diffuse objects such as a distant galaxy or to reconstruct point-like objects such as a star field over a uniform background. Moreover, since the regularization algorithms contain a free parameter, namely the regularization parameter, once the algorithm has been chosen, one still has to test the various methods proposed for setting the value of the parameter.

For all the reasons mentioned above, the use of numerical simulations is vital when one has to solve a practical inverse problem and hence to decide which method is best suitable for the given problem.

A program of numerical simulations consists of the following steps.

- Step 1: Forward problem: establish a mathematical model of the imaging system and compute the matrix which approximates the imaging operator.
- Step 2: Choice of typical representatives of the class or classes of objects to be reconstructed.
- Step 3: Generation of noise-free images by applying the imaging matrix computed in Step 1 to the objects selected in Step 2.
- Step 4: Generation of noisy images by perturbing the noise-free images by means of a realistic model of the noise due to the detection system; it is important to point out that, for each representative, it could be useful to generate a set of different noisy images using different noise realizations (an example is provided in Section 10.5.1).
- Step 5: Selection of appropriate reconstruction methods and computation of the optimal solutions (defined by a suitable estimate of the regularization parameter or of the number of iterations, in the case of semi-convergence, etc.) for each of the different realizations of the noisy image. To this purpose it could be important to select the FOM, i.e. the measure of the quality of the reconstruction, which looks most appropriate to the particular problem one is considering.

The final result of such a program of numerical simulations should be a decision about the strategy to be used in practice for each class of objects. These simulations should also indicate what kind of information can be obtained and what kind of information cannot be obtained from the reconstruction of the blurred and noisy images.

We now briefly discuss the various steps mentioned above.

Step 1 is obvious: it is the basic point for the formulation of the problem of image formation and it can be called the mathematical modeling of the forward problem. In some cases, as in X-ray tomography, this step may be rather easy while it may be very difficult in others (for instance, in the case of emission tomography). In image deconvolution this step is equivalent to the identification of the PSF by means of measurements or to its computation based on a sufficiently accurate model of the imaging system. If one has only a poor knowledge of the PSF one can attempt to improve it by means of methods for *blind deconvolution*, a term introduced by Stockham, Cannon and Ingebretsen [270] for denoting problems where both the PSF and the object are unknown. One has now at one's disposal many methods of blind deconvolution: spectral methods [270], iterative methods [14] or maximum likelihood methods [186]. The discussion of such methods is beyond the scope of this book.

When the PSF is known, some important quantities obtainable from the PSF must be estimated. The first is the effective bandwidth Ω which provides the resolution achievable by means of simple methods such as Tikhonov and CG. Moreover, the values of $\hat{K}_{max}$ and $\hat{K}_{min}$, i.e. the maximum and the minimum value of the modulus of the TF, must be computed. If the maximum value is reached at $\omega = 0$, then it may be convenient to normalize the TF (and therefore the PSF) in such a way that $\hat{K}_{max} = 1$. This is the normalization we used in all simulations presented in this book and which is used in most practical applications of deconvolution methods. From $\hat{K}_{max}$ and $\hat{K}_{min}$ we obtain the condition number which provides an estimate of the noise propagation from the noisy image to the inverse-filter solution. Finally the value of $\hat{K}_{max}$ can be used for the choice of the relaxation parameter when the Landweber or the projected Landweber method is used.

Step 2 is the true starting point of a program of numerical simulations. Indeed, the objects imaged by the imaging system under consideration are not completely unknown. We have at least some kind of *a priori* information about their qualitative features. Even if these features cannot be expressed in a quantitative mathematical form (such as a constraint, for instance), they can be used for generating typical examples of the possible objects to be imaged. These representatives or test objects can be digital or physical. By digital we mean a computer-generated test object, i.e. an array of numbers providing a representation of a possible object; by physical we mean a manufactured and perfectly known test object. While in the first case the image will be computer-generated, in the second case one can obtain a true image of the test object. It is obvious that it is not always possible to proceed in this way. It is certainly not possible in the case of astronomical images but it is possible, for instance, in the case of medical imaging, where the use of the so-called *phantoms* is very common.

The previous remarks can be clarified by means of a few examples. The first example is provided by the problem of deconvolving the HST images before the installation of the corrective optics (see Section 2.6). In order to test and compare various reconstruction methods, several different brightness distributions were generated corresponding to typical celestial objects [145] and made available to the scientific community. In Figure 14.6, we show two of these distributions: one corresponds to a cluster of 470 stars (already used in Section 5.2) while the other one corresponds to a galaxy with a simple elliptical shape and no inner structure.

Nowadays, some standard images of natural objects or of people (preferably girls, how strange...) are frequently used as test objects by different authors with the purpose of providing a comparison of their own method with those proposed by other authors. An example is the Barbara image used in this book. The reader is invited to have a look at papers on image reconstruction and from there to prepare a list of the test objects which are the most frequently used for a given problem.

Tomography is another problem where test objects are frequently used. A well-known example is the Shepp-Logan phantom [252], also used in this book, which provides a model of a head section consisting of ten ellipses with different densities. This model has been extensively used for testing the accuracy of the reconstruction algorithms in X-ray tomography because it was believed that especially the reconstruction of the human head requires great numerical accuracy and should be free from artifacts. A different test object is used for emission tomography. It is the so-called Hoffman phantom [160] which simulates cerebral blood-flow and metabolic features of the human brain. Two sections of this phantom are

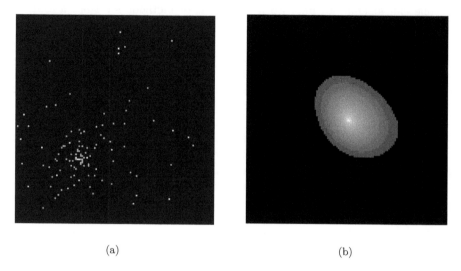

(a)   (b)

FIGURE 14.6   Two test objects used for comparing reconstruction methods in the case of HST imaging: (a) a star cluster; (b) an elliptic galaxy (images produced by AURA/STScI).

shown in Figure 14.7. This is the digital version of the phantom. A physical version has also been manufactured. It can be used for data acquisition by PET or SPECT machines.

The third step of the program is the computation of the noise-free image $g^{(0)}$, which is obtained by computing a matrix-vector multiplication or a convolution product, using the results of Step 1 and Step 2.

The fourth step is the generation of a realistic noise model to corrupt the noise-free image $g^{(0)}$. In this book only two noise models are discussed in detail and used in our simulations, namely additive Gaussian noise (with zero mean and given variance σ^2) and Poisson noise. In Section 11.1 other noise models are briefly introduced. Sometimes a noise model is not even available. For instance, both in X-ray tomography and in emission tomography, it is assumed that the noise is mainly due to fluctuations in the counting process so that it can be accurately described by a Poisson distribution. However, in the case of X-ray tomography this is the noise affecting the raw data; since it is their logarithm which is used for image reconstruction, the noise of the resulting data will be certainly different and its properties are usually not taken into account. The reconstruction problem with raw data is in fact nonlinear [191]. Poisson noise due to the counting process is also the main type of noise in the case of HST images where two other kinds of noise are also present: the so-called background noise, which is also Poisson distributed, and the read-out noise which can be described by an additive Gaussian process [259]. Therefore, the total noise is a mixture of Gaussian and Poisson noise and is discussed in Section 11.1.

In general, codes for the generation of Gaussian distributed r.v. as well as of Poisson distributed random numbers are sufficient to cover the most important cases of noise. The Gauss distribution can be obtained from the uniform distribution (which is the usual one for the generators of sequences of random numbers) by means of the transformation method, while the Poisson distribution can be obtained by means of the rejection method [236]. In the case of Gaussian additive noise, one has to choose the variance of the distribution which determines the amount of noise affecting the noise-free data (the expectation value, in general, is zero); next, one generates a sequence of random numbers according to this distribution and adds the numbers of the sequence to the various pixel values of the noise-free image. In the case of Poisson noise, one has first to normalize the noise-free image in such a way

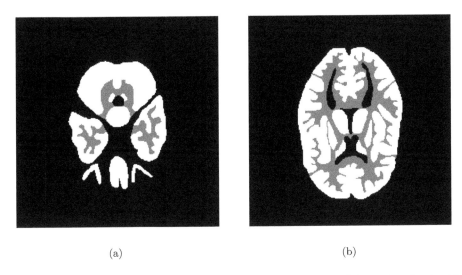

(a)   (b)

FIGURE 14.7 Two slices of the 3D Hoffman phantom which can be used for testing the algorithms of emission tomography.

that the sum of the values of the image corresponds to the expected total number of counts. Then the value of the image at any given pixel must be considered as the expectation value of a Poisson process in that pixel. The value obtained from the generator of Poisson random numbers, with that expectation value, must be taken as the value of the noisy image at that sampling point.

Once the noisy image g has been generated, the difference $g - g^{(0)}$ provides the realization of the noise term w. This image of the noise looks completely random if the noise is independent of the object; otherwise, as in the case of Poisson noise, it still keeps a 'memory' of the noise-free image. In any case the Euclidean norm of the noise term w provides the exact value of the discrepancy which is basic for estimating the regularization parameter or the number of iterations.

We point out that in many circumstances it may be useful to generate several noisy images corresponding to the same noise-free image, i.e. images which correspond to different realizations of the noise process. Indeed, by applying the same reconstruction algorithm to these images one can obtain a clear picture of the noise propagation from the image to the restored object and possibly identify parts of the object domain where the error due to the noise is more important than in others.

At this point we have all elements we need to perform the numerical experiments of Step 5.

This is essentially the story of all numerical experiments we did to illustrate this book. However, since we used the same matrix for generating the images and for reconstructing the objects, we are guilty of what some authors call an *inverse crime*.

Anyhow, let us remark that solving an inverse problem is very similar to finding the culprit of a crime. Sir Conan Doyle expressed this concept into his novel '*A Study in Scarlet*', through the words of Sherlock Holmes: '*Most people, if you describe a train of events to them, will tell you what the result would be. They can put those events together in their minds, and argue from them that something will come to pass. There are few people, however, who, if you told them a result, would be able to evolve from their own inner consciousness what the steps were which led up to that result. This power is what I mean when I talk of reasoning backwards, or analytically.*'

There is however a noticeable difference between solving a crime in a novel and in real life. It is also much more difficult to solve real inverse problems without knowing exactly the corresponding direct problem. Therefore, while pleading guilty to some inverse crimes, we nevertheless hope to be absolved from them by the reader.

References

1. F. Abramovich and B. W. Silverman. Wavelet Decomposition Approaches to Statistical Inverse Problems. *Biometrika*, 85(1):115–129, 1998.
2. M. Abramovitz and I. A. Stegun. *Handbook of Mathematical Functions*. Dover, 1965.
3. R. Acar and C. R. Vogel. Analysis of Bounded Variation Penalty Methods for Ill-Posed Problems. *Inverse Problems*, 10:1217–1229, 1994.
4. H. M. Adorf. Hubble Space Telescope Image Restoration in Its Fourth Year. *Inverse Problems*, 11:639–653, 1995.
5. H. M. Adorf, R. N. Hook, L. B. Lucy, and F. D. Murtagh. Accelerating the Richardson-Lucy Restoration Algorithm. In *Proceedings of 4th ESO/ST-ECF Data Analysis Workshop*, pages 99–103. Space Telescope Science Institute, 1992.
6. L. Alvarez, P. L. Lions, and J. M. Morel. Image Selective Smoothing and Edge Detection by Nonlinear Diffusion. II. *SIAM Journal on Numerical Analysis*, 29:845–866, 1992.
7. A. W. Anderson. Measurement of Fiber Orientation Distributions Using Angular Resolution Diffusion Imaging. *Magnetic Resonance in Medicine*, 54:1194–1206, 2005.
8. F. J. Anscombe. The Transformation of Poisson, Binomial and Negative-Binomial Data. *Biometrika*, 35:246–254, 1948.
9. A. Antoniadis and J. Fan. Regularization of Wavelet Approximations. *Journal of the American Statistical Association*, 96(455):939–967, 2001.
10. S. R. Arridge. Optical Tomography in Medical Imaging. *Inverse Problems*, 15(2):R41–R93, 1999.
11. S. R. Arridge and J. C. Schotland. Optical Tomography: Forward and Inverse Problems. *Inverse Problems*, 25(12):123010, 2009.
12. M. Aubailly, M. C. Roggemann, and T. J. Schulz. Approach for Reconstructing Anisoplanatic Adaptive Optics Images. *Applied Optics*, 46(24):6055–6063, 2007.
13. G. Aubert and J. F. Aujol. A Variational Approach to Removing Multiplicative Noise. *SIAM Journal on Applied Mathematics*, 68:925–946, 2008.
14. G. R. Ayres and J. C. Dainty. Iterative Blind Deconvolution Method and its Applications. *Optics Letters*, 13:547–549, 1988.
15. H. W. Babcock. The Possibility of Compensating Astronomical Seeing. *Publications of the Astronomical Society of the Pacific*, 65:229–236, 1953.
16. F. Bach, R. Jenatton, J. Mairal, and G. Obozinski. Structured Sparsity through Convex Optimization. *Statistical Science*, 27(4):450–468, 2012.
17. A. V. Balakrishnan. *Applied Functional Analysis*, volume 3 of *Applications of Mathematics*. Springer, 1976.
18. H. P. Baltes, editor. *Inverse Source Problems in Optics*, volume 20 of *Topics in Current Physics*. Springer, 1978.
19. L. Bar, N. Sochen, and N. Kiryati. Image Deblurring in the Presence of Salt-and-Pepper Noise. In R. Kimmel, N. A. Sochen, and J. Weickert, editors, *Scale Space and PDE Methods in Computer Vision. Scale-Space 2005*, volume 3459 of *Lecture Notes in Computer Science*, pages 107–118. Springer, 2005.
20. H. H. Barrett and K. J. Meyers. *Foundations of Image Science*. Wiley and Sons, 2003.
21. F. Bauer, S. Pereverzev, and L. Rosasco. On Regularization Algorithms in Learning Theory. *Journal of Complexity*, 23(1):52–72, 2007.
22. A. Beck. *First-Order Methods in Optimization*. SIAM, 2017.
23. A. Beck and M. Teboulle. A Fast Iterative Shrinkage-Thresholding Algorithm for Linear Inverse Problems. *SIAM Journal on Imaging Sciences*, 2(1):183–202, 2009.

24. J. M. Beckers. Adaptive Optics for Astronomy: Principles, Performance and Applications. *Annual Review of Astronomy and Astrophysics*, 31:13–62, 1993.

25. M. Beister, D. Kolditz, and W. A. Kalender. Iterative Reconstruction Methods in X-ray CT. *Physica Medica*, 28:94–108, 2012.

26. E. Beltrami. Sulle Funzioni Bilineari. *Giornale di Matematiche ad Uso degli Studenti delle Università*, 11:98–106, 1873.

27. S. Ben Hadj and L. Blanc-Féraud. Modeling and Removing Depth Variant Blur in 3D Fluorescence Microscopy. In *Proceedings IEEE International Conference ICASSP 2012*, pages 689–692, 2012.

28. F. Benvenuto, A. La Camera, C. Theys, A. Ferrari, H. Lantèri, and M. Bertero. The Study of an Iterative Method for the Reconstruction of Images Corrupted by Poisson and Gaussian Noise. *Inverse Problems*, 24:035016, 2008.

29. F. Benvenuto, A. La Camera, C. Theys, A. Ferrari, H. Lantèri, and M. Bertero. Corrigendum - the Study of an Iterative Method for the Reconstruction of Images Corrupted by Poisson and Gaussian Noise. *Inverse Problems*, 28:069502, 2012.

30. F. Benvenuto, R. Zanella, L. Zanni, and M. Bertero. Nonnegative Least-Squares Image Deblurring: Improved Gradient Projection Approaches. *Inverse Problems*, 26:025004, 2010.

31. M. Bertero. Regularization Methods for Linear Inverse Problems. In G. Talenti, editor, *Inverse Problems*, volume 1225 of *Lecture Notes in Mathematics*, pages 52–112. Springer, 1986.

32. M. Bertero. Linear Inverse and Ill-Posed Problems. In P. W. Hawkes, editor, *Series on Advances in Electronics and Electron Physics*, volume 75, pages 1–120. Academic Press, 1989.

33. M. Bertero, P. Boccacci, and V. Ruggiero. *Inverse Imaging with Poisson Data - From Cells to Galaxies*. IOP Publishing, 2018.

34. M. Bertero, P. Boccacci, G. Talenti, R. Zanella, and L. Zanni. A Discrepancy Principle for Poisson Data. *Inverse Problems*, 26:105004, 2010.

35. M. Bertero and C. De Mol. Super-Resolution by Data Inversion. In E. Wolf, editor, *Progress in Optics*, volume 36, pages 129–178. Elsevier, 1996.

36. M. Bertero, C. De Mol, and E. R. Pike. Linear Inverse Problems with Discrete Data. I: General Formulation and Singular System Analysis. *Inverse Problems*, 1:301–330, 1985.

37. M. Bertero and E. R. Pike. Signal Processing for Linear Instrumental Systems with Noise: A General Theory with Illustrations from Optical Imaging and Light Scattering Problems. In N. K. Bose and C. R. Rao, editors, *Signal Processing and its Applications*, volume 10 of *Handbook of Statistics*, pages 1–46. Elsevier, 1993.

38. M. Bertero, T. A. Poggio, and V. Torre. Ill-Posed Problems in Early Vision. *Proceedings of the IEEE*, 76(8):869–889, 1988.

39. E. Betzig and J. K. Trautman. Near-field Optics: Microscopy, Spectroscopy, and Surface Modification Beyond the Diffraction Limit. *Science*, 257(5067):189–195, 1992.

40. H. Bialy. Iterative Behandlung Linearer Funktionalgleichungen. *Archive for Rational Mechanics and Analysis*, 4:166–176, 1959.

41. R. L. Biemond, J. Lagendijk and R. M. Mersereau. Iterative Methods for Image Deblurring. *Proceedings of the IEEE*, 78(5):856–883, 1990.

42. J. Biretta. WFPC and WFPC 2 Instrumental Characteristics. In R. J. Hanish and R. L. White, editors, *The Restoration of HST Images and Spectra-II*, pages 224–235. Space Telescope Science Institute, 1994.

43. N. Bleistein, J. K. Cohen, and J. W. Stockwell. *Mathematics of Multidimensional Seismic Imaging, Migration, and Inversion*, IAM volume 54. Springer, 2001.

44. T. Blumensath and M. E. Davies. Iterative Thresholding for Sparse Approximations. *Journal of Fourier Analysis and Applications*, 14:629–654, 2008.

45. N. Bojarski. A Survey of the Physical Optics Inverse Scattering Identity. *IEEE Transactions on Antennas and Propagation*, 30:980–989, 1982.

46. T. Bonesky. Morozov's Discrepancy Principle and Tikhonov-type Functionals. *Inverse Problems*, 25(1):015015, 2009.

47. S. Bonettini and V. Ruggiero. An Alternating Extragradient Method for Total Variation Based Image Restoraton from Poisson Data. *Inverse Problems*, 27:095001, 2011.

48. S. Bonettini, R. Zanella, and L. Zanni. A Scaled Gradient Projection Method for Constrained Image Deblurring. *Inverse Problems*, 25:015002, 2008.

49. L. Borcea. Electrical Impedance Tomography. *Inverse Problems*, 18(6):R99–R136, 2002.

50. R. N. Bracewell. *The Fourier Transform and its Applications*. McGraw-Hill, 1986.

51. G. J. Brakenhoff, H. T. M. van der Voort, E. A. van Spronsen, W. A. M. Linnemans, and N. Nanninga. Three-dimensional Chromatin Distribution in Neuroblastoma Nuclei Shown by Confocal Scanning Laser Microscopy. *Nature*, 317:748–749, 1985.

52. K. Bredies, K. Kunish, and T. Pock. Total Generalized Variation. *SIAM Journal on Imaging Sciences*, 3:492–526, 2010.

53. L. M. Briceño-Arias, P. L. Combettes, J. C. Pesquet, and N. Pustelnik. Proximal Algorithms for Multicomponent Image Recovery Problems. *Journal of Mathematical Imaging and Vision*, 41(1-2):3–22, 2011.

54. W. L. Briggs and V. E. Henson. *The DFT: An Owner's Manual for the Discrete Fourier Transform*. SIAM, 1995.

55. A. Buades, B. Coll, and J. M. Morel. A Review of Image Denoising Algorithms, with a New One. *Multiscale Modelling and Simulation*, 4:490–530, 2005.

56. M. Burger and S. Osher. A Guide to the TV Zoo. In M. Burger and S. Osher, editors, *Level Set and PDE Based Reconstruction Methods in Imaging*, volume 2090 of *Lecture Notes in Mathematics*, pages 1–70. Springer, 2013.

57. P. T. Callaghan. *Principles of Nuclear Magnetic Resonance Microscopy*. Clarendon Press, 1993.

58. E. J. Candès, J. Romberg, and T. Tao. Stable Signal Recovery from Incomplete and Inaccurate Measurements. *Communications on Pure and Applied Mathematics*, 59(8):1207–1223, 2006.

59. E. J. Candès and T. Tao. Decoding by Linear Programming. *IEEE Transactions on Information Theory*, 51(12):4203–4215, 2005.

60. A. Caponnetto and M. Bertero. Tomography with a Finite Set of Projections: Singular Value Decomposition and Resolution. *Inverse Problems*, 13:1191–1205, 1997.

61. K. Chadan and P. C. Sabatier. *Inverse Problems in Quantum Scattering Theory*. Springer, 2nd edition, 2011.

62. A. Chambolle, V. Caselles, D. Cremers, M. Novaga, and T. Pock. An Introduction to Total Variation for Image Analysis. In M. Fornasier, editor, *Theoretical Foundations and Numerical Methods for Sparse Recovery*, volume 9 of *Radon Series on Computational and Applied Mathematics*, pages 263–340. de Gruyter, Berlin, 2010.

63. A. Chambolle, R. A. DeVore, N. Y. Lee, and B. J. Lucier. Nonlinear Wavelet Image Processing: Variational Problems, Compression, and Noise Removal Through Wavelet Shrinkage. *IEEE Transactions on Image Processing*, 7(3):319–335, 1998.

64. A. Chambolle and P. L. Lions. Image Recovery via Total Variation Minimization and Related Problems. *Numerische Mathematik*, 76:167–188, 1997.

65. A. Chambolle and T. Pock. A First-Order Primal-Dual Algorithm for Convex Problems with Applications to Imaging. *Journal of Mathematical Imaging and Vision*, 40:120–145, 2010.

66. T. F. Chan and J. Shen. *Image Processing and Analysis*. SIAM, 2005.

67. P. Charbonnier, L. Blanc-Féraud, G. Aubert, and M. Barlaud. Deterministic Edge-preserving Regularization in Computed Imaging. *IEEE Transactions on Image Processing*, 6:298–311, 1997.

68. M. Cheney, D. Isaacson, and J. C. Newell. Electrical Impedance Tomography. *SIAM Review*, 41(1):85–101, 1999.

69. R. R. Coifman and D. L. Donoho. *Translation-Invariant De-noising*, pages 125–150. Springer-Verlag, 1995.

70. L. Colin, editor. Mathematics of Profile Inversion. In *Proceedings of a Workshop held at Ames Research Center, Moffett Field, California, July 12-16, 1971*. NASA Technical Memorandum, TMX-62-150, 1972.

71. P. Colli Franzone, B. Taccardi, and C. Viganotti. An Approach to Inverse Calculation of Epicardial Potentials from Body Surface Maps. *Advances in Cardiology*, 21:167–170, 1977.

72. D. L. Colton and R. Kress. *Integral Equation Methods in Scattering Theory*. Pure and Applied Mathematics. Wiley, 1983.

73. D. L. Colton and R. Kress. *Inverse Acoustic and Electromagnetic Scattering Theory*, volume 93 of *Applied Mathematical Sciences*. Springer, 1992.

74. P. L. Combettes and V. R. Wajs. Signal Recovery by Proximal Forward-Backward Splitting. *Multiscale Modeling & Simulation*, 4(4):1168–1200, 2005.

75. R. Courant and D. Hilbert. *Methods of Mathematical Physics*, volume 1. Wiley, 1989.

76. I. J. D. Craig and J. C. Brown. *Inverse Problems in Astronomy*. Adam Hilger, 1986.

77. P. Craven and G. Wahba. Smoothing Noisy Data with Spline Functions. *Numerische Mathematik*, 31:377–403, 1978.

78. M. E. Daube-Witherspoon and G. Muehllehner. An Iterative Image Space Reconstruction Algorithm Suitable for Volume ECT. *IEEE Transactions on Medical Imaging*, 5:61–66, 1986.

79. I. Daubechies. *Ten Lectures on Wavelets*. SIAM, 1992.

80. I. Daubechies, M. Defrise, and C. De Mol. An Iterative Thresholding Algorithm for Linear Inverse Problems with a Sparsity Constraint. *Communications on Pure and Applied Mathematics*, 57(11):1413–1457, 2004.

81. I. Daubechies, M. Defrise, and C. De Mol. Sparsity-enforcing Regularisation and ISTA Revisited. *Inverse Problems*, 32(10):104001, 2016.

82. I. Daubechies and G. Teschke. Variational Image Restoration by Means of Wavelets: Simultaneous Decomposition, Deblurring, and Denoising. *Applied and Computational Harmonic Analysis*, 19(1):1–16, 2005.

83. M. E. Davison. A Singular Value Decomposition for the Radon Transform in n-Dimensional Euclidean Space. *Numerical Functional Analysis and Optimization*, 3:321–340, 1981.

84. M. E. Davison and F. A. Grunbaum. Tomographic Reconstructions with Arbitrary Directions. *Communications on Pure and Applied Mathematics*, 34:77–119, 1981.

85. C. De Mol, E. De Vito, and L. Rosasco. Elastic-net Regularization in Learning Theory. *Journal of Complexity*, 25(2):201–230, 2009.

86. C. De Mol and M. Defrise. A Note on Wavelet-based Inversion Algorithms. In M. Z. Nashed and O. Scherzer, editors, *Inverse Problems, Image Analysis, and Medical Imaging*, volume 313 of *Contemporary Mathematics*, pages 85–96. American Mathematical Society, 2002.

87. C. De Mol and M. Defrise. Inverse Imaging with Mixed Penalties. In *Proceedings URSI EMTS 2004*, pages 798–800. Ed. PLUS, 2004.

88. I. de Pater. Time Evolution of Io's Volcanoes Pele and Pillan from 1996 - 2015, as Derived from GALILEO, NIMS, Keck, Gemini, IRTF, and LBTI Observations. In *AAS/Division for Planetary Sciences Meeting Abstracts*, volume 47, page 409.06, nov 2015.

89. I. de Pater, C. Laver, A. G. Davies, K. de Kleer, D. A. Williams, R. R. Howell, J. A. Rathbun, and J. R. Spencer. Io: Eruptions at Pillan, and the Time Evolution of Pele and Pillan from 1996 to 2015. *Icarus*, 264:198–212, 2016.

90. A. R. De Pierro. On the Convergence of the Iterative Image Space Reconstruction Algorithm for Volume ECT. *IEEE Transactions on Medical Imaging*, 6:174–175, 1987.

91. A. R. De Pierro. On the Relation Between the ISRA and the EM Algorithm for Positron Emission Tomography. *IEEE Transactions on Medical Imaging*, 12(2):328–333, 1993.

92. A. R. De Pierro. A Modified Expectation Maximization Algorithm for Penalized Likelihood Estimation in Emission Tomography. *IEEE Transactions on Medical Imaging*, 14(1):132–137, 1995.

93. G. de Villiers and E. R. Pike. *The Limits of Resolution*. CRC Press, 2019.

94. E. De Vito, L. Rosasco, A. Caponnetto, U. De Giovannini, and F. Odone. Learning from Examples as an Inverse Problem. *Journal of Machine Learning Research*, 6(30):883–904, 2005.

95. M. Defrise and C. De Mol. A Note on Stopping Rules for Iterative Regularization Methods and Filtered SVD. In P. C. Sabatier, editor, *Inverse Problems: An Interdisciplinary Study*, volume Supplement 19 of *Advances in Electronics and Electron Physics*, pages 261–258. Academic Press, 1987.

96. M. Defrise, P. E. Kinahan, and C. J. Michel. Image Reconstruction Algorithms in PET. In D. L. Bailey et al., editors, *Positron Emission Tomography*, pages 63–91. Springer, 2004.

97. M. Defrise, C. Vanhove, and X. Liu. An Algorithm for Total Variation Regularization in High-dimensional Linear Problems. *Inverse Problems*, 27(6):065002, 2011.

98. F. Dell'Acqua, G. Rizzo, F. Scifo, R. A. Clarke, G. Scotti, and F. Fazio. A Model-Based Deconvolution Approach to Solve Fiber Crossing in Diffusion-Weighted MR Imaging. *IEEE Transactions on Biomedical Engineering*, 54:462–472, 2007.

99. A. P. Dempster, N. M. Laird, and D. B. Rubin. Maximum Likelihood for Incomplete Data via the EM Algorithm. *Journal of Royal Statistical Society*, B 39:1–38, 1977.

100. L. Denis, E. Thiébaut, and F. Soulez. Fast Model of Space-Variant Blurring and its Application to Deconvolution in Astronomy. In *Proceedings 18th IEEE Int. Conf. on Image Proc.*, pages 2817–2820, 2011.

101. A. J. Devaney. Inverse-Scattering Theory within the Rytov Approximation. *Optics Letters*, 6:374–376, 1981.

102. A. Diaspro, editor. *Confocal and Two-Photon Microscopy: Foundations, Applications, and Advances*. Wiley-Liss, New York, 2002.

103. J. T. Dobbins III and D. J. Godfrey. Digital X-Ray Tomosynthesis: Current State of the Art and Clinical Potential. *Physics in Medicine and Biology*, 48:R65–R106, 2003.

104. D. L. Donoho. De-noising by Soft-Thresholding. *IEEE Transactions on Information Theory*, 41(3):613–627, 1995.

105. D. L. Donoho. Nonlinear Solution of Linear Inverse Problems by Wavelet-Vaguelette Decomposition. *Applied and Computational Harmonic Analysis*, 2(2):101–126, 1995.

106. D. L. Donoho. Compressed Sensing. *IEEE Transactions on Information Theory*, 52(4):1289–1306, 2006.

107. D. L. Donoho. From Blackboard to Bedside. *Notices of the AMS*, 65(1):40–44, 2018.

108. D. L. Donoho and I. M. Johnstone. Ideal Spatial Adaptation by Wavelet Shrinkage. *Biometrika*, 81(3):425–455, 1994.

109. D. L. Donoho and B. F. Logan. Signal Recovery and the Large Sieve. *SIAM Journal on Applied Mathematics*, 52(2):577–591, 1992.

110. B. Efron, T. Hastie, I. Johnstone, and R. Tibshirani. Least Angle Regression. *The Annals of Statistics*, 32(2):407–499, 2004.

111. B. Eicke. Iterative Methods for Convexly Constrained Ill-Posed Problems in Hilbert Space. *Numerical Functional Analysis and Optimization*, 13:413–429, 1992.

112. L. Elden. Algorithms for the Regularization of Ill-Conditioned Least Squares Problems. *BIT Numerical Mathematics*, 17:134–145, 1977.

113. H. W. Engl and W. Grever. Using the L–curve for Determining Optimal Regularization Parameters. *Numerische Mathematik*, 69:25–31, 1994.

114. H. W. Engl, M. Hanke, and A. Neubauer. *Regularization of Inverse Problems*. Kluwer, Dordrecht, 1996.

115. S. Esposito, A. Riccardi, L. Fini, A. T. Puglisi, E. Pinna, M. Xompero, R. Briguglio, F. Quires-Pacheco, P. Stefanini, J. C. Guerra, L. Busoni, A. Tozzi, F. Pieralli, G. Agapito, G. Brusa-Zappellini, R. Demers, J. Brynnel, C. Arcidiacono, and P. Salinari. First Light AO (FLAO) System for LBT: Final Integration, Acceptance Test in Europe, and Preliminary On-Sky Commissioning Results. In *SPIE Astronomical Telescope+Instrumentation*, volume 7736, pages 107–118, 2010.

116. T. Evgeniou, T. Poggio, M. Pontil, and A. Verri. Regularization and Statistical Learning Theory for Data Analysis. *Computational Statistics and Data Analysis*, 38(4):421–432, 2002.

117. T. Evgeniou, M. Pontil, and T. A. Poggio. Regularization Networks and Support Vector Machines. *Advances in Computational Mathematics*, 13:1–50, 04 2000.

118. M. A. T. Figueiredo and R. D. Nowak. An EM Algorithm for Wavelet-based Image Restoration. *IEEE Transactions on Image Processing*, 12(8):906–916, 2003.

119. M. Fornasier and H. Rauhut. Recovery Algorithms for Vector-valued Data with Joint Sparsity Constraints. *SIAM Journal on Numerical Analysis*, 46(2):577–613, 2008.

120. S. Foucart and H. Rauhut. *A Mathematical Introduction to Compressive Sensing*. Birkhauser Basel, 2013.

121. I. E. Frank and J. H. Friedman. A Statistical View of Some Chemometrics Regression Tools. *Technometrics*, 35(2):109–135, 1993.

122. J. N. Franklin. Well-Posed Stochastic Extensions of Ill-Posed Linear Problems. *Journal of Mathematical Analysis and Applications*, 31:682–716, 1970.

123. V. Fridman. Methods of Successive Approximations for Fredholm Integral Equations of the First Kind. *Uspekhi Matematicheskikh Nauk*, 11:233–234, 1956.

124. B. R. Frieden. Evaluation, Design and Extrapolation Methods for Optical Signals, Based on Use of the Prolate Functions. In E. Wolf, editor, *Progress in Optics*, volume 9, pages 311–407. Elsevier, 1971.

125. J. Frikel and E. T. Quinto. Characterization and Reduction of Artifacts in Limited Angle Tomography. *Inverse Problems*, 20:125007, 2013.

126. W. J. Fu. Penalized Regressions: The Bridge Versus the Lasso. *Journal of Computational and Graphical Statistics*, 7(3):397–416, 1998.

127. S. Geman and D. Geman. Stochastic Relaxation, Gibbs distributions, and the Bayesian Restoration of Images. *IEEE Transactions on Pattern Analysis and Machine Intelligence*, 6:721–741, 1984.

128. S. Geman and D. E. McClure. Bayesian Image Analysis: An Application to Single Photon Emission Tomography. *Proceedings of the Statistical Computational Section, American Statistical Association*, pages 12–18, 1985.

129. S. Geman and D. E. McClure. Statistical Methods for Tomographic Image Reconstruction. *Bulletin of International Statistical Institute*, LII-4:5–21, 1987.

130. R. W. Gerchberg. Super-Resolution through Error Energy Reduction. *Optica Acta*, 21(9):709–720, 1974.

131. E. Gilad and J. von Hardenberg. A Fast Algorithm for Convolution Integrals with Space and Time Variant Kernels. *Journal of Computational Physics*, 216:326–336, 2006.

132. S. F. Gilyasov. Iterative Solution Methods for Inconsistent Operator Equations. *Moscow University Computational Mathematics and Cybernetics*, 3:78–84, 1977.

133. E. Giusti. *Minimal Surfaces and Functions of Bounded Variation*. Springer, 1984.

134. G. H. Golub, M. Heath, and G. Wahba. Generalized Cross-Validation as a Method for Choosing a Good Ridge Parameter. *Technometrics*, 21:215–223, 1979.

135. G. H. Golub and C. Reinsch. Singular Value Decomposition and Least Squares Solutions. *Numerische Mathematik*, 14:403–420, 1970.

136. G. H. Golub and C. F. Van Loan. *Matrix Computations*. Johns Hopkins University Press, 1989.

137. J. W. Goodman. *Introduction to Fourier Optics*. McGraw-Hill, 1996.

138. P. J. Green. Bayesian Reconstructions from Emission Tomography Data using a Modified EM Algorithm. *IEEE Transactions on Medical Imaging*, 9:84–93, 1990.

139. U. Grenander. *Tutorial in Pattern Theory*. Brown University, 1984.

140. C. W. Groetsch. *The Theory of Tikhonov Regularization for Fredholm Equations of the First Kind*, volume 105 of *Research Notes in Mathematics*. Pitman, Boston, 1984.

141. C. W. Groetsch. *Inverse Problems in the Mathematical Sciences*. Springer, 1993.

142. H. Gudbjartsson and S. Patz. The Rician Distribution of Noisy MRI Data. *Magnetic Resonance in Medicine*, 34:910–914, 1995.

143. J. Hadamard. Sur les Problèmes aux Dérivées Partielles et Leur Signification Physique. *Princeton University Bulletin*, 13:49–52, 1902.

144. J. Hadamard. *Lectures on Cauchy's Problem in Linear Partial Differential Equations*. Dover Publications, 1952.

145. R. J. Hanisch. WFPC Simulation Data Sets Available. *Newsletter of STScI's Image Restoration Project*, 1:76–77, 1993.

146. M. Hanke. *A Taste of Inverse Problems*. SIAM, 2017.

147. M. Hanke and P. C. Hansen. Regularization Methods for Large-scale Problems. *Surveys on Mathematics for Industry*, 3:253–315, 1993.

148. P. C. Hansen. Analysis of Discrete Ill-Posed Problems by Means of the L-Curve. *SIAM Review*, 34(4):561–580, 1992.

149. P. C. Hansen. *Discrete Inverse Problems*. SIAM, 2010.

150. T. Hastie, R. Tibshirani, and J. Friedman. *The Elements of Statistical Learning: Data Mining, Inference, and Prediction*. Springer Series in Statistics. Springer New York, 2013.

151. T. Hastie, R. Tibshirani, and M. Wainwright. *Statistical Learning with Sparsity: The Lasso and Generalizations*. Chapman & Hall/CRC Monographs on Statistics & Applied Probability. CRC Press, 2015.

152. T. Hebert and R. Leahy. A Generalized EM Algorithm for 3-D Bayesian Reconstruction from Poisson Data using Gibbs Priors. *IEEE Transactions on Medical Imaging*, 8:194–202, 1989.

153. S. W. Hell and J. Wichmann. Breaking the Diffraction Resolution Limit by Stimulated Emission: Stimulated-Emission-Depletion Fluorescence Microscopy. *Optics Letters*, 19:780–782, 1994.

154. R. M. Henkelman. Measurement of Signal Intensities in the Presence of Noise in MR Images. *Medical Physics*, 12:232–233, 1985.

155. G. T. Herman, editor. *Image Reconstruction from Projections, Implementation and Applications*, volume 32 of *Topics in Applied Physics*. Springer, 1979.

156. G. T. Herman. *Fundamentals of Computerized Tomography: Image Reconstruction from Projections.* Springer, 2009.

157. M. R. Hestenes and E. Stiefel. Methods of Conjugate Gradient for Solving Linear Systems. *Journal of Research of the National Bureau of Standards,* 49:409–439, 1952.

158. M. Hirsch, S. Sra, B. Scholkopf, and S. Harmeling. Efficient Filter Flow for Space-Variant Multiframe Blind Deconvolution. In *Proceedings IEEE CVPR 2010,* pages 607–614, 2010.

159. A. E. Hoerl and R. W. Kennard. Ridge Regression: Biased Estimation for Nonorthogonal Problems. *Technometrics,* 12(1):55–67, 1970.

160. E. Hoffman, P. Cutler, W. Digby, and J. Mazziotta. 3-D Phantom to Simulate Cerebral Blood Flow and Metabolic Images for PET. *IEEE Transactions on Nuclear Science,* 37:616–620, 1990.

161. T. Hohage and F. Werner. Inverse Problems with Poisson Data: Statistical Regularization Theory, Applications and Algorithms. *Inverse Problems,* 32:09300, 2016.

162. T. J. Holmes and Y. H. Liu. Acceleration of Maximum-Likelihood Image Restoration for Fluorescence Microscopy. *Journal of Optical Society of America,* A-8:893–907, 1991.

163. G. N. Hounsfield. Computerized Transverse Axial Scanning (Tomography). Part I. Description of the System. *The British Journal of Radiology,* 46:1016–1022, 1973.

164. H. M. Hudson and R. S. Larkin. Accelerated Image Reconstruction using Ordered Subsets of Projection Data. *IEEE Transactions on Medical Imaging,* 13:601–609, 1994.

165. H. S. Hundal. An Alternating Projection that Does Not Converge in Norm. *Nonlinear Analysis,* 57:35–61, 2004.

166. V. Isakov. *Inverse Problems for Partial Differential Equations.* Springer, 1998.

167. A. N. Iusem. Convergence Analysis for a Multiplicatively Relaxed EM Algorithm. *Mathematical Methods in the Applied Sciences,* 14(8):573–593, 1991.

168. A. N. Iusem. A Short Convergence Proof of the EM Algorithm for a Specific Poisson Model. *Revista Brasileira de Probabilidade e Estatistica,* 6:57–67, 1992.

169. V. K. Ivanov. On Linear Problems Which Are Not Well-Posed. *Soviet Mathematics Doklady,* 145:270–272, 1962.

170. A. J. Jerri. The Shannon Sampling Theorem — Its Various Extensions and Applications: A Tutorial Review. *Proceedings of the IEEE,* 65:1565–1596, 1977.

171. F. John. Numerical Solution of the Equation of Heat Conduction for Preceding Times. *Annali di Matematica Pura ed Applicata,* 40:129–142, 1955.

172. D. K. Jones, M. A. Horsfield, and A. Simmons. Optimal Strategies for Measuring Diffusion in Anisotropic Systems by Magnetic Resonance Imaging. *Magnetic Resonance in Medicine,* 42:515–525, 1999.

173. C. Jordan. Mémoire sur les Formes Bilinéaires. *Journal de Mathématiques Pures et Appliquées,* 19:35–54, 1874.

174. M. Kac. Can One Hear the Shape of a Drum? *The American Mathematical Monthly,* 73:1–23, 1966.

175. J. Kaipio and E. Somersalo. *Statistical and Computational Inverse Problems.* Springer, Berlin, 2005.

176. A. C. Kak and M. Slaney. *Principles of Computerized Tomographic Imaging.* IEEE Press, 1988.

177. B. Kaltenbacher, A. Neubauer, and O. Scherzer. *Iterative Regularization Methods for Nonlinear Ill-Posed Problems,* volume 6 of *Radon Series on Computational and Applied Mathematics.* Walter de Gruyter, 2008.

178. W. J. Kammerer and M. Z. Nashed. Iterative Methods for Best Approximate Solutions of Linear Integral Equations of the First and Second Kinds. *Journal of Mathematical Analysis and Applications,* 40:547–573, 1972.

179. J. B. Keller. Inverse Problems. *The American Mathematical Monthly,* 83:107–118, 1976.

180. A. Kirsch. *An Introduction to the Mathematical Theory of Inverse Problems*. Springer, 2011.

181. T. A. Klar and S. W. Hell. Subdiffraction Resolution in Far-Field Fluorescence Microscopy. *Optics Letters*, pages 954–956, 1999.

182. S. Kullback and R. A. Leibler. On Information and Sufficiency. *The Annals of Mathematical Statistics*, pages 79–86, 1951.

183. M. Kunt. *Digital Signal Processing*. Artech House Communication and Electronic Defense Library. Norwood, MA : Artech House, 1986.

184. G. Kutyniok, W. Q . Lim, and R. Reisenhofer. Shearlab 3D: Faithful Digital Shearlet Transforms Based on Compactly Supported Shearlets. *ACM Transactions on Mathematical Software*, 42(1), 2016.

185. A. La Camera, L. Schreiber, E. Diolaiti, P. Boccacci, M. Bertero, M. Bellazzini, and P. Ciliegi. A Method for Space-variant Deblurring with Application to Adaptive Optics Imaging in Astronomy. *Astronomy and Astrophysics*, 579:A1, 2015.

186. R. L. Lagendijk and J. Biemond. *Iterative Identification and Restoration of Images*. Boston: Kluwer, 1991.

187. R. L. Lagendijk, J. Biemond, and D. E. Boekee. Regularized Iterative Image Restoration with Ringing Reduction. *IEEE Transactions on Acoustics Speech and Signal Processing*, 32(12):1874–1888, 1988.

188. C. Lanczos. *Linear Differential Operators*. Van Nostrand, 1961.

189. L. Landweber. An Iteration Formula for Fredholm Integral Equations of the First Kind. *American Journal of Mathematics*, 73(3):615–624, 1951.

190. K. Lange. Convergence of EM Image Reconstruction Algorithms with Gibbs Smoothing. *IEEE Transactions on Medical Imaging*, 9:439–446, 1990.

191. K. Lange and R. Carson. EM Reconstruction Algorithms for Emission and Transmission Tomography. *Journal of Computer Assisted Tomography*, 8:306–316, 1984.

192. K. Lange, D. R. Hunter, and I. Yang. Optimization Transfer Using Surrogate Objective Functions. *Journal of Computational and Graphical Statistics*, 9(1):1–20, 2000.

193. H. Lantéri, M. Roche, and C. Aime. Penalized Maximum Likelihood Image Restoration with Positivity Constraints: Multiplicative Algorithms. *Inverse Problems*, 18:1397–1419, 2002.

194. H. Lantéri, M. Roche, O. Cuevas, and C. Aime. A General Method to Devise Maximum-Likelihood Signal Restoration Multiplicative Algorithms with Nonnegativity Constraints. *Signal Processing*, 81:945–974, 2001.

195. P. Lauterbur. Image Formation by Induced Local Interaction: Examples Employing Nuclear Magnetic Resonance. *Nature*, 242:190–191, 1973.

196. D. Le Bihan, J. F. Mangin, C. Poupon, C. A. Clark, S. Pappata, N. Molko, and H. Chabriat. Diffusion Tensor Imaging: Concepts and Applications. *Journal of Magnetic Resonance Imaging*, 13:534–546, 2001.

197. M. Lindenbaum, M. Fischer, and A. M. Bruckstein. On Gabor Contribution to Image Enhancement. *Pattern Recognition*, 27:1–8, 1994.

198. J. Llacer. On the Validity of Hypothesis Testing for Feasibility of Image Reconstruction. *IEEE Transactions on Medical Imaging*, 9:226–230, 1990.

199. J. Llacer and E. Veklerov. Feasible Images and Practical Stopping Rules for Iterative Algorithms in Emission Tomography. *IEEE Transactions on Medical Imaging*, 8:186–193, 1989.

200. I. Loris, M. Bertero, C. De Mol, R. Zanella, and L. Zanni. Accelerating Gradient Projection Methods for Constrained Signal Recovery by Steplength Selection Rules. *Applied and Computational Harmonic Analysis*, 27(2):247–254, 2009.

201. I. Loris and C. Verhoeven. On a Generalization of the Iterative Soft-Thresholding Algorithm for the Case of Non-separable Penalty. *Inverse Problems*, 27(12):125007, 2011.

202. A. K. Louis. Orthogonal Function Series Expansions and the Null Space of the Radon Transform. *SIAM Journal on Mathematical Analysis*, 15:621–633, 1984.

203. L. B. Lucy. An Iterative Technique for the Rectification of Observed Distribution. *Astronomical Journal*, 79:745–754, 1974.

204. R. K. Luneburg. *Mathematical Theory of Optics*. University of California Press, 1966.

205. M. Lustig, D. L. Donoho, and J. M. Pauly. Sparse MRI: The Application of Compressed Sensing for Rapid MR Imaging. *Magnetic Resonance in Medicine*, 58(6):1182–1195, 2007.

206. S. Mallat. *A Wavelet Tour of Signal Processing, Third Edition: The Sparse Way*. Academic Press, Inc., 3rd edition, 2008.

207. P. Mansfield. Multi-Planar Image Formation using NMR Spin Echoes. *Journal of Physics C: Solid State Physics*, 10:L55–L58, 1977.

208. S. F. Mc Cormick and G. H. Rodrigue. A Uniform Approach to Gradients Methods for Linear Operator Equations. *Journal of Mathematical Analysis and Applications*, 49:275–285, 1975.

209. S. G. Mikhlin. *Integral Equations*. Pergamon Press, 1957.

210. K. Miller. Least Squares Methods for Ill-Posed Problems with a Prescribed Bound. *SIAM Journal on Mathematical Analysis*, 1:52–74, 1970.

211. M. Moeller and D. Cremers. Image Denoising Old and New. In M. Bertalmio, editor, *Denoising of Photographic Images and Video*, pages 63–91. Springer, 2018.

212. V. A. Morozov. The Error Principle in the Solution of Operational Equations by the Regularization Method. *USSR Computational Mathematics and Mathematical Physics*, 8:63–87, 1968.

213. V. A. Morozov. *Methods for Solving Incorrectly Posed Problems*. Springer, 1984.

214. J. L. Mueller and S. Siltanen. *Linear and Nonlinear Inverse Problems with Practical Applications*. SIAM, 2012.

215. H. N. Multhei and B. Schorr. On Properties of the Iterative Maximum Likelihood Reconstruction Method. *Mathematical Methods in Applied Science*, 11:331–342, 1989.

216. K. J. Myers, H.H. Barrett, M. C. Borgstrom, D. D. Patton, and G. W. Seeley. Effect of Noise Correlation on Detectability of Disk Signals in Medical Imaging. *Journal of Optical Society of America A*, 2:1752–1759, 1985.

217. J. G. Nagy and D. P. O'Leary. Restoring Images Degraded by Spatially Variant Blur. *SIAM Journal on Scientific Computing*, 19:1063–1082, 1998.

218. F. Natterer. *The Mathematics of Computerized Tomography*. Wiley, New York, 1986.

219. F. Natterer and F. Wübbeling. *Mathematical Methods in Image Reconstruction*. SIAM, 2001.

220. Y. Nesterov. A Method of Solving a Convex Programming Problem with Convergence Rate $O(1/k^2)$. *Soviet Mathematics Doklady*, 27(2):372–376, 1983.

221. A. Neubauer. Tikhonov-Regularization of Ill-posed Linear Operator Equations on Closed Convex Sets. *Journal of Approximation Theory*, 53(3):304–320, 1988.

222. M. Nikolova. A Variational Approach to Remove Outliers and Impulse Noise. *Journal of Mathematical Imaging and Vision*, 20:99–120, 2004.

223. M. R. Osborne, B. Presnell, and B. Turlach. A New Approach to Variable Selection in Least Squares Problems. *IMA Journal of Numerical Analysis*, 20(3):389–403, 2000.

224. S. Osher, M. Burger, D. Goldfarb, J. Xu, and W. Yin. An Iterative Regularization Method for Total Variation-Based Image Restoration. *Multiscale Modelling and Simulation*, 4:460–489, 2005.

225. A. Papoulis. *Probability, Random Variables and Stochastic Processes*. Mc Graw-Hill, 1965.

226. A. Papoulis. *The Fourier Integral and its Applications.*, Mc Graw-Hill, 1962.

227. A. Papoulis. A New Algorithm in Spectral Analysis and Band-Limited Extrapolation. *IEEE Transactions on Circuits and Systems*, 22(9):735–742, 1975.

228. J. B. Pawley, editor. *Handbook of Biological Confocal Microscopy*. Plenum, New York, 2005.

229. P. Perona and J. Malik. Scale Space and Edge Detection using Anisotropic Diffusion. *IEEE Transactions on Pattern Analysis and Machine Intelligence*, 12:629–639, 1990.

230. D. L. Phillips. A Technique for the Numerical Solution of Certain Integral Equations of the First Kind. *Journal of ACM*, 9:84–97, 1962.

231. M. Piana and M. Bertero. Projected Landweber Method and Preconditioning. *Inverse Problems*, 13:441–463, 1997.

232. C. Pierpaoli, P. Jezzard, P. Bassers, A. Barnett, and G. Di Chiro. Diffusion Tensor MR Imaging of the Human Brain. *Radiology*, 201:637–648, 1996.

233. T. A. Poggio, V. Torre, and C. Koch. Computational Vision and Regularization Theory. *Nature*, 317:314–319, 1985.

234. M. Potmesil and I. Chakravarty. Synthetic Image Generation with a Lens and Aperture Camera Model. *ACM Transactions on Graphics*, 1:85–108, 1982.

235. M. Prato, R. Cavicchioli, L. Zanni, P. Boccacci, and M. Bertero. Efficient Deconvolution Methods for Astronomical Imaging: Algorithms and IDL-GPU Codes. *Astronomy and Astrophysics*, 539:A133, 2012.

236. W. H. Press, S. A. Teukolsky, W. T. Vetterling, and B. P. Flannery. *Numerical Recipes*. Cambridge University Press, 1992.

237. C. Pucci. Sui Problemi di Cauchy non Ben Posti. *Atti della Accademia Nazionale dei Lincei*, 18:473–477, 1955.

238. J. Radon. On the Determination of Functions From Their Integrals Along Certain Manifolds. *Berichte über die Verhandlungen der Königlich-Sächsischen Akademie der Wissenschaften zu Leipzig, Mathematisch-Physische Klasse*, 69:262–277, 1917.

239. N. A. Rahman. *A Course in Theoretical Statistics*. London: Griffin, 1968.

240. R. Ramlau and G. Teschke. Tikhonov Replacement Functionals for Iteratively Solving Nonlinear Operator Equations. *Inverse Problems*, 21(5):1571–1592, 2005.

241. W. H. Richardson. Bayesian-Based Iterative Method of Image Restoration. *Journal of Optical Society of America*, 62:55–59, 1972.

242. F. Rigaut. Astronomical Adaptive Optics. *Publications of the Astronomical Society of the Pacific*, 127(958):1197–1203, 2015.

243. F. Roddier. The Effects of Atmospheric Turbulence in Optical Astronomy. *Progress in Optics*, 19:281–376, 1981.

244. L. I. Rudin, S. Osher, and E. Fatemi. Nonlinear Total Variation based Noise Removal Algorithms. *Physica D*, pages 259–268, 1992.

245. K. Salvo and M. Defrise. A Convergence Proof of MLEM with Fixed Background. *IEEE Transactions on Medical Imaging*, 38:721–729, 2019.

246. F. Santosa and W. W. Symes. Linear Inversion of Band-Limited Reflection Seismograms. *SIAM Journal on Scientific and Statistical Computing*, 7(4):1307–1330, 1986.

247. J. L. Sanz and T. S. Huang. Unified Hilbert Space Approach to Iterative Least-Squares Linear Signal Restoration. *Journal of Optical Society of America*, 73:1455–1465, 1983.

248. O. Scherzer, editor. *Handbook of Mathematical Methods in Imaging*. Springer, New York, 2 edition, 2015.

249. O. Scherzer, M. Grasmair, H. Grossauer, M. Haltmeier, and F. Lenzen. *Variational Methods in Imaging*. Springer, New York, 2009.

250. R. Scholte, I. Lopez, N. B. Roozen, and H. Nijmeijer. Truncated Aperture Extrapolation for Fourier-based Near-Field Acoustic Holography by means of Border Padding. *Journal of Acoustic Society of America*, 125:3844, 2009.

251. I. Sechopoulos. A Review of Breast Tomosynthesis. Part II. Image Reconstruction, Processing and Analysis, and Advanced Applications. *Medical Physics*, 40:014302, 2013.

252. L. A. Shepp and B. F. Logan. The Fourier Reconstruction of a Head Section. *IEEE Transactions on Nuclear Science*, 21:21–43, 1974.

253. L. A. Shepp and Y. Vardi. Maximum Likelihood Reconstruction for Emission Tomography. *IEEE Transactions on Medical Imaging*, 1:113–122, 1982.

254. G. C. Sherman. Application of the Convolution Theorem to Rayleigh's Integral Formulas. *Journal of Optical Society of America*, 57(4):546–547, 1967.

255. J. R. Shewell and E. Wolf. Inverse Diffraction and a New Reciprocity Theorem. *Journal of Optical Society of America*, 58:1596–1603, 1968.

256. D. Slepian. Prolate Spheroidal Wave functions. Fourier Analysis and Uncertainty - IV: Extensions to Many Dimensions, Generalized Prolate Spheroidal Functions. *Bell System Technical Journal*, 43:3009–3057, 1964.

257. D. Slepian and H. O. Pollak. Prolate Spheroidal Wave Functions, Fourier Analysis and Uncertainty - I. *Bell System Technical Journal*, 40(1):43–63, 1961.

258. D. L. Snyder. Modifications of the Lucy-Richardson Iteration for Restoring Hubble Space-Telescope Imagery. In R L White and R J Allen, editors, *The Restoration of HST Images and Spectra*, pages 56–61. The Space Telescope Science Institute, Baltimore, 1990.

259. D. L. Snyder, A. M. Hammond, and R. L. White. Image Recovery from Data Acquired with a Charge-Coupled-Device Camera. *Journal of Optical Society of America A*, A-10:1014–1023, 1993.

260. D. L. Snyder, C. W. Helstrom, A. D. Lanterman, M. Faisal, and R. L. White. Compensation for Read-Out Noise in HST Image Restoration. In R. J. Hanish and R. L. White, editors, *The Restoration of HST Images and Spectra II*, pages 139–154. The Space Telescope Science Institute, Baltimore, 1994.

261. D. L. Snyder and M. I. Miller. The Use of Sieves to Stabilize Images Produced with the EM Algorithm for Emission Tomography. *IEEE Transactions on Nuclear Science*, 32:3864–3872, 1985.

262. A. Staglianò, P. Boccacci, and M. Bertero. Analysis of an Approximate Model for Poisson Data Reconstruction and a Related Discrepancy Principle. *Inverse Problems*, 27:125003, 2011.

263. J. L. Starck, D. L. Donoho, and E. J. Candès. Astronomical Image Representation by the Curvelet Transform. *Astronomy and Astrophysics*, 398(2):785–800, 2003.

264. J. L. Starck, M. Elad, and D. L. Donoho. Image Decomposition via the Combination of Sparse Representations and a Variational Approach. *IEEE Transactions on Image Processing*, 14(10):1570–1582, 2005.

265. J. L. Starck and F. Murtagh. *Astronomical Image and Data Analysis*. Springer, 2002.

266. J. L. Starck, F. Murtagh, and A. Bijaoui. *Image Processing and Data Analysis: The Multiscale Approach*. Cambridge University Press, 1998.

267. J. L. Starck, F. Murtagh, and J. M. Fadili. *Sparse Image and Signal Processing - Wavelets, Curvelets, Morphological Diversity. 2nd edition.* Cambridge University Press, 2015.

268. H. Stark, editor. *Image Recovery: Theory and Applications*. Academic Press, 1987.

269. G. W. Stewart. On the Early History of the Singular Value Decomposition. *SIAM Review*, 35:551–566, 1993.

270. T. G. Stockham, T. M. Cannon, and R. B. Ingebretsen. Blind Deconvolution through Digital Signal Processing. *Proceedings of the IEEE*, 63:678–693, 1975.

271. P. A. Stokseth. Properties of a Defocused Optical System. *Journal of Optical Society of America*, 59:1314–1321, 1969.

272. O. N. Strand. Theory and Methods Related to the Singular-Function Expansion and Landweber's Iteration for Integral Equations of the First Kind. *SIAM Journal on Numerical Analysis*, 11:798–825, 1974.

273. O. N. Strand and E. R. Westwater. Statistical Estimation of the Numerical Solution of a Fredholm Integral Equation of the First Kind. *Journal of the Association Computing Machinery*, 15:100–124, 1968.

274. D. Strong and T. Chan. Spatially and Scale Adaptive Total Variation Based Regularization and Anisotropic Diffusion in Image Processing. Technical report, Diffusion in Image Processing, UCLA Math Department CAM Report, 1996.

275. D. Strong and T. Chan. Edge-Preserving and Scale-Dependent Properties of Total Variation Regularization. *Inverse Problems*, 19:S165–S187, 2003.

276. A. Tarantola. *Inverse Problem Theory and Methods for Model Parameter Estimation*. SIAM, 2005.

277. H. L. Taylor, S. C. Banks, and J. F. McCoy. Deconvolution With the $\ell 1$ Norm. *Geophysics*, 44(1):39–52, 1979.

278. L. Tenorio. *An Introduction to Data Analysis and Uncertainty Quantification for Inverse Problems*. SIAM, 2017.

279. G. Teschke. Multi-Frame Representations in Linear Inverse Problems with Mixed Multi-Constraints. *Applied and Computational Harmonic Analysis*, 22(1):43–60, 2007.

280. J. H. Thomas, V. Grulier, S. Paillasseur, and J. C. Pascal. Real-Time Near-Field Acoustic Holography for Continuously Visualizing Nonstationary Acoustic Field. *Journal of Acoustic Society of America*, page 3554, 2010.

281. R. Tibshirani. Regression Shrinkage and Selection via the Lasso. *Journal of the Royal Statistical Society. Series B (Methodological)*, 58:267–288, 1996.

282. A. N. Tikhonov. On the Stability of Inverse Problems. *Doklady Academii Nauk*, 39:195–198, 1943.

283. A. N. Tikhonov. Solution of Incorrectly Formulated Problems and the Regularization Method. *Soviet Mathematics Doklady*, 4:1035–1038, 1963.

284. A. N. Tikhonov and V. Y. Arsenin. *Solution of Ill-Posed Problems*. Wiley (Russian edition in 1974), 1977.

285. G. Toraldo di Francia. Degrees of Freedom of an Image. *Journal of the Optical Society of America*, 59(7):799–804, 1969.

286. H. J. Trussell and B. R. Hunt. Image Restoration of Space Variant Blurs by Sectioned Methods. *IEEE Transactions on Acoustics Speech and Signal Processing*, 26:608–609, 1978.

287. H. J. Trussell and B. R. Hunt. Sectioned Methods for Image Restoration. *IEEE Transactions on Acoustics Speech and Signal Processing*, 26:157–164, 1978.

288. D. S. Tuch, T. Reese, M. Wiegell, N. Makris, J. Belliveau, and V. Weeden. High Angular Resolution Diffusion Imaging Reveals Intravoxel White Matter Fiber Heterogeneity. *Magnetic Resonance in Medicine*, 48:577–582, 2002.

289. V. F. Turchin, V. P. Kozlov, and M. S. Malkevich. The Use of Mathematical-Statistics Methods in the Solution of Incorrectly Posed Problems. *Soviet Physics Uspekhi*, 13:681–703, 1971.

290. S. Twomey. *Introduction to the Mathematics of Inversion in Remote Sensing and Indirect Measurements*, volume 3 of *Developments in Geomathematics*. Elsevier, 1977.

291. G. Uhlmann. Electrical Impedance Tomography and Calderon's Problem. *Inverse Problems*, 25(12):123011, 2009.

292. P. H. van Cittert. Zum Einfluss der Spaltbreite auf die Intensitaetsverteilung in Spektrallinien. II. *Zeitschrift für Physik*, pages 298–308, 1931.

293. A. van der Sluis and H. A. van der Vorst. Numerical Solution of Large Sparse Linear Algebraic Systems Arising from Tomographic Problems. In G. Nolet, editor, *Seismic Tomography*, chapter 3, pages 49–83. Springer, 1987.

294. H. T. M. van der Voort and K. C. Strasters. Restoration of Confocal Images for Quantitative Image Analysis. *Journal of Microscopy*, 178:165–181, 1995.

295. Y. Vardi, L. A. Shepp, and L. Kaufman. A Statistical Model for Positron Emission Tomography. *Journal of American Statistical Association*, 80:8–37, 1984.

296. G. Vicidomini, P. Bianchini, and A. Diaspro. STED Super-Resolved Microscopy. *Nature Methods*, 15:173–182, 2018.

297. G. Vicidomini, P. Boccacci, A. Diaspro, and M. Bertero. Application of the Split-Gradient Method to 3D Image Deconvolution in Fluorescence Microscopy. *Journal of Microscopy*, 234:47–61, 2009.

298. C. R. Vogel. Non-convergence of the L-curve Regularization Parameter Selection Method. *Inverse Problems*, 12:535–547, 1996.

299. C. R. Vogel. *Computational Method for Inverse Problems*. SIAM, 2002.

300. S. Voronin and I. Daubechies. An Iteratively Reweighted Least Squares Algorithm for Sparse Regularization. In M.Z. Cwickel and M. Milman, editors, *Functional Analysis, Harmonic Analysis, and Image Processing*, volume 693 of *Contemporary Mathematics*, pages 391–411. American Mathematical Society, 2017.

301. G. Wahba. Practical Approximate Solutions to Linear Operator Equations when the Data are Noisy. *SIAM Journal on Numerical Analysis*, 14(4):651–667, 1977.

302. Z. Wang, A. C. Bovik, H. R. Sheikh, and E. P. Simoncelli. Image Quality Assessment: From Error Visibility to Structural Similarity. *IEEE Transactions on Image Processing*, 13:600–612, 2004.

303. S. Warach, D. Chien, W. Li, M. Ronthal, and R. R. Edelman. Fast Magnetic Resonance Diffusion Weighted Imaging of Acute Human Stroke. *Neurology*, 42:1717–1723, 1992.

304. S. Webb editor. *The Physics of Medical Imaging*. IOP Publishing, 1988.

305. Y. W. Wen and R. H. Chan. Parameter Selection for Total-Variation-Based Image Restoration using Discrepancy Principle. *IEEE Transactions on Image Processing*, 21(4):1770–1781, 2012.

306. E. G. Williams. *Fourier Acoustics: Sound Radiation and Near-field Acoustic Holography*. Academic, London, 1999.

307. E. G. Williams and J. D. Maynard. Holographic Imaging without the Wavelength Resolution Limit. *Physics Review Letters*, 45:554–557, 1980.

308. T. Wilson, editor. *Confocal Microscopy*. Academic Press: London, 1990.

309. E. Wolf. Three-Dimensional Structure Determination of Semi-Transparent Objects from Holographic Data. *Optics Communications*, 1:153–156, 1969.

310. L. P. Yaroslavsky. *Digital Picture Processing. An Introduction*. Springer-Verlag, 1985.

311. M. Yuan and Y. Lin. Model Selection and Estimation in Regression with Grouped Variables. *Journal of the Royal Statistical Society: Series B (Statistical Methodology)*, 68(1):49–67, 2006.

312. R. Zanella, P. Boccacci, L. Zanni, and M. Bertero. Corrigendum: Efficient Gradient Projection Methods for Edge-Preserving Removal of Poisson Noise. *Inverse Problems*, 29:119501, 2013.

313. W. I. Zangwill. *Nonlinear Programming*. New York: Englewood Cliffs, 1969.

Index

Milton Keynes UK
Ingram Content Group UK Ltd.
UKHW050449071024
449327UK00014B/305